Metal Recovery and Separation from Wastes

Metal Recovery and Separation from Wastes

Editors

Lijun Wang
Shiyuan Liu

Basel • Beijing • Wuhan • Barcelona • Belgrade • Novi Sad • Cluj • Manchester

Editors

Lijun Wang
University of Science and
Technology Beijing
Beijing, China

Shiyuan Liu
University of Science and
Technology Beijing
Beijing, China

Editorial Office
MDPI
St. Alban-Anlage 66
4052 Basel, Switzerland

This is a reprint of articles from the Special Issue published online in the open access journal *Metals* (ISSN 2075-4701) (available at: https://www.mdpi.com/journal/metals/special_issues/Recovery_Separation_Wastes).

For citation purposes, cite each article independently as indicated on the article page online and as indicated below:

Lastname, A.A.; Lastname, B.B. Article Title. *Journal Name* **Year**, *Volume Number*, Page Range.

ISBN 978-3-0365-8718-9 (Hbk)
ISBN 978-3-0365-8719-6 (PDF)
doi.org/10.3390/books978-3-0365-8719-6

© 2023 by the authors. Articles in this book are Open Access and distributed under the Creative Commons Attribution (CC BY) license. The book as a whole is distributed by MDPI under the terms and conditions of the Creative Commons Attribution-NonCommercial-NoDerivs (CC BY-NC-ND) license.

Contents

Lijun Wang and Shiyuan Liu
Metal Recovery and Separation from Wastes
Reprinted from: *Metals* **2023**, *13*, 1411, doi:10.3390/met13081411 1

Aleksey Vishnyakov
Vanadium and Nickel Recovery from the Products of Heavy Petroleum Feedstock Processing: A Review
Reprinted from: *Metals* **2023**, *13*, 1031, doi:10.3390/met13061031 5

Wenjing Chen, Manning Li and Jiancheng Tang
Efficient Recovery of Cu from Wasted CPU Sockets by Slurry Electrolysis
Reprinted from: *Metals* **2023**, *13*, 643, doi:10.3390/met13040643 25

Junyan Du, Yiyu Xiao, Shiyuan Liu, Lijun Wang and Kuo-Chih Chou
Mechanism of Selective Chlorination of Fe from Fe_2SiO_4 and FeV_2O_4 Based on Density Functional Theory
Reprinted from: *Metals* **2023**, *13*, 139, doi:10.3390/met13010139 41

Qingdong Miao, Ming Li, Guanjin Gao, Wenbo Zhang, Jie Zhang and Baijun Yan
Improved Process for Separating TiO_2 from an Oxalic-Acid Hydrothermal Leachate of Vanadium Slag
Reprinted from: *Metals* **2023**, *13*, 20, doi:10.3390/met13010020 55

Zaure Karshyga, Almagul Ultarakova, Nina Lokhova, Azamat Yessengaziyev, Kaisar Kassymzhanov and Maxat Myrzakulov
Technology for Complex Processing of Electric Smelting Dusts of Ilmenite Concentrates to Produce Titanium Dioxide and Amorphous Silica
Reprinted from: *Metals* **2022**, *12*, 2129, doi:10.3390/met12122129 65

Fang Zhang, Jun Peng, Hongtao Chang and Yongbin Wang
Vacuum Carbon Reducing Iron Oxide Scale to Prepare Porous 316 Stainless Steel
Reprinted from: *Metals* **2022**, *12*, 2118, doi:10.3390/met12122118 81

Lei Li, Zhipeng Xu and Shiding Wang
Tin Removal from Tin-Bearing Iron Concentrate with a Roasting in an Atmosphere of SO_2 and CO
Reprinted from: *Metals* **2022**, *12*, 1974, doi:10.3390/met12111974 99

Praphaphan Wongsawan, Weerayut Srichaisiriwech and Somyote Kongkarat
Synthesis of Ferroalloys via Mill Scale-Dross-Graphite Interaction: Implication for Industrial Wastes Upcycling
Reprinted from: *Metals* **2022**, *12*, 1909, doi:10.3390/met12111909 111

Seyedreza Hosseinipour, Eskandar Keshavarz Alamdari and Nima Sadeghi
Selenium and Tellurium Separation: Copper Cementation Evaluation Using Response Surface Methodology
Reprinted from: *Metals* **2022**, *12*, 1851, doi:10.3390/met12111851 125

Ervins Blumbergs, Vera Serga, Andrei Shishkin, Dmitri Goljandin, Andrej Shishko, Vjaceslavs Zemcenkovs, et al.
Selective Disintegration–Milling to Obtain Metal-Rich Particle Fractions from E-Waste
Reprinted from: *Metals* **2022**, *12*, 1468, doi:10.3390/met12091468 143

Pavel Yudaev and Evgeniy Chistyakov
Chelating Extractants for Metals
Reprinted from: *Metals* **2022**, *12*, 1275, doi:10.3390/met12081275 159

Mo Lan, Zhanwei He and Xiaojun Hu
Optimization of Iron Recovery from BOF Slag by Oxidation and Magnetic Separation
Reprinted from: *Metals* **2022**, *12*, 742, doi:10.3390/met12050742 219

Shiyuan Liu, Weihua Xue and Lijun Wang
Extraction of the Rare Element Vanadium from Vanadium-Containing Materials by Chlorination Method: A Critical Review
Reprinted from: *Metals* **2021**, *11*, 1301, doi:10.3390/met11081301 229

Fuqiang Zheng, Yufeng Guo, Feng Chen, Shuai Wang, Jinlai Zhang, Lingzhi Yang and Guanzhou Qiu
Fluoride Leaching of Titanium from Ti-Bearing Electric Furnace Slag in $[NH_4^+]$-$[F^-]$ Solution
Reprinted from: *Metals* **2021**, *11*, 1176, doi:10.3390/met11081176 247

Editorial

Metal Recovery and Separation from Wastes

Lijun Wang * and Shiyuan Liu *

Collaborative Innovation Center of Steel Technology, University of Science and Technology Beijing, Beijing 100083, China
* Correspondence: lijunwang@ustb.edu.cn (L.W.); shiyuanliu@ustb.edu.cn (S.L.)

1. Introduction and Scope

With the development of society, large amounts of solid waste (slag, sludge, tailing, electronic waste, etc.) are generated every year. Each type of waste contains specific metals, such as As, Cr, V, Cu, Pb, and Zn, which are valuable resources and are also harmful to the environment. Currently, problems regarding the environment have increasingly attracted widespread attention as the global interest in these issues increases. If the metals in waste are not recovered effectively, not only are the resources wasted, but the environment is also seriously polluted.

The current processes for recovering metals (V, Cr, Ti, Fe, Mn, Pb, Zn, Cu, Ni, Co, Al, As, Nb, Mg, Au, etc.) from wastes (slag, sludge, tailing, electronic waste, etc.) include gravimetric, magnetic, floatation, pyrometallurgical, hydrometallurgical, bioleaching, chlorination, and electrolysis methods, etc. [1–3].

2. Contributions

Eleven research articles and three review articles were published in this Special Issue of *Metals*. The main topics covered include:

Cu from wasted CPU sockets was efficient recovered via slurry electrolysis [4]. The valuable metals (Ti, Fe, Mn, etc.) were extracted from vanadium slag by means of chlorination or an oxalic-acid hydrothermal leachate [5,6]. Karshyga et al. report the development of a technology intended to process electric smelting dusts of ilmenite concentrate with the extraction of silicon and titanium and the production of products in the form of their dioxides [7]. Fang et al. investigate the vacuum carbon reducing iron oxide scale to prepare porous 316 stainless steel [8]. Tin from Tin-bearing iron concentrate was removed via roasting in an atmosphere containing SO_2 and CO [9]. Mill scale and aluminum dross are the industrial wastes from steel and aluminum industries, which have high concentrations of Fe_2O_3 and Al_2O_3, respectively. Wongsawan reports the synthesis of ferroalloys via mill scale-dross-graphite [10]. Hosseinipour et al. investigate the significant factors of Se and/or Te recovery in the copper cementation process using the response surface methodology [11]. Current state-of-the-art milling methods also lead to the presence of significantly more reactive polymers still adhered to milled target metal particles. Blumbergs et al. find a novel and double-step disintegration–milling approach that can obtain metal-rich particle fractions from e-waste [12]. In order to solve the problem of solid waste pollution from basic oxygen furnace (BOF) slag, Lan et al. investigated oxidation reconstruction of BOF slag and alcohol wet magnetic separation recovery of iron. Compared with the initial steel slag, the iron grade increased by 8.22%, and the iron recovery increased by 46.38% compared with direct magnetic separation without oxidation [13]. Ti from Ti-bearing electric furnaces slag was leached by a $[NH_4^+]$-$[F^-]$ solution, providing the foundation for industrialization [14].

Vishnyakov reviews the recent developments in the recovery of vanadium and nickel from the heavy petroleum feedstock (HPF) as a raw source of metals [15]. Yudaev and Chistyakov review the efficiency and selectivity of the extractants in the recovery of metals

Citation: Wang, L.; Liu, S. Metal Recovery and Separation from Wastes. *Metals* **2023**, *13*, 1411. https://doi.org/10.3390/met13081411

Received: 19 July 2023
Accepted: 26 July 2023
Published: 7 August 2023

Copyright: © 2023 by the authors. Licensee MDPI, Basel, Switzerland. This article is an open access article distributed under the terms and conditions of the Creative Commons Attribution (CC BY) license (https://creativecommons.org/licenses/by/4.0/).

from industrial wastewater, soil, spent raw materials, and the separation of metal mixtures [16]. Liu et al. reviews the research progress of chlorination in the treatment of vanadium-containing materials [17].

3. Conclusions and Outlook

The purpose of this Special Issue is to focus on the current state-of-the-art ideas, methods, technologies, etc., for utilizing waste. This Special Issue provides a very good reference for the effective utilization of solid wastes. To minimize production costs and environmental impacts, it will be more and more necessary to use cleaner and more economical methods to recover metals from wastes.

The Guest Editor would like to thank the authors who contributed the manuscripts, the reviewers who took time out from their busy schedules to review the papers, and the journal.

Funding: The authors are grateful for the financial support of this work from the National Natural Science Foundation of China (No. 52274406, 51904286, 51922003), the Fundamental Research Funds for the Central Universities (FRF-TP-19-004C1) and Interdisciplinary Research Project for Young Teachers of USTB (Fundamental Research Funds for the Central Universities FRF-IDRY-21-015).

Data Availability Statement: Not applicable.

Conflicts of Interest: The authors declare no conflict of interest.

References

1. Liu, S.; Wang, L.; He, X.; Chou, K. Insight into the oxidation mechanisms of vanadium slag and its application in the separation of V and Cr. *J. Clean. Prod.* **2023**, *405*, 136981. [CrossRef]
2. Liu, S.; Ye, L.; Wang, L.; Chou, K. Selective oxidation of vanadium from vanadium slag by CO_2 during $CaCO_3$ roasting treatment. *Sep. Purif. Technol.* **2023**, *312*, 123407. [CrossRef]
3. Liu, S.; He, X.; Wang, Y.; Wang, L. Cleaner and effective extraction and separation of iron from vanadium slag by carbothermic reduction-chlorination-molten salt electrolysis. *J. Clean. Prod.* **2021**, *284*, 124674. [CrossRef]
4. Chen, W.; Li, M.; Tang, J. Efficient Recovery of Cu from Wasted CPU Sockets by Slurry Electrolysis. *Metals* **2023**, *13*, 643. [CrossRef]
5. Du, J.; Xiao, Y.; Liu, S.; Wang, L.; Chou, K.-C. Mechanism of Selective Chlorination of Fe from Fe_2SiO_4 and FeV_2O_4 Based on Density Functional Theory. *Metals* **2023**, *13*, 139. [CrossRef]
6. Miao, Q.; Li, M.; Gao, G.; Zhang, W.; Zhang, J.; Yan, B. Improved Process for Separating TiO_2 from an Oxalic-Acid Hydrothermal Leachate of Vanadium Slag. *Metals* **2023**, *13*, 20. [CrossRef]
7. Karshyga, Z.; Ultarakova, A.; Lokhova, N.; Yessengaziyev, A.; Kassymzhanov, K.; Myrzakulov, M. Technology for Complex Processing of Electric Smelting Dusts of Ilmenite Concentrates to Produce Titanium Dioxide and Amorphous Silica. *Metals* **2022**, *12*, 2129. [CrossRef]
8. Zhang, F.; Peng, J.; Chang, H.; Wang, Y. Vacuum Carbon Reducing Iron Oxide Scale to Prepare Porous 316 Stainless Steel. *Metals* **2022**, *12*, 2118. [CrossRef]
9. Li, L.; Xu, Z.; Wang, S. Tin Removal from Tin-Bearing Iron Concentrate with a Roasting in an Atmosphere of SO_2 and CO. *Metals* **2022**, *12*, 1974. [CrossRef]
10. Wongsawan, P.; Srichaisiriwech, W.; Kongkarat, S. Synthesis of Ferroalloys via Mill Scale-Dross-Graphite Interaction: Implication for Industrial Wastes Upcycling. *Metals* **2022**, *12*, 1909. [CrossRef]
11. Hosseinipour, S.; Keshavarz Alamdari, E.; Sadeghi, N. Selenium and Tellurium Separation: Copper Cementation Evaluation Using Response Surface Methodology. *Metals* **2022**, *12*, 1851. [CrossRef]
12. Blumbergs, E.; Serga, V.; Shishkin, A.; Goljandin, D.; Shishko, A.; Zemcenkovs, V.; Markus, K.; Baronins, J.; Pankratov, V. Selective Disintegration–Milling to Obtain Metal-Rich Particle Fractions from E-Waste. *Metals* **2022**, *12*, 1468. [CrossRef]
13. Lan, M.; He, Z.; Hu, X. Optimization of Iron Recovery from BOF Slag by Oxidation and Magnetic Separation. *Metals* **2022**, *12*, 742. [CrossRef]
14. Zheng, F.; Guo, Y.; Chen, F.; Wang, S.; Zhang, J.; Yang, L.; Qiu, G. Fluoride Leaching of Titanium from Ti-Bearing Electric Furnace Slag in $[NH_4^+]$-$[F^-]$ Solution. *Metals* **2021**, *11*, 1176. [CrossRef]
15. Vishnyakov, A. Vanadium and Nickel Recovery from the Products of Heavy Petroleum Feedstock Processing: A Review. *Metals* **2023**, *13*, 1031. [CrossRef]

16. Yudaev, P.; Chistyakov, E. Chelating Extractants for Metals. *Metals* **2022**, *12*, 1275. [CrossRef]
17. Liu, S.; Xue, W.; Wang, L. Extraction of the Rare Element Vanadium from Vanadium-Containing Materials by Chlorination Method: A Critical Review. *Metals* **2021**, *11*, 1301. [CrossRef]

Disclaimer/Publisher's Note: The statements, opinions and data contained in all publications are solely those of the individual author(s) and contributor(s) and not of MDPI and/or the editor(s). MDPI and/or the editor(s) disclaim responsibility for any injury to people or property resulting from any ideas, methods, instructions or products referred to in the content.

Review

Vanadium and Nickel Recovery from the Products of Heavy Petroleum Feedstock Processing: A Review

Aleksey Vishnyakov [1,2,†,‡]

[1] Skolkovo Institute of Science and Technology, Moscow 121205, Russia; avishnja@polly.phys.msu.ru
[2] Department of Physics, Moscow State University, Moscow 119991, Russia
[†] Organization where the project was carried out.
[‡] Current addresses: Institute of Solutions Chemistry RAS, Ivanovo 153038, Russia.

Abstract: The steadily growing demand for non-ferrous metals, a shift to heavier crude oil recovery and tightened environmental standards have increased the importance of heavy petroleum feedstock (HPF) as a raw source of metals. This paper reviews the recent developments in the recovery of vanadium and nickel from HPF. During crude oil processing and the application of its products, HPF is converted to various metal-enriched byproducts ("heavy oil", petcoke, ashes and slags) from which the metals can be recovered. This paper briefly describes the sources and recovery pathways (both mainstream and exotic), and discusses the economic viability and possible future directions. Particular attention is paid to (i) the electrochemical recovery of metals from petrofluids and alternative approaches; (ii) pre-combustion metal recovery from petcoke; and (iii) metal reclamation from fly ash from heavy fuel oil or petroleum coke combustion: hydro- and pyro-metallurgical and bio-based techniques. The current stage of development and prospects for the future are evaluated for each method and summarized in the conclusion. Increasing research activity is mostly observed in traditional areas: metal extraction from fly ash and the reduction of metals from the ash to V–Fe and Ni–Fe alloys. Bioengineering approaches to recover vanadium from ashes are also actively developed and have the potential to become commercially viable in the future.

Keywords: vanadium; nickel; heavy oil; reclamation

1. Introduction

Metal ions, which are abundant in heavy petroleum feedstock (HPF), are pollutants, and research on the techniques for petroleum demetallization never stops. Metal ions irreversibly poison the catalysts employed in oil processing and increase equipment corrosion. Furthermore, many metal compounds, including nickel oxides and, possibly, vanadium oxides, are carcinogens [1]. At the same time, the non-ferrous metals contained in HPF are a valuable resource.

The most abundant metal in HPF is vanadium. In Russia, the vanadium content of heavy petroleum recovered every year amounts to one third of the current annual production using traditional methods. Nickel is a distant second, and cobalt is found in sizable concentrations. Vanadium is of special interest: although it is ubiquitous in the Earth's crust in the forms of oxides, sulfides and phosphates, and is associated with other metals (iron, titanium, uranium, etc.), the vanadium grade is typically too low for direct production. Almost 70% the vanadium produced worldwide is extracted from slag formed in the process of steel production. The world demand for vanadium amounted to 120,067 mt in 2021, and only 8% of that originated from HPF. In the US, vanadium produced from petroleum feedstock amounted to 20% of the total in 2007 [2]. It should be noted that vanadium supply is key for the transition to a circular economy: vanadium redox batteries are regarded as the best option in stationary, long duration, high power applications for their ability to sustain nearly unlimited charge–discharge cycles, safety features and the

ease of reuse/recycling of the components and the electrolyte. The factors limiting their application are vanadium toxicity and high vanadium costs. The transition to a circular economy and the steady demand from the steel and biomedical industries are expected to drive the vanadium market up at an annual rate of 8.5% through to 2033 [3]. The demand for nickel is expected to grow by 44% by 2030 [4].

Metal recovery from crude oil demetallization waste is by no means a new field. The efforts to recover metals started in the 1940s and have been progressing ever since. Whilst it cannot be said that the metal recovery from HPF processing waste is rapidly growing, the situation is changing slowly, but surely. What was considered as waste in the 1940s has now become treasure. The three main factors in that evolving landscape are:

(i) Low-hanging fruits have mostly been reached: the depletion of fields with light and medium oils forces the recovery of heavy oil, which constitutes 70% of all resources that are available now [5]; the metal concentration generally grows with the asphaltene content in the crude, which makes metal recovery more attractive;

(ii) Improvements in oil processing technologies increase the share of the crude that is converted to fluid products, thus increasing the metal content in the residues;

(iii) The ever-tightening environmental regulations drive the search for safer and environmentally benign techniques of metal production.

As a result, both academic and industrial research has substantially intensified over the last 5–8 years. Remarkably, more efforts are invested in the traditional areas, e.g., the hydrometallurgic recovery of metals from the ashes. It is worth noting that the methods originally developed for metal recovery from mineral oils are now extended to spent refinery catalysts, municipal waste and other sorts of refuse.

Over the last several years, several comprehensive reviews were published on neighboring topics. In particular, Magomedov et al. [6] reviewed the methods of petroleum demetallization. Kurniawan et al. [7] considered the chemistry, origins and removal of metalloporphyrins; Yuan et al. [8] considered vanadium extraction techniques from all possible sources. Yet, none of them actually cover the metals recovery from HPF demetallization residues in reasonable detail. The recovery of metals from spent catalysts and ashes was reviewed eight years ago by Akcil et al. [9]. Spent catalysts recovery has certainly progressed towards full-scale commercial implementation. A very recent work by Baritto et al. [10] presented an economic evaluation of the recovery of vanadium from a spent catalyst employed in the catalytic reforming of bitumen. Data intensive process models developed for mass and energy balances allowed for the estimation of capital and operating costs. It was concluded that the recovery of vanadium from a spent catalyst obtained from bitumen upgrading the operations is potentially profitable considering the current vanadium market price. Spent catalysts from oil processing will not be considered here in detail, because hydrometallurgical pathways are similar to those applied for ashes and have not drastically changed since 2015, while novel biomethods were reviewed by Pathak et al. [11]. Wastewater deserves mentioning as well, because oil conditioning and desalting leaches substantial amounts of metals. Some additives, e.g., phosphoric acid, facilitate demetallization during oil pre-treatments, resulting in a high metal content in the wastewaters [12]. The methods applied to wastewater are drastically different from those reviewed in this paper and were recently described elsewhere [13].

This paper aims at providing a critical review of the existing methodologies of metal extraction from HPF, describing the current trends in the academic and industrial literature, and providing at outlook for further developments in the market, which is expected to grow [3,4]. The focus of this review is the recovery of vanadium and nickel, as they are most abundant in HPF, and most of the research efforts focus on them. The approaches to recovery are divided into two main categories: pre-combustion and post-combustion (recovery from ashes). We try to cover both major pathways and alternative ideas.

2. Vanadium and Nickel in Crude Oil

Vanadium and nickel compounds in crude oil are called "molecular fossils" due to their origin from heme and chlorophyll. Just as in living cells, in petroleum they are found in various coordination complex compounds, with metal porphyrins being the most known (Figure 1). The polyaromatic structures are flat and interact favorably with other polyaromatics via dispersion and π–π forces. Polyaromatic asphaltenes form the most hydrophilic fraction of crude (e.g., [14]), and the metal-containing complexes shown in Figure 1a are more hydrophilic in comparison with typical asphaltene molecules. In oil–water emulsions, they are located near the oil–water interface, which is important in some prospective recovery processes. In general, the metal content in crude oils increases with the density (and with the asphaltene content, correspondingly) [15]. The vanadium content grows with density faster than the nickel content (Figure 1b). Upon de-asphalting with heptane, most metals remain with the asphaltene fraction: vanadium content reached 1% of the total asphaltene mass precipitated from the heaviest oil samples (Figure 1c) [15]. As the asphaltene content in the petroleum feedstock increases, more effort has to be invested in demetallization, and in turn the demetallization waste becomes more valuable as a source of metals.

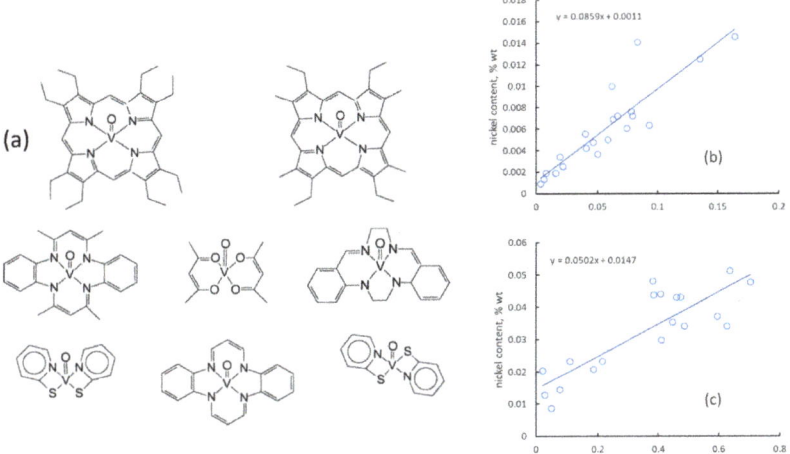

Figure 1. (a) Chemical structures of vanadium compounds in crude oil [16] (reproduced from ref. [16] with a permission from Elsevier), (b) vanadium to nickel ratio as a function of total vanadium content in the crude oil for different oil fields in Russia, (c) vanadium to nickel ratio in asphaltene fraction for the same oil samples upon standard extraction with heptane. Vanadium content in the asphaltene fraction reaches 1% mass, nickel mass is about twenty times as low (redrawn using data from ref. [17]).

The metal-containing molecules with strong coordination bonds between the metal and the surrounding organics are separated from the lighter components by fluid/critical/supercritical extraction and broken in the processes of thermal and catalytic cracking of the HPF, hydrogenation, coking and gasification. Each process listed above generally yields a lighter fraction, is more or less conveniently usable (as fuel, in organic synthesis, etc.), and has a denser "residue", enriched with carbon and heteroatoms. Metals always tend to stay with the residue. Each of the stages of oil processing can therefore be regarded as a demetallization process, which produces the desired product and "waste", which is often the raw material for the next stage. The heavier products, which are of interest as sources of metals, are:

(i) Heavy fuel oil. Although attempts to recover metals from heavy oil without combustion exist, there are few and these methods are not well established.

(ii) Asphalt, a solid or semi-solid bituminous residue from mostly critical and supercritical extraction. Asphalt can be coked and gasified, or can be used for metal recovery as is.
(iii) Cokes from delayed coking and flexicoking; also, coke gasification produces soot enriched with metals, which can also be used as a source of metals.
(iv) Ashes from heavy oil and coke combustion.

Below, we describe the techniques for metal recovery at each stage. The conclusion gives a summary comparison of the approaches.

3. Pre-Combustion Metal Recovery from Fluid Oil

3.1. Electrochemical Approaches

The electrochemical methods described here attempt to break the metal complexes in the liquid phase, and thus do away with any thermal-based demetallization treatment. Despite the hydrophobicity of the complexes, they can be either dissolved in a non-aqueous medium of a reasonable conductivity or dispersed in water (say, as a fine emulsion or a micellar solution), where they can be subjected to electrochemical oxidation or reduction. It is important that the organometallic compounds are among the most polar in the crude.

Electrochemical approaches are versatile, eco-benign and the metals can potentially be obtained in pure, easily usable forms. The fundamental possibility of the electrochemically assisted demetallization of porphyrins has been explored since the 1960s [18–20]. The efficiency depends on the composition of the electrolytic medium, the electrode material and the applied voltage. Ovales et al. [21] systematically studied the electrochemically assisted processing of HPF, including the reduction of polyaromatic compounds, demetallization and desulfurization. Demetallization and desulfurization of the real crudes could not be achieved simultaneously (high metal yields corresponded to low sulfur yields). Up to 81% of metals were successfully removed from bitumen residues, but the currents were as low as 0.01–0.02 A/cm^2 of the electrode surface, with an approximate maximum of 0.002 eq. of metals per m^2 s.

Jorge et al. [22] investigated the electrochemical decomposition of vanadyl tetraphenyl-porphyrins dissolved in xylene and dispersed in aqueous solutions of K_2SO_4. The electrolysis of the two-phase system was carried out potentiostatically. A steady-state regime with a constant current density was reached. The demetallization mechanism included vanadyl oxidation followed by the formation of cationic and zwitterionic radicals which led to the destruction of the porphyrin cycle. This work demonstrated a possibility for vanadium extraction via the electrolysis of a fine liquid emulsion. Welter et al. [23] attempted the demetallization of two synthetic compounds, vanadyl (IV) meso-tetraphenylporphyrin and vanadyl (IV) octaethylporphyrin, as well as Ayacucho Venezuelan crude oil samples by electrochemical techniques. They found that a protonated medium was essential for metal removal. Metal extraction resulted in the formation of free porphyrins that were not destroyed by the electrolysis. The cycling voltammograms were measured in a wide range of applied voltages (−2.3 to 0 V vs. Ag/Ag+) and were quite complex (Figure 2). The best metal extraction from commercial petroporphyrins (84%) was obtained on the glassy carbon electrode at −2.3 V in 0.1M $HClO_4$ solution in 4/1 vol. mixture of tetrahydrofuran and CH_3OH. For crude oil samples, the best metal yield was 66.4% of graphite, while for direct electrolysis on crude, the yields were 7.5% of vanadium, 8.2% of nickel and 79.6% of iron in charge efficiency. The ambitious goal of precipitating different metals in relatively pure forms with a successive step-by-step reduction at different voltages was not achieved because of very slow reduction kinetics. To increase the current, the authors had to increase the overpotential, which resulted in different reduction reactions running concurrently. According to the paper, "the results of commercial interest related to direct demetallization in crude oil still remains as an open challenge for electrochemists." As far as we understand, this challenge still remains open as of today. A somewhat similar attempt was made by Kurbanova and co-authors [24]. The authors used ethanol as a solvent, and the currents were significantly higher than in previous studies, 0.035 A/m^2 (assuming we understood the text correctly). It should be noted that as soon as the crude oil (rather than the extracted porphyrins) is diluted with expensive solvents [24], the process becomes economically non-viable.

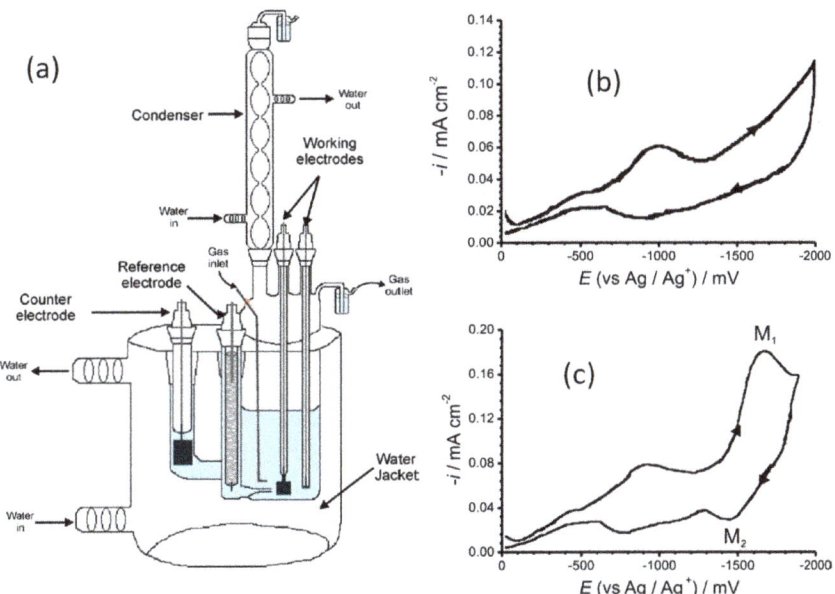

Figure 2. (**a**) Electrochemical cell used for electrolysis of metal–porphyrin compounds in polar non-aqueous solvents. (**b**) Cyclic voltammograms of mesotetraphenyl in THF/LiClO4/MeOH/HClO4 solution on platinum, voltage change rate of 50 mV/s, ambient conditions. (**c**) Cyclic voltammograms of vanadyl mesotetraphenyl at the same conditions. M1 reduction wave at −1 V is associated with direct electron transference to the porphyrin nucleus and is reversible. M2 is related to metal oxidation and is not fully reversible. Reproduced from ref [23] with permission from Elsevier.

Afanasieva et al. [25] applied electrolysis to a micellar solution of metal-containing asphaltenes. The authors experimented with both synthetic porphyrins and Castilla crude oil. The oil was mixed with either 0.2M LiClO$_4$ solution in THF and methanol (70/20 vol) or 0.5% H$_3$PO$_4$ + 0.1 M LiClO$_4$ in the mixture of THF and acetonitrile, and then the solution was electrolyzed. Cyclic voltammograms were measured at potentiostatic conditions between −2 V and 0 V vs. Ag/Ag$^+$. As a result, 80%+ of the metals (V, Ni and Fe) precipitated on the carbon electrode. Theoretically, such a method can separate the different metals contained in the asphaltene mixture. The kinetics, however, was also a serious problem: the currents were so low that metal separation was not achieved even in the laboratory experiment, and the conclusion made by the authors indicated that as of 2015, the electrochemical route was not viable. This was not surprising since the electrolysis was performed with a non-aqueous micellar solution. Increasing the overpotential in order to improve the kinetics was hardly reasonable due to numerous side reactions.

Electrolysis was also applied to the products of asphaltene reduction in an effort to extract vanadium and nickel [26]. Portions of the Boscan asphaltenes were treated with Raney nickel (a fine-grained solid composed mostly of nickel derived from a nickel–aluminum alloy [27]) and electrolyzed in LiCl solution in ethylenediamine. According to the study, the loss in vanadium was proportional to the loss of metalloporphyrins. While interesting, this technique does not seem economically promising: the process is quite complex, requires the re-solubilization of asphaltenes and does not promise much in terms of kinetics.

Electrochemical techniques can also assist metal extraction by hydrogenation. For example, Acevedo et al. [28] deposited vanadium and nickel from tetraphenylporphyrins dissolved in CH₃Cl. Hydrogen was generated by electrolysis and diffused through a Pd electrode which served as a catalyst in non-aqueous hydrogenation.

In summary, electrochemical methods are attractive for several reasons, and efforts to make them viable have not ceased. The hydrophobicity of porphyrin structures still makes electrochemical deposition very slow for the necessity to use non-aqueous solvents or colloidal solutions. As of today, electrochemical extraction provides moderate to high yields, but the kinetics are always questionable. The slow kinetics problem has not been resolved, and electrochemical methods, often called "most promising" are in fact not close to commercial applications.

3.2. Bioengineering Approaches

This section describes interesting ideas, each of which was reported in a single or in very few research papers and/or patents. These methods are apparently far from any commercial implementation, but deserve attention as interesting concepts for the future. Bioengineering approaches are inspired by the vigorous efforts invested into the biodegradation of crude oil (e.g., [29]). Metalloporphyrins are specifically targeted in a number of studies [7]. As a result of the biodegradation of the metalloporphyrins, metal ions are leached from the extremely stubborn asphaltene precipitate to forms that are more or less accessible to electrochemical and other methods. Preliminary studies were carried out on pure substances—surrogates of petroleum organometallic compounds. For example, in ref. [30], fungal cultures of aspergillus were used to decompose vanadyloxide octaethylporphyrin. The protoporphyrins of nickel were decomposed [31] by a very complex process using an enzyme obtained from P. azelaica YA-1 cultures. As means to obtain metals, bacterial and yeast cultures are hardly promising as of today: it is not even exactly clear as to what raw source they should be applied to. Yet, it is not impossible that crude oil biotransformation will also lead to metals in reasonable forms and quantities.

3.3. Other Alternative Approaches

Shiraishi et al. [32] proposed a photochemical process to destroy the most stable bonds between metal and porphyrin rings. The authors first considered a simultaneous photoreaction and extraction in a two-phase oil/water system (Figure 3). The results obtained for vanadium (IV) and nickel (II) tetraphenyl porphyrins dissolved in tetralin were compared with the results obtained for the residues from the atmospheric distillation of a crude oil samples. It was found that the first process was able to demetallize the "free" metalloporphyrins, but experienced difficulties in demetallizing "bound" metalloporphyrins, which are strongly associated with asphaltene molecules in the oil residue. In order to weaken this association and thus convert bound metalloporphyrins into free ones, a protonating solvent, 2-propanol, was added to the residual oil and subjected to photoemission. Then, the 2-propanol was evaporated, and the resulting oil residue was processed with an aqueous HCl solution, into which the vanadium and nickel successfully dissolved. In total, 93% of vanadium and 98% of nickel were extracted from the atmospheric residue; 73% of vanadium and 85% of nickel were extracted from the vacuum residue, respectively [32]. This method looks interesting, but suffers from the same problems as the electrochemical methods: the photochemical process is rather slow and the procedure requires relatively expensive solvents.

Lebedev et al. [33] proposed metal extraction using a microwave discharge. The focusing lens was placed against the side wall of the oil tank, and an antenna wire was inserted into the tank. As a result of the expansion and formation of plasma around the antenna, a solid residue of a fractal structure was formed on the wire. The residue contained metals in high concentrations (order of magnitude greater than in the parent bitumen); the metals were easy to leach with an acid. The practical significance of the idea is unclear, since: (i) metals still have to be extracted from the residue; (ii) the scalability of the process is dubious; and (iii) the process leads to the formation of conductive carbon particles in the liquid phase, thus preventing the process from being repeated. Garyfzyanova [34] also proposed the application of a plasma arch to cause the pyrolysis of heavy oil products into lighter hydrocarbons with the deposition of a residue containing soot and metal oxides. Trutnev et al. [35] proposed sonication to facilitate the recovery of metals (V, Ni, Cr and W) during the thermal cracking of crude oil. Finely dispersed metal oxide particles were introduced to the liquid-like residue obtained by thermal cracking and heated to a temperature of 380–420 °C. The application of an acoustic field to the dispersion is supposed to facilitate the precipitation of the particles and separation of the metal-containing fraction, which is periodically removed and subjected to calcination and purification, thus recovering both metal particles and porphyrins. We are not aware of any practical application of this idea.

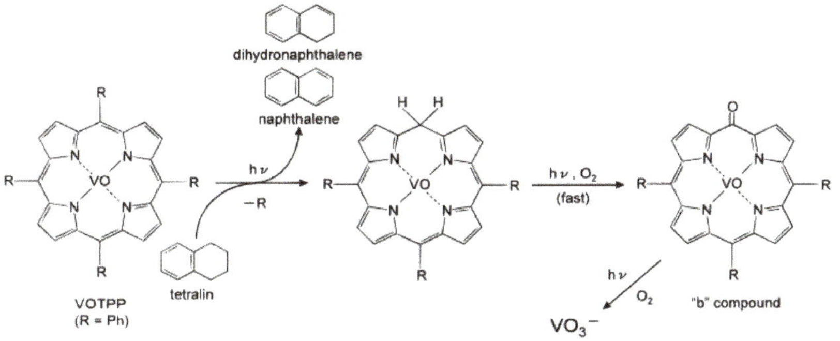

Figure 3. The proposed photodecomposition pathway for vanadyl (IV) tetraphenylporphyrin by photoirradiation to the tetralin + vanadyl (IV) tetraphenylporphyrin/water two-phase system [32] (reproduced with permission).

4. Pre-Combustion Metal Recovery from Petroleum Cokes

4.1. Metal Content in Petcokes

Petcoke results from the thermal cracking of heavy petroleum feedstock and is used as a fuel and in metallurgy. As always, in coking processes the coke is a "residue" enriched with metals, which are undesirable pollutants in metallurgy and energy production. A DTE petcoke technical data sheet (according to ref. [36]) estimated the vanadium content in petcokes at 0.12% and nickel content as 0.025% wt. in 2009. A recent thesis [37] cites a vanadium content of 0.1 to 0.2% and nickel content from 0.0035% to 0.06%. The current standard for the needle coke used in metallurgical electrode production is <0.025% vanadium, and the actual content is even lower [38]. ExxonMobile advertises the coke from flexicoking units for metal reclamation [39]. This section reviews the techniques of pre-combustion metal removal. Processing with acids, bases, sintering with salts and oxidation are mainstream methods in the published efforts. It is worth noting that similar methods are applied in metal extraction from stone coal, spend catalysts and other industrial waste materials [40–43].

4.2. Pre-Combustion Metal Leaching from Petcokes

Hepworth and Slimane [44] extracted vanadium and nickel from flexicoke obtained from Orinoco crude by leaching with acids at atmospheric pressure followed by extraction and crystallization. They obtained 99.6% pure V_2O_5. Acid leaching is more effective for a lower valence state of vanadium (V_2O_3) than for a higher valence (V_2O_5). The effect of vanadium re-leaching following the heat treatment using H_2SO_4 or NaOH was also studied. Vanadium recovery increased up to 98% by repeated leaching with 2M NaOH. Ultrasound also facilitated vanadium recovery. Alvarado et al. [45] treated coal and petroleum coke from Venezuelan oil with HNO_3 solutions (!) and subjected the result to a heat treatment in a microwave oven. In total, 93% to 98% of vanadium was recovered from the different samples. Sitnikova at al. [46] extracted vanadium from cokes produced by delayed coking and flexicoking from West Siberian HPF samples. Similarly to ref. [45], leaching was carried out with an oxidative solution (H_2SO_4 + NaClO) for 4 h at 70–80 °C. Here, 99% of vanadium was extracted from flexicoke, and only 40% of the available vanadium was extracted from the cokes produced by delayed coking. Ryumin et al. [47] leached vanadium from petroleum coke with H_2SO_4. The coke was pulverized to <100 μm particles, heated at 380–420 °C in air for 2–6 h, then processed with a H_2SO_4 solution for 2–3 h at 90–100 °C. Depending on the acid concentration, temperature and duration, 69–92% of the available vanadium was extracted with 40–70% (!) carbon mass loss. Rudko et al. [48,49] treated a lab sample of coke obtained from asphaltenes of West Siberian crudes with H_2SO_4 and H_2SO_4–HNO_3 mixtures of different concentrations. Coke that was manually crushed to 100 μm particles was exposed to leaching agents for 1–2 h at 100 °C. The degree of extraction reached 90%, which looks good, but no data on the carbon losses were provided.

The kinetics of alkaline leaching of vanadium from flexicoking residues was also explored. For example, in ref [50], flexicoke pulverized to 0.09–0.106 μm-sized particles was exposed to concentrated solutions of Na_2CO_3 and NaOH in the presence of a H_2O_2 oxidant over 5 h at 100 °C. The ratio of liquid and solid phases was 6 to 1. A 72.7% degree of demetallization was achieved.

Vanadium can be extracted from petrocokes by sintering with alkali metal salts at temperatures below salt melting, and then leaching in aqueous media with acids or bases. Patent [51] considers flexicokes, although does not limit the application to this specific source. The disadvantage of this method is the loss of a significant part of the carbon mass of the coke as a result of the process. Paper [52] applies a similar method for concurrent sulfur removal and vanadium extraction. Only 60% of vanadium was extracted.

In general, the attempts to "have your cake and eat it" (that is, to extract metals and efficiently use the carbon as a fuel or otherwise) have a long history and have not ceased. The problem is the high carbon loss, which increases the cost and, in many cases, low degree of extraction. A more subtle route to do away with the oxidation step is leaching metals from coke or asphalt gasification residue, rather than from the coke itself [53,54]. Gasification is the processing of a dried residue (coal, coke or asphalt) with steam and oxygen at high temperatures leading to a syngas (CO, H_2 or CO_2) and carbon soot enriched with metals (review [55]). By thermodynamic and kinetic modeling, the authors [53,54] attempted to find conditions that would lead to easy-to-extract metal compounds in the resulting soot residue. The authors found that only a small fraction of molybdenum and nickel would be reduced to pure metals; rather they would form carbides. Vanadium would be reduced to V_2O_3, especially at lower gasification temperatures. The resulting compounds are rather difficult to dissolve (vanadium dissolves in water in vanadate forms only), and thus the process would require an extra oxidation step. As an alternative, the authors explored controlled soot burnout to an ash and concluded that burnout was superior as a method for metal recovery.

4.3. Conversion of Metals Contained in Petcokes to Other Products without Prior Leaching

There are exotic ideas that target metal conversion to useful products without the isolation and purification of vanadium and nickel compounds. Abdrabo and Husein [56] proposed the conversion of oil demetallization waste (namely, the gasification residue) to the dispersion of metal oxide nanoparticles in heavy oil. This interesting idea does not however fully qualify as a technique for metal recovery. Finally, Zhan et al. [57] synthesized a vanadium-based metal organic framework (MOF) from vanadium-containing waste: carbon black from an oil refinery. This idea is demonstrated in Figure 4. No matter how interesting this idea is (it is!) as a method of *vanadium recovery*, this work is not very convincing, because vanadium has to be converted in a reasonably usable form before MOF synthesis.

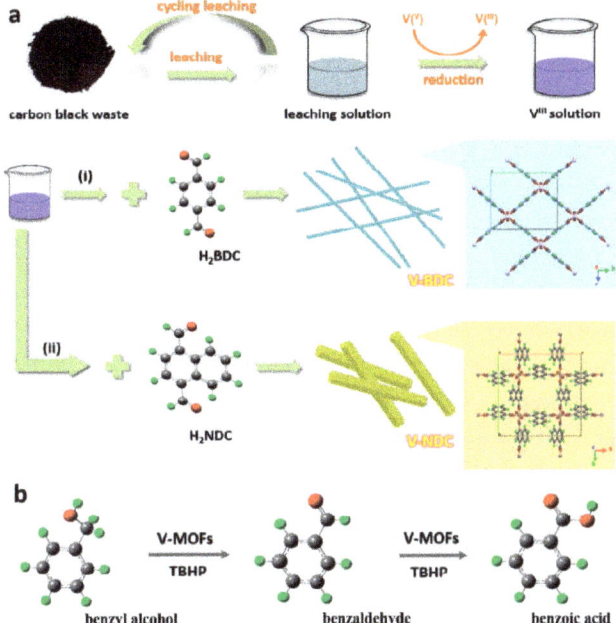

Figure 4. (a) Schematic illustrations of vanadium recovery from carbon black waste into V-MOFs: (i) synthesis of vanadium–benzene dicarboxylate, and (ii) synthesis of vanadium–naphtalenedicarboxylate nanorods. The cycling leaching process represents the usage of the first leachate as a leaching agent in order to increase the concentration of vanadium in the second leachate. (b) The reaction route of catalytic benzyl alcohol oxidation over the vanadium-MOF catalyst. Color codes in molecular structures and ball–stick models: green balls (hydrogen atoms), gray balls (carbon atoms), red balls (oxygen atoms) and pink balls (vanadium atoms). (Reproduced from ref. [57] with permission from ACS).

5. Ashes from Heavy Fuel Oil, Petroleum Coke and Asphalt as Sources of Metals

5.1. Fly Ash from Heavy Fuel Oil and Coke Combustion

Unlike CO_2 and water, metal oxides are relatively non-volatile. During the combustion of hydrocarbon fuel, they concentrate in the solid residue, aka ashes (Figure 5). The recovery of metals from combustion products attracts intensive efforts and is (as of today) the only path that is industrially implemented. The implementation started back in the 1970s. The literature review in thesis [58] already described a long research history, and as of 1988, eight units of vanadium recovery from combustion residue operated in Canada and Venezuela (according to review [6]).

As sources of metals, heavy oil (or "heavy fuel") fly ash (HOFA) and ashes from petcoke combustion are of special interest. Since most metals (V, Ni, Co and Fe) are chemically bonded to the polyaromatic fragments of the asphaltenes, they are found in the fly ash rather than the bottom ash of the pulverized or fluidized petcoke furnaces. Metal-containing ash is a waste and substantial efforts are being invested into its utilization (see special issue [59] and review [60]). The most common applications for fly ash are composites [61], waste stabilization [62] and construction materials [63]. Still, about half of the fly ash in the US is currently landfilled [64]. Using fine heavy oil and petcoke fly ash in concrete and asphalt concrete is safer compared to landfill depositing, as metal emissions to the environment are much slower when the metal-containing particles remain in a stabilized form [65]. The prospective applications of fly ashes include even soil improvement [66].

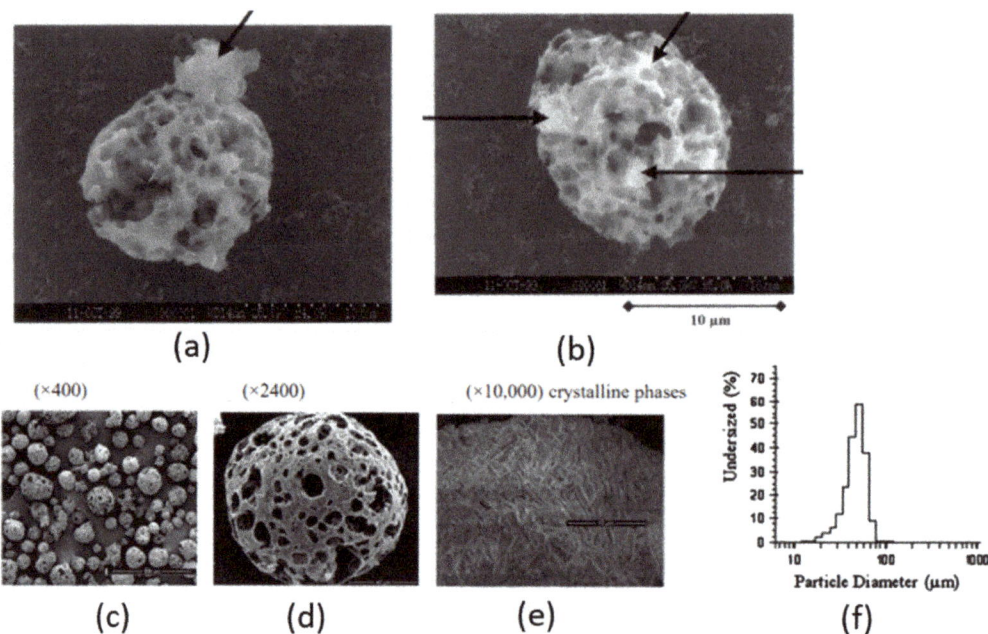

Figure 5. Top row (**a**,**b**) particles of fly ash from semi-solid asphalt combustion. The arrows indicate metal-rich inorganic inclusions; the carbon phase is darker gray (reproduced from ref. [67] with permission from T&F). Bottom row (**c**–**e**) ash particles formed during the combustion of heavy fuel oil in boilers and (**f**) the particle size distribution (reproduced from ref [68] with permission from Elsevier).

5.2. Controlled Combustion and Gasification of Petcoke to Obtain Ash Enriched with Metals

Although power plants are the main sources of fly ash, sometimes it makes sense to burn refuse polluted coke or solid asphalt resulted from demetallization via critical or supercritical extraction. The controlled oxidation generates ash that is easy to collect and transport.

In 1945, a patent was granted [69] in the US for a method of vanadium recovery from oil, which involved separation of the high boiling solid or semi-solid fraction (asphalt) from the feed oil, burning and further oxidation of the product in a controlled manner and leaching the metal from the resulting ash. The patent schematically presented the construction of the furnace for the controlled burning of the asphalt. Note that metal recovery is the specific target of asphalt oxidation here, rather than a byproduct of power generation.

Jack and co-authors [70] studied the possibility of the extraction of iron, vanadium, nickel and titanium by hydrometallurgical methods from petroleum coke obtained during flexicoking from the Athabasca oil bitumen sands. When treated with strong acids, the extraction of vanadium, iron and nickel was more than 50%, while only 20% of titanium was extracted. By means of weak acid leaching, nickel extraction was about 30%, but no transition to a solution of other metals was achieved. If coke obtained from the bitumen of Athabasca oil sands by flexicoking or the usual method of delayed coking is oxidized by air at temperatures below 500 °C, then the vanadium and nickel remaining in the ash are easily leached by acid: the solid residue is treated with acids, as a result of which the metals pass into the solution. Nickel and cobalt are extracted as cations $[Co(H_2O)_6]^{2+}$, $[Ni(H_2O)_6]^{2+}$, and vanadium is extracted in the form of isopolymers; for example, $(V_2O_7)^{4-}$, $(V_3O_9)^{3-}$, etc.

A straightforward method of extracting vanadium from petroleum coke proposed by Gardner [71] is based on the complete gasification of coke, the production of ash and combustible gas and the extraction of vanadium from the ash. For the flexicoking residue specifically mentioned by the author, such an approach may be fully justified. For example, two Canadian companies, MGX Minerals and Highbury Energy, announced a venture to recover metals from excessively vast petcoke stockpiles close to the Athabasca region in Alberta, Canada [72]. Gasification units were developed specially for the project. According to the report [73], MGX studied thirteen coke samples from two different stockpiles. The two stockpiles contained 0.042% and 0.046% of vanadium (wt% of dry petcoke). The ashes contained 6.6% and 45% of V_2O_5, correspondingly.

5.3. Hydrometallurgical Approaches for Metal Extraction from Ash

Heavy oil fly ash (HOFA) is a by-product generated in power plants by the burning of heavy fuel oil. The main constituent of HOFA is unburned carbon; it also contains other elements such as As, Cd, Co, Cr, Hg, Ni, Pb Cu, Zn, Se, Ca, Mg, Na and Si. The composition and physico-chemical characteristics of HOFA were studied quite comprehensively. HOFA particles have a typical size of 10–100 μm [74], although the size depends heavily on the process. The bulk density of HOFA varies from 0.50 to 1.50 g/cm^3 and the porosity is estimated as 10.31% [75]. A comparison of the BET [76] surface area and the particle size reported in ref. [77] also suggests a substantial porosity. The metal content of HOFA varies widely (depending on the raw material and the combustion process), but is always significant. HOFA consists mostly of carbon [78] and contains large amounts of sulfur. Besides vanadium and nickel, HOFA samples can contain substantial amounts of zinc and iron, and traces of cobalt, chromium, lead and copper [68]. Carbon soot is the most significant constituent of HOFA; the total metal content can approach 5% [68,79–82]. For example, Jung and Mishra measured the metal content by inductively coupled plasma atomic emission spectroscopy and X-ray fluorescence. Their samples of HOFA contained 2.2% wt vanadium, 1.9% germanium and 0.4% nickel. Carbon constituted >93% wt of the total HOFA mass. Vanadium is mostly present in HOFA in the $Mg_3V_2O_8$ form. It is worth mentioning that the V/Ni ratio does not obey the relationships common to heavy oil [83], possibly suggesting metal loss.

The metal content in ashes depends on the combustion process and on the ability to collect the fly ash (by electromagnetic traps). Linak and Miller [84] burned gasified coals and fuel oil in laboratory installations simulating power plants. Coals with various sulfur and metal contents were used: the vanadium content ranged from 2.25 to 13 μg/g, and the nickel content ranged from 0 to 6.4 μg/g of the raw materials. The highest content was in a high grade fuel oil sample, 220 μg/g. Next, the metal content in the trapped particles formed during combustion was measured. The type of combustion equipment played an important role: during the combustion of fuel oil in narrow tubes surrounded by water, which led to the rapid cooling of the combustion products, a large number of very small non-trappable particles formed, and the metal content in the trapped ash was lower. The metal content in the smaller particles was always higher than in the larger ones. When combustion occurred in a larger volume (water was supplied in the tubes, burning was conducted in the surrounding space), the combustion products cooled more slowly, which led to larger particles and a vanadium content up to 13.6%. Since the processes of burning coal and petroleum coke are similar, it is expected that in ash with a particle size smaller than 2.5 μm, the metal content in the fine ash fraction is increased by 25–100 times compared to the original coke. For HOFA, we might expect the ratio of 250+ times.

The main approaches to metal extraction are based on solvent extraction/leaching, roasting/calcination with salts also followed by leaching and controlled oxidation to burn out the carbon and obtain metal oxide mixture to an almost-pure form. The degree of recovery and the selectivity are two main targets. The straightforward approach is a "hydrometallurgical" recovery of vanadium and nickel directly from HOFA by acids [85–89], bases [90–93] or water [94] leaching. Vanadium is oxidized to vanadate forms (by O_2 [85], H_2O_2 [90,95], HClO [8,96], etc. [86,87]) and precipitated with non-alkali cations or an ion exchange [97]. Then, it can be isolated and purified with recrystallization or solvent extraction [88,89]. We will not describe all these studies here; a comprehensive review can be found in [98] with the focus on the removal rather than the recovery. The process can be assisted by sonication [99]. The straightforward procedure promises moderate success. Al-Degs et al. [68] used various solvents (HNO_3, NaOH, EDTA, etc.), as well as acid and base solutions on a set of HOFA samples to extract different metals (Ni, V, Mo, Cr, Zn, Cu and Mg) and recovered up to 95% of the total magnesium and up to 15% of the total vanadium, for which the yields were notably low. Other published studies report a somewhat higher recovery [100]. For example, in paper [101] leaching HOFA with 0.5 N sulfuric acid resulted in the extraction of 65% vanadium, 60% nickel and 42% iron. During leaching in a 2M NaOH, vanadium recovery was 80%, and nickel recovery was insignificant, which allows for the selective recovery of metals. Selective nickel extraction was achieved with a mixture of ammonium water and ammonium sulfate. Paper [102] followed similar pathways of vanadium extraction from HOFA, but explored in finer details the influence of process parameters, such as the solvent composition, time, particle size and mixing rate. Navarro and co-authors [103] treated HOFA with alkalis to selectively separate vanadium in the form of soluble vanadates, from which insoluble forms were precipitated with non-alkaline metal ions. Generally, leaching at a high pH is selective to vanadium, while other metals, including nickel and cobalt, are leached by acids. A three-step process of metal extraction from Orimulsion fly ashes was presented by Vitolo et al. [80].

5.4. Roasting with Salts and Controlled Oxidation

Fly ash sintering (roasting, calcination) with salts in order to convert metals (vanadium and nickel, first of all) to forms better suitable to leaching makes another group of traditional methods that can be traced back to at least early 1980 [58]. Gomez-Bueno et al. [104] roasted the ashes from Athabasca HPF combustion with NaCl at 875–950 °C prior to leaching with an alkali at boiling temperatures. The authors recovered 85% of the available vanadium. Holloway et al. applied a very similar approach to oil sands fly ash [105]. The same group examined the selectivity of vanadium extraction and found that roasting with high amounts of Na_2CO_3 allowed for the selective recovery of vanadium, and somewhat later, they proposed the production technology [106]. Recent developments include roasting oxidation for the selective recovery of vanadium that could be leached by water, and nickel that could only be recovered with acids [107]. More than 80% of the available vanadium was recovered as a result.

The approaches to fly ash from petcoke combustion are similar. Vasilyeva et al. [108] studied the elemental composition of ash produced by the combustion of petcokes of Syrian origin. The very scheme of leaching vanadium compounds dates back to a very old patent [109]. Most attempts to isolate metals involve the roast-and-leach approach [110]. For example, Rezai et al. [111] recently applied roasting with Na_2CO_3 and nitric acid leaching to extract a wide group of metals from coal coke. Ziyadanoğullari [112] applied the roast-and-leach approach to asphalt residue after critical solvent extraction and was able to recover 70%+ of the available vanadium. Petcokes are more often subjected to controlled oxidation prior to calcination with salts in order to remove most of the remaining carbon [113,114]. Kadhim [115] applied a two-stage process: first, the ash was treated with air at 650 °C and 850 °C in order to reduce the carbon content. Then, the product was processed with NaOH to extract the vanadium. Vitolo et al. [80] also conducted preliminary oxidation with air at 650 to 1150 °C, below the initial deformation temperature of the fly ash. The temperature of the preliminary burning step substantially influences the outcome. Above 950 °C, the volatilization of vanadium and the formation of V–Ni refractory compounds adversely affected the recovery of vanadium. The burning temperature of 850 °C was found to be the optimum as a result of the trade-off between the overall vanadium recovery yield (83%) and the V_2O_5 weight percentage in the precipitate (84.8%). For coke from oil obtained from the oil sands of Western Canada, a complex process was proposed [116], including: (i) the burning of ash in the presence of oxygen and NaCl (5 to 35% wt of the petcoke) at temperatures ranging from 700 °C to 950 °C; (ii) leaching of the residue with aqueous solutions at pH = 5–12 and temperatures ranging from room to boiling until a significant portion of the vanadium dissolves; and (iii) precipitation of metavanadate ammonium with a solution of ammonium chloride or sulfate acidified to a pH = 2–3.

Volkov et al. [82] published a detailed analysis of HOFA roasting with Na_2CO_3 followed by aluminothermic melting. The process leads to a "slag" enriched with V, Ni and Fe in various forms ($NaMg_4(VO_4)_3$, $NaVO_3$, $Ca_xMg_yNa_z(VO_4)_6$; (FeV_2O_4), $V_{2-x-y-z}Fe_xAl_yCr_zO_3$, $Mg_{1-x-y-z}Ni_xFe_yV_zO$). The metals were leached with either acid or base solutions. Leaching from the slug was far more effective compared to the direct leaching from the HOFA source (72.3–96.2% V and about 90% Ni was recovered with H_2SO_4). A somewhat similar method was previously applied to vanadium-containing sludge, which is a byproduct of vanadium pentoxide obtained by hydrometallurgical methods. Vanadium was mostly in the $FeO \cdot V_2O_3$ form. The authors explored various oxidation roasting methods for a sludge treatment to facilitate vanadium extraction. Oxidation roasting of the sludge at 1000 °C with 1% CaCO3 increased the acid-soluble V_2O_5 from 1.5% to 3.7% and lowered the content of $FeO \cdot V_2O_3$ from 3% to 0.4% [117]. Vanadium and nickel from flexicoking waste were isolated by Queneau et al. [118] by pressure oxidation: steam with a pH = 9.5 was supplied under pressure in an industrial-type facility. Exothermic reactions produced enough steam to generate enough electricity to make the process self-sustainable. Vanadium and nickel are concentrated in some kind of ash in carbonate forms and converted to high purity salts and then oxides with extraction and crystallization. The general scheme of vanadium and nickel extraction from fly ash is shown in Figure 6.

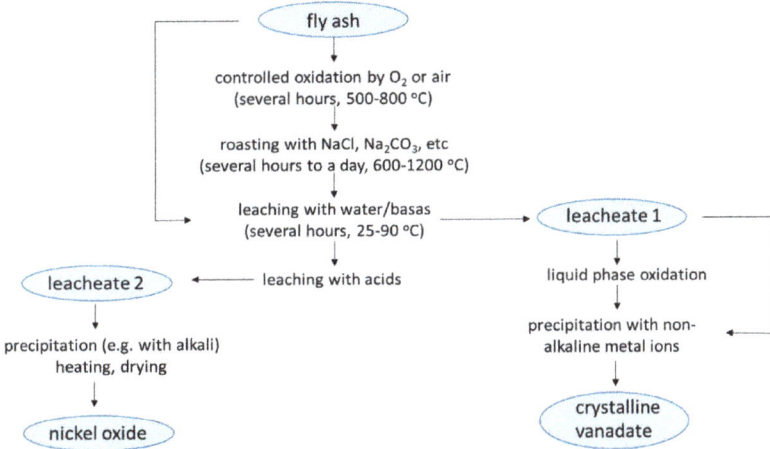

Figure 6. Composite scheme of hydrometallurgical vanadium and nickel recovery from fly ash, according to the literature. There are many modifications of this process; for example, Liu et al. [8] describe chlorination-based methods.

5.5. Pyrometallurgical Methods

Pyrometallurgical approaches [119–122] are qualitatively different from sintering with salts or controlled oxidation; their product are alloys of the target metals with a less expensive metal, usually Fe. For example, Xiao et al. [120] suggested an interesting pyrometallurgical method to recover vanadium by obtaining a ferrovanadium alloy from two industrial waste resources: petcoke fly ash and flue dust. A ferrovanadium alloy with about 20 wt% vanadium was obtained at 1550 °C. On average, about 30% of the metal yield was obtained during smelting of the mixture of petroleum fly ash to flue dust in a 4/5 weight ratio. Sun et al. [122] proposed a pyrometallurgical process to recover nickel in the form of an Fe–Ni alloy from coal fly ash, spent petroleum catalyst, CaO, Fe powder and H_3BO_3.

Finally, Tectonics' DC operates a plasma arc melting furnace (PAF) to recover metals from the ash that is used just as any ore [123]. In PAF, the content is melted under an inert gas atmosphere at an ambient pressure with the plasma arc torch column which provides temperatures sufficient for the production of metals. Wet materials containing substantial amounts of water can be heated safely and effectively in a plasma arc smelting furnace. The recovery of nickel as an Fe–Ni alloy and a vanadium-rich slag is achieved with a carbothermic reduction process common in pyrometallurgy: nickel and iron oxides are reduced by carbon. Then, vanadium as an Fe–V alloy from the first stage slag is obtained via reduction with aluminum. Nickel and vanadium are thus obtained separately in this two-stage process. Further processing of the alloys (especially the first stage Fe–Ni alloy) may be required to meet the standard ferroalloy specifications.

5.6. Bioengineering-Based Approaches to Post-Combustion Extraction Metals

Bioengineering approaches to vanadium and nickel separation from ashes appear to be far from production but potentially promising [124]. Li and co-authors [125] studied the metabolism of vanadium in technologies where coke was mixed with an organic matter of biological origin. It must be said that the bio-demetallization of coal combustion products has been studied for quite a long time, and it is based on the consumption of carbon residues by microorganisms. Rasulnia and Mousavi [126] applied fungal microorganisms, including ordinary penicillin, to produce vanadium and nickel from ashes (Figure 7). The utilization of carbon for the vital activity of fungi leads to the release of metals, because microorganisms do not absorb metal ions, which in high concentrations are harmful to them. Fungal cultures are introduced to fairly dense suspensions of ash in water. Metal ions can be precipitated into insoluble forms. At a temperature of 60 °C, 90% of vanadium and about half of nickel were extracted within a week. The process is long but environmentally benign. Generally, the idea of the biological extraction of metals received wide interest, including applications to metallurgical slag [127], the ash from municipal waste [128–130], wastewater treatment and the extraction of metals from the wreckage of electronic devices (review [131]).

 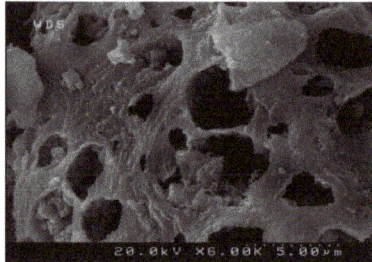

Figure 7. The effect of bioleaching in FE-SEM images of the surface morphology of the original HOFA particles (**left**) and the before and after bioleaching at ×6.00 K magnification (**right**). Reproduced from ref. [126] with permission from RSC.

6. Conclusions

Table 1 lists the main approaches to metal reclamation from the products of HPF processing, their advantages and disadvantages, the stages of development and commercialization. The successful commercial projects of metal extraction from HPF have existed for decades. Nevertheless, they remain relatively small and "local": metal reclamation seems to have not reached the scale comparable with traditional sources—the slag from steel production and the direct production from vanadium-bearing ores. Post-combustion metal extraction from power plant ashes and gasification residues is the only route of metal recovery from the HPF commercially explored on a regular basis as of today. Metal-polluted ash is really a waste that needs to be safely disposed of. In general, the approaches to recovery are somewhat similar to those applied to the metal extraction from slag. The important difference is the high carbon content of the fly ashes, while the slag consists primarily of metal oxides, silicates and phosphates, [132] with V_2O_5 content up to 20 wt% [133] (we did not review the vanadium extraction from the slag since it is mostly of non-petroleum origin and was described extensively [133,134]). The environmental challenges specific to the fly ashes are related to the ash capture (as we showed, metals are contained in a small particle fraction that needs to be captured), storage and transportation. Pyrometallurgy is energy-intensive, but relies on relatively traditional smelting technologies. The hydrometallurgical route is the most studied in the literature and can be easily adjusted to the various types of raw materials. Most research efforts are also concentrated in these traditional and mature areas. The progress is mostly incremental and follows the prospective industrial demand.

Table 1. Summary of the main methods of vanadium and nickel extraction from HPF demetallization residues.

Source	Method	Advantages	Disadvantages	
Heavy oil	Electrochemical: Emulsification/solvation in non-aqueous solvent, electrolysis	Convenient, environmentally benign, potentially selective	Very slow	Very unlikely to reach industrial implementation
Coke	Leaching with acids, solvents	Simple	Loss of reagents on carbon oxidation	In perspective, inferior to other methods
Coke or solid asphalt	Oxidation or gasification to ash	Produces syngas		Good way to utilize low quality or refuse material Implementation underway
Ash from heavy oil and coke combustion	Hydrometallurgy: oxidation–roasting–leaching–precipitation	Source is really a waste; Extraction is a low energy process	Complex, multi-stage; fly ash collection, storage and transportation is difficult	Most used and studied, will remain for a while
	Pyrometallurgy: reduction to V–Fe and Ni–Fe alloys	Metals obtained in convenient form	Energy consumption, expensive equipment	Ready technology, will remain for a while
	Bioengineering	Environmentally benign	Slow, not studied enough	Years of research needed, but potentially viable

Where we may expect serious developments is metal reclamation from the residues resulted from the on-site supercritical extraction of the lighter components from heavy oil. The on-site extraction and hydrocracking of super heavy oils are aimed at resolving the transportation problem: currently, very viscous heavy oil is often diluted with lighter oils just to be transported to refineries. The on-site extraction yields liquid and solid products that are much easier to deal with. Whether the semi-solid asphalt obtained on-site is worth further cracking processing depends on the original crude and the extraction conditions, which can be tailored to deeper extractions (less saturates and lower aromatics on the residue; more organometallic compounds in the deasphaltate) or shallower extractions (deeper demetallization; more SAR in the residue). In the case of a deeper extraction, the asphalt might likely be treated similarly to petcoke with gasification and hydrometallurgical extraction, but further research to optimize the extraction and metal reclamation is needed.

The other area of actively increasing research activity is bio-extraction from ashes with fungi species known to accumulate vanadium and/or consume the carbon fraction of the ashes. How long it may take to develop into ready technologies is hard to predict; it is possible that the bioengineering methods will never become commercially viable. As we may see from the review, the pre-combustion extraction of metals from petcoke, electrochemical extraction and exotic methods are very far from commercial application and are unlikely to develop into real technologies anytime soon.

Funding: This research received no external funding.

Institutional Review Board Statement: Not applicable.

Informed Consent Statement: Not applicable.

Data Availability Statement: Not applicable.

Conflicts of Interest: The authors declare no conflict of interest.

References

1. Luz, A.L.; Wu, X.; Tokar, E.J. Chapter One—Toxicology of Inorganic Carcinogens. In *Advances in Molecular Toxicology*; Fishbein, J.C., Heilman, J.M., Eds.; Elsevier: Amsterdam, The Netherlands, 2018; Volume 12, pp. 1–46.
2. Raja, B.V.R. Vanadium market in the world. *Steelworld* **2007**, *13*, 19–22.
3. FacrMR. Vanadium Market Growth Outlook (2023–2033). Available online: https://www.factmr.com/report/vanadium-market (accessed on 20 January 2023).
4. Samora, R.; Bell, A. Vale Sees 44% Increase in Global Nickel Demand by 2030. Available online: https://www.reuters.com/markets/commodities/vale-sees-44-increase-global-nickel-demand-by-2030-2022-09-07/ (accessed on 20 January 2023).
5. Kornienko, V.; Avtonomov, P. Application of Neutron Activation Analysis for Heavy Oil Production Control. *Procedia Soc. Behav. Sci.* **2015**, *195*, 2451–2456. [CrossRef]
6. Magomedov, R.; Popova, A.; Maryutina, T.; Kadiev, K.M.; Khadzhiev, S. Current status and prospects of demetallization of heavy petroleum feedstock. *Pet. Chem.* **2015**, *55*, 423–443. [CrossRef]
7. Kurniawan, Y.S. The Origin, Physicochemical Properties, and Removal Technology of Metallic Porphyrins from Crude Oils. *Indonesian J. Chem.* **2021**, *21*, 1023–1038.
8. Liu, S.; Xue, W.; Wang, L. Extraction of the Rare Element Vanadium from Vanadium-Containing Materials by Chlorination Method: A Critical Review. *Metals* **2021**, *11*, 1301.
9. Akcil, A.; Vegliò, F.; Ferella, F.; Okudan, M.D.; Tuncuk, A. A review of metal recovery from spent petroleum catalysts and ash. *Waste Manag.* **2015**, *45*, 420–433. [CrossRef]
10. Baritto, M.; Oni, A.O.; Kumar, A. The development of a techno-economic model for the assessment of vanadium recovery from bitumen upgrading spent catalyst. *J. Clean. Prod.* **2022**, *363*, 132376. [CrossRef]
11. Pathak, A.; Kothari, R.; Vinoba, M.; Habibi, N.; Tyagi, V.V. Fungal bioleaching of metals from refinery spent catalysts: A critical review of current research, challenges, and future directions. *J. Environ. Manag.* **2021**, *280*, 111789. [CrossRef]
12. Kukes, S.G.; Battiste, D.R. Demetallization of Heavy Oils with Phosphorous Acid. U.S. Patent 6,655,113, 11 June 1985.
13. Taghvaie Nakhjiri, A.; Sanaeepur, H.; Ebadi Amooghin, A.; Shirazi, M.M.A. Recovery of precious metals from industrial wastewater towards resource recovery and environmental sustainability: A critical review. *Desalination* **2022**, *527*, 115510. [CrossRef]
14. Zheng, F.; Shi, Q.; Vallverdu, G.S.; Giusti, P.; Bouyssiere, B. Fractionation and Characterization of Petroleum Asphaltene: Focus on Metalopetroleomics. *Processes* **2020**, *8*, 1504. [CrossRef]
15. Yakubov, M.R.; Milordov, D.V.; Yakubova, S.G.; Borisov, D.N.; Ivanov, V.T.; Sinyashin, K.O. Concentrations of vanadium and nickel and their ratio in heavy oil asphaltenes. *Pet. Chem.* **2016**, *56*, 16–20. [CrossRef]
16. Amorim, F.A.C.; Welz, B.; Costa, A.C.S.; Lepri, F.G.; Vale, M.G.R.; Ferreira, S.L.C. Determination of vanadium in petroleum and petroleum products using atomic spectrometric techniques. *Talanta* **2007**, *72*, 349–359. [CrossRef]
17. Yakubov, M.R. Composition and Properties of Asphaltenes from Heavy Oils with High V and Ni Content. Ph.D. Thesis, Kazan Research Center RAS, Ufa, Russia, 2019. [CrossRef]
18. Sugihara, J.M.; Okada, T.; Branthaver, J.F. Reductive Desulfuration on Vanadium and Metalloporphyrin Contents of Fractions from Boscan Asphaltenes. *J. Chem. Eng. Data* **1965**, *10*, 190–194. [CrossRef]
19. Lanese, J.G.; Wilson, G.S. Electrochemical studies of zinc tetraphenylporphin. *J. Electrochem. Soc.* **1972**, *119*, 1039–1043. [CrossRef]
20. Khanova, L.A.; Lafi, L.F. Adsorption characteristics of tetraphenylporphyrins on an amalgamated platinum electrode. *J. Electroanal. Chem.* **1993**, *345*, 393–411. [CrossRef]
21. Ovalles, C.; Rojas, I.; Acevedo, S.; Escobar, G.; Jorge, G.; Gutierrez, L.B.; Rincon, A.; Scharifker, B. Upgrading of Orinoco Belt crude oil and its fractions by an electrochemical system in the presence of protonating agents. *Fuel Process. Technol.* **1996**, *48*, 159–172. [CrossRef]
22. Jorge, G.A.; Garcia, E.; Scott, C.E. Electrolysis of V(IV) and free meso-tetraphenyl porphyrin in two immiscible liquids. *J. Appl. Electrochem.* **2002**, *32*, 569–572. [CrossRef]
23. Welter, K.; Salazar, E.; Balladores, Y.; Marquez, O.P.; Marquez, J.; Martinez, Y. Electrochemical removal of metals from crude oil samples. *Fuel Process. Technol.* **2009**, *90*, 212–221. [CrossRef]
24. Kurbanova, A.N.; Akhmetov, N.K.; Yeshmuratov, A.; Zulkharnay, R.N.; Sugurbekov, Y.T.; Demeuova, G.; Baisariyev, M.; Sugurbekova, G.K. Removal of nickel and vanadium from crude oil by using solvent extraction and electrochemical process. *Phys. Sci. Technol.* **2017**, *4*, 74–80. [CrossRef]
25. Afanasjeva, N.; Lizcano-Valbuena, W.H.; Aristizabal, N.; Mañozca, I. V and Ni electrochemical deposition from asphaltenes in heavy oils. *Ing. Compet.* **2015**, *17*, 9–17. [CrossRef]
26. Farag, A.S.; Sif El-Din, O.I.; Youssef, M.H.; Hassan, S.I.; Farmawy, S. Solvent demetalization of heavy oil residue. *Hung. J. Ind. Chem.* **1989**, *17*, 289–294.

27. Billica, H.; Adkins, H. Cataylst, Raney Nickel, W6 (with high contents of aluminum and adsorbed hydrogen). *Org. Synth.* **1949**, *29*, 24. [CrossRef]
28. Acevedo, D.; Camacho, L.F.E.; Moncada, J.; Puentes, Z. Electrochemically assisted demetallisation of model metalloporphyrins and crude oil porphyrinic extracts in emulsified media, by using active permeated atomic hydrogen. *Fuel* **2012**, *92*, 264–270. [CrossRef]
29. Khandelwal, A.; Singh, S.B.; Sharma, A.; Nain, L.; Varghese, E.; Singh, N. Effect of surfactant on degradation of *Aspergillus* sp. and *Trichoderma* sp. mediated crude oil. *Int. J. Environ. Anal. Chem.* **2021**, *103*, 1667–1680. [CrossRef]
30. Salehizadeh, H.; Mousavi, M.; Hatamipour, S.; Kermanshahi, K. Microbial Demetallization of Crude Oil Using *Aspergillus* sp.: Vanadium Oxide Octaethyl Porphyrin (VOOEP) as a Model of Metallic Petroporphyrins. *Iran. J. Biotechnol.* **2007**, *5*, 31.
31. Dedeles, G.R.; Abe, A.; Saito, K.; Asano, K.; Saito, K.; Yokota, A.; Tomita, F. Microbial demetallization of crude oil: Nickel protoporphyrin disodium as a model organo-metallic substrate. *J. Biosci. Bioeng.* **2000**, *90*, 515–521. [CrossRef]
32. Shiraishi, Y.; Hirai, T.; Komasawa, I. A novel demetalation process for vanadyl- and nickelporphyrins from petroleum residue by photochemical reaction and liquid-liquid extraction. *Ind. Eng. Chem. Res.* **2000**, *39*, 1345–1355. [CrossRef]
33. Lebedev, Y.A.; Averin, K.A. Extraction of valuable metals by microwave discharge in crude oil. *J. Phys. D-Appl. Phys.* **2018**, *51*, 5. [CrossRef]
34. Garifzyanova, G.G.; Garifzyanov, G.G. Pyrolysis of vacuum resid by the plasma chemical method. *Chem. Tech. Fuels Oils* **2006**, *42*, 172–175. [CrossRef]
35. Trutnev, J.A.; Mufazalov, R.S.; Mukhtorov, N.J.; Zaripov, R.K. Method and Apparatus for Demetallization of Crude Oil. RU Patent 2133766C1, 27 July 1999.
36. Fan, X. The Fates of Vanadium and Sulfur Introduced with Petcoke to Lime Kilns. Ph.D. Thesis, University of Toronto, Toronto, ON, Canada, 2010.
37. Abdolahnezhad, M. Metal Leaching from Oil Sands Fluid Petroleum Coke under Different Geochemical Conditions. Master's Thesis, University of Saskatchewan, Saskatoon, SK, Canada, 2020.
38. Lukoil. Coke. Available online: https://lukoil.ru/Products/business/petroleumproducts/coke (accessed on 20 January 2023).
39. ExxonMobile. Resid Conversion: Flexicoking. Available online: https://www.exxonmobilchemical.com/en/catalysts-and-technology-licensing/resid-conversion (accessed on 20 January 2023).
40. Cai, Z.; Zhang, Y.; Liu, T.; Huang, J. Mechanisms of vanadium recovery from stone coal by novel $BaCO_3$/CaO composite additive roasting and acid leaching technology. *Minerals* **2016**, *6*, 26. [CrossRef]
41. Mishra, B.; Jung, M. Recovery and Recycling of Valuable Metals from Fine Industrial Waste Materials. *Int. J. Soc. Mater. Eng. Resour.* **2018**, *23*, 105–108. [CrossRef]
42. Szymczycha-Madeja, A. Kinetics of Mo, Ni, V and Al leaching from a spent hydrodesulphurization catalyst in a solution containing oxalic acid and hydrogen peroxide. *J. Hazard. Mater.* **2011**, *186*, 2157–2161. [CrossRef]
43. Kinoshita, T.; Yamaguchi, K.; Akita, S.; Nii, S.; Kawaizumi, F.; Takahashi, K. Hydrometallurgical recovery of zinc from ashes of automobile tire wastes. *Chemosphere* **2005**, *59*, 1105–1111. [CrossRef]
44. Hepworth, M.; Slimane, R. Recovery of vanadium, nickel, and carbon from Orinoco crudes via flexicoking. *J. Solid Waste Techn. Manag.* **1997**, *24*, 104–112.
45. Alvarado, J.; Alvarez, M.; Cristiano, A.R.; Marcó, L. Extraction of vanadium from petroleum coke samples by means of microwave wet acid digestion. *Fuel* **1990**, *69*, 128–130. [CrossRef]
46. Sitnikova, G.Y.; Rastova, N.V.; Davydova, S.L. Role of thermal oxidation in the process of vanadium extraction from refinery coke (solid oil residues). *Petrol. Chem. USSR* **1990**, *30*, 172–175.
47. Ryumin, A.; Belonin, A.; Gribkov, V. Method of Extraction of Vanadium from Petroleum Coke (Способ Извлечения Ванадия из Нефтяного Кокса). RU Patent 2070940C1, 27 December 1996.
48. Rudko, V.A.; Kondrasheva, N.K.; Lukonin, R.E. Influence of acid leaching parameters on the extraction of vanadium from petroleum coke. *Bullet. S. Petersb. Technol. Inst.* **2018**, *42*, 43–48.
49. Kondrasheva, N.K.; Rudko, V.A.; Lukonin, R.E.; Derkunskii, I.O. The influence of leaching parameters on the extraction of vanadium from petroleum coke. *Pet. Sci. Technol.* **2019**, *37*, 1455–1462. [CrossRef]
50. Zhang, Y.; Yang, L. Alkali leaching of vanadium from petroleum coke and kinetics analysis. *Int. J. Environ. Eng.* **2015**, *7*, 90–100. [CrossRef]
51. McCorriston, L.L. Process Using Sulphate Reagent for Recovering Vanadium from Cokes Derived from Heavy Oils. U.S. Patent 4,389,378, 21 June 1983.
52. Shlewit, H.; Alibrahim, M. Extraction of sulfur and vanadium from petroleum coke by means of salt-roasting treatment. *Fuel* **2006**, *85*, 878–880. [CrossRef]
53. Visaliev, M.Y.; Shpirt, M.Y.; Kadiev, K.M.; Dvorkin, V.I.; Magomadov, E.E.; Khadzhiev, S.N. Integrated conversion of extra-heavy crude oil and petroleum residue with the recovery of vanadium, nickel, and molybdenum. *Solid Fuel Chem.* **2012**, *46*, 100–107. [CrossRef]
54. Shpirt, M.; Nukenov, D.; Punanova, S.; Visaliev, M. Principles of the Production of Valuable Metal Compounds from Fossil Fuels. *Solid Fuel Chem.* **2013**, *47*, 71–82. [CrossRef]
55. Murthy, B.N.; Sawarkar, A.; Deshmukh, N.; Mathew, T.; Joshi, J. Petroleum coke gasification: A review. *Can. J. Chem. Eng.* **2014**, *92*, 441–468. [CrossRef]

56. Abdrabo, A.E.; Husein, M.M. Method for Converting Demetallization Products into Dispersed Metal Oxide Nanoparticles in Heavy Oil. *Energy Fuels* **2012**, *26*, 810–815. [CrossRef]
57. Zhan, G.W.; Ng, W.C.; Lin, W.Y.; Koh, S.N.; Wang, C.H. Effective Recovery of Vanadium from Oil Refinery Waste into Vanadium-Based Metal-Organic Frameworks. *Environ. Sci. Technol.* **2018**, *52*, 3008–3015. [CrossRef]
58. Griffin, P.J. Extraction of Vanadium and Nickel from Athabasca Oil Sands Fly Ash. Ph.D. Thesis, University of Alberta, Edmonton, AB, Canada, 1981.
59. Valeev, D.; Kondratiev, A. Current State of Coal Fly Ash Utilization: Characterization and Application. *Materials* **2023**, *16*, 27. [CrossRef]
60. Basha, S.I.; Aziz, A.; Maslehuddin, M.; Ahmad, S.; Hakeem, A.S.; Rahman, M.M. Characterization, Processing, and Application of Heavy Fuel Oil Ash, an Industrial Waste Material—A Review. *Chem. Rec.* **2020**, *20*, 1568–1595. [CrossRef]
61. Qaidi, S.; Najm, H.M.; Abed, S.M.; Ahmed, H.U.; Al Dughaishi, H.; Al Lawati, J.; Sabri, M.M.; Alkhatib, F.; Milad, A. Fly ash-based geopolymer composites: A review of the compressive strength and microstructure analysis. *Materials* **2022**, *15*, 7098. [CrossRef]
62. Kanhar, A.H.; Chen, S.; Wang, F. Incineration Fly Ash and Its Treatment to Possible Utilization: A Review. *Energies* **2020**, *13*, 6681. [CrossRef]
63. Herath, C.; Gunasekara, C.; Law, D.W.; Setunge, S. Performance of high volume fly ash concrete incorporating additives: A systematic literature review. *Constr. Build. Mater.* **2020**, *258*, 120606. [CrossRef]
64. UnionProcess. Coal, Coke & Fly Ash Applications for Particle Size Reduction Equipment. Available online: https://www.unionprocess.com/industries-applications/energy/coal-coke-fly-ash/ (accessed on 20 January 2023).
65. Mofarrah, A.; Husain, T.; Danish, E.Y. Investigation of the Potential Use of Heavy Oil Fly Ash as Stabilized Fill Material for Construction. *J. Mater. Civ. Eng.* **2012**, *24*, 684–690. [CrossRef]
66. Sett, P. Flyash: Characteristics, Problems and Possible Utilization. *Adv. Appl. Sci. Res.* **2017**, *8*, 32–50.
67. Huffman, G.P.; Huggins, F.E.; Shah, N.; Huggins, R.; Linak, W.P.; Miller, C.A.; Pugmire, R.J.; Meuzelaar, H.L.C.; Seehra, M.S.; Manivannan, A. Characterization of fine particulate matter produced by combustion of residual fuel oil. *J. Air Waste Manag. Assoc.* **2000**, *50*, 1106–1114. [CrossRef]
68. Al-Degs, Y.S.; Ghrir, A.; Khoury, H.; Walker, G.M.; Sunjuk, M.; Al-Ghouti, M.A. Characterization and utilization of fly ash of heavy fuel oil generated in power stations. *Fuel Process. Technol.* **2014**, *123*, 41–46. [CrossRef]
69. Noel, H.M. Recovery of Vanadium. U.S. Patent 2,372,109, 20 March 1945.
70. Jack, T.R.; Sullivan, E.A.; Zajic, J.E. Leaching of vanadium and other metals from Athabasca Oil Sands coke and coke ash. *Fuel* **1979**, *58*, 589–594. [CrossRef]
71. Gardner, H.E. Recovery of Vanadium and Nickel from Petroleum Residues. U.S. Patent 4816236A, 28 March 1989.
72. Junior_Gold_Report. MGX Minerals Partners with Highbury Energy to Extract Nickel, Vanadium, and Cobalt from Petroleum Coke. Available online: https://juniorgoldreport.com/mgx-minerals-partners-with-highbury-energy-to-extract-nickel-vanadium-and-cobalt-from-petroleum-coke/ (accessed on 20 January 2023).
73. Kikauka, A. *MGX Minerals and Highbury Energy Produce 45% Vanadium Concentrate from Petroleum Coke Ash*; MGX Minerals: Vancouver, BC, Canada, 2017.
74. Saeed, A.; Banoqitah, E.; Alaqab, A.; Alshahrie, A.; Saleh, A.; Alhawsawi, A.; Damoom, M.; Salah, N. Epoxy Flooring/Oil Fly Ash as a Multifunctional Composite with Enhanced Mechanical Performance for Neutron Shielding and Chemical-Resistance Applications. *ACS Omega* **2022**, *8*, 747–760. [CrossRef]
75. Kwon, W.-T.; Kim, D.-H.; Kim, Y.-P. Characterization of heavy oil fly ash generated from a power plant. *Adv. Technol. Mater. Mater. Process. J.* **2004**, *6*, 260–263.
76. Brunauer, S.; Emmett, P.H.; Teller, E. Adsorption of gases in multimolecular layers. *J. Am. Chem. Soc.* **1938**, *60*, 309–319. [CrossRef]
77. Kamil, F.H.; Abood, W.M.; Mohammed, Y.; Khedair, Z. Heavy Oil Fly Ash as an Adsorbent for Methylene Blue Removal from Aqueous Solution. *Iraqi J. Ind. Res.* **2022**, *9*, 48–56. [CrossRef]
78. Dalhat, M.A.; Al-Abdul Wahhab, H.I. Sulfur extended heavy oil fly ash and cement waste asphalt mastic for roofing and waterproofing. *Mater. Struct.* **2015**, *48*, 205–216. [CrossRef]
79. Akita, S.; Maeda, T.; Takeuchi, H. Recovery of vanadium and nickel in fly ash from heavy oil. *J. Chem. Technol. Biotechnol. Int. Res. Process Environ. Clean Technol.* **1995**, *62*, 345–350. [CrossRef]
80. Vitolo, S.; Seggiani, M.; Filippi, S.; Brocchini, C. Recovery of vanadium from heavy oil and Orimulsion fly ashes. *Hydrometallurgy* **2000**, *57*, 141–149. [CrossRef]
81. Al-Ghouti, M.A.; Al-Degs, Y.S.; Ghrair, A.; Khoury, H.; Ziedan, M. Extraction and separation of vanadium and nickel from fly ash produced in heavy fuel power plants. *Chem. Eng. J.* **2011**, *173*, 191–197. [CrossRef]
82. Volkov, A.; Kologrieva, U.; Stulov, P. Study of Forms of Compounds of Vanadium and Other Elements in Samples of Pyrometallurgical Enrichment of Ash from Burning Oil Combustion at Thermal Power Plants. *Materials* **2022**, *15*, 8596. [CrossRef]
83. Bakirova, S.; Kotova, A.; Yag'yaeva, S.; Fedorova, N.; Nadirov, N. Structural features of vanadyl porphyrins of petroleum of West Kazakhstan. *Pet. Chem. USSR* **1984**, *24*, 733–738. [CrossRef]
84. Linak, W.P.; Miller, C.A. Comparison of particle size distributions and elemental partitioning from the combustion of pulverized coal and residual fuel oil. *J. Air Waste Manag. Assoc.* **2000**, *50*, 1532–1544. [CrossRef]
85. Amer, A.M. Processing of Egyptian boiler-ash for extraction of vanadium and nickel. *Waste Manag.* **2002**, *22*, 515–520. [CrossRef]
86. Whigham, W. New in extraction: Vanadium from petroleum. *Chem. Eng.* **1965**, *72*, 64.

87. Park, K. Recovery of vanadium and nickel from heavy oil fly ash. In Proceedings of the 2nd International Symposium on East-Asian Resources Recycling Technology, Seoul, Republic of Korea, 30 June 1993; p. 211.
88. Tsuboi, I.; Kasai, S.; Kunugita, E.; Komasawa, I. Recovery of Gallium and Vanadium from Coal Fly Ash. *J. Chem. Eng. Jpn.* **1991**, *24*, 15–20. [CrossRef]
89. Tsuboi, I.; Kasai, S.; Yamamoto, T.; Komasawa, I.; Kunugita, E. Recovery of Rare Metals from Coal Fly Ash. In *Process Metallurgy*; Sekine, T., Ed.; Elsevier: Amsterdam, The Netherlands, 1992; Volume 7, pp. 1199–1204.
90. Chmielewski, A.G.; Urbański, T.S.; Migdał, W. Separation technologies for metals recovery from industrial wastes. *Hydrometallurgy* **1997**, *45*, 333–344. [CrossRef]
91. Edwards, C.R. The recovery of metal values from process residues. *JOM* **1991**, *43*, 32–33. [CrossRef]
92. Lakshmanan, V.I.; McQueen, N. Recovery of Vanadium from Suncor Flyash—Flowsheet Development. *Proc. Second. Int. Conf. Sep. Sci. Technol.* **1989**, *2*, 525–531.
93. Aydin, A. Recovery of vanadium compounds from the slags and fly ashes of power plants. *Chim. Acta Turc.* **1988**, *16*, 153–182.
94. Akaboshi, T.; Kaneko, N.; Sakuma, A.; Sugiyama, T. Recovery of ammonium metavanadate from petroleum-combustion residues. *Jpn. Kokai Tokkio Koho JP* **1987**, *62*, 298.
95. Tsygankova, M.V.; Bukin, V.I.; Lysakova, E.I.; Smirnova, A.G.; Reznik, A.M. The recovery of vanadium from ash obtained during the combustion of fuel oil at thermal power stations. *Russ. J. Non Ferr. Met.* **2011**, *52*, 19–23. [CrossRef]
96. Murase, K.; Nishikawa, K.-i.; Ozaki, T.; Machida, K.-i.; Adachi, G.-y.; Suda, T. Recovery of vanadium, nickel and magnesium from a fly ash of bitumen-in-water emulsion by chlorination and chemical transport. *J. Alloys Compd.* **1998**, *264*, 151–156. [CrossRef]
97. Tokuyama, H.; Nii, S.; Kawaizumi, F.; Takahashi, K. Process development for recovery of vanadium and nickel from heavy oil fly ash by leaching and ion exchange. *Sep. Sci. Technol.* **2003**, *38*, 1329–1344. [CrossRef]
98. Meer, I.; Nazir, R. Removal techniques for heavy metals from fly ash. *J. Mater. Cycles Waste Manag.* **2018**, *20*, 703–722. [CrossRef]
99. Rahimi, G.; Rastegar, S.; Chianeh, F.R.; Gu, T. Ultrasound-assisted leaching of vanadium from fly ash using lemon juice organic acids. *RSC Adv.* **2020**, *10*, 1685–1696. [CrossRef]
100. Liu, J. Recovery of Vanadium and Nickel from Oil Fly Ash. Ph.D. Thesis, Memorial University of Newfoundland, St. John's, NL, Canada, 2017.
101. Tsai, S.L.; Tsai, M.S. A study of the extraction of vanadium and nickel in oil-fired fly ash. *Resour. Conserv. Recycl.* **1998**, *22*, 163–176. [CrossRef]
102. Parvizi, R.; Khaki, J.V.; Moayed, M.H.; Ardani, M.R. Hydrometallurgical Extraction of Vanadium from Mechanically Milled Oil-Fired Fly Ash: Analytical Process Optimization by Using Taguchi Design Method. *Metall. Mater. Trans. B Process Metall. Mater. Process. Sci.* **2012**, *43*, 1269–1276. [CrossRef]
103. Navarro, R.; Guzman, J.; Saucedo, I.; Revilla, J.; Guibal, E. Vanadium recovery from oil fly ash by leaching, precipitation and solvent extraction processes. *Waste Manag.* **2007**, *27*, 425–438. [CrossRef] [PubMed]
104. Gomez-Bueno, C.; Spink, D.; Rempel, G. Extraction of vanadium from Athabasca tar sands fly ash. *Metall. Trans. B* **1981**, *12*, 341–352. [CrossRef]
105. Holloway, P.C.; Etsell, T. Salt roasting of suncor oil sands fly ash. *Metall. Mater. Trans. B* **2004**, *35*, 1051–1058. [CrossRef]
106. Holloway, P.; Etsell, T. Process for the complete utilization of oil sands fly ash. *Can. Metall. Q.* **2006**, *45*, 25–32. [CrossRef]
107. Goncharov, K.; Kashekov, D.Y.; Sadykhov, G.; Olyunina, T. Processing of fuel oil ash from thermal power plant with extraction of vanadium and nickel. *Non Ferr. Metals* **2020**, *1*, 3–7. [CrossRef]
108. Vassileva, C.; Vassilev, S.; Daher, D. Preliminary results on chemical and phase-mineral composition of Syrian petroleum coke and ash. *Comptes Rendus L'acad. Bulg. Sci.* **2010**, *63*, 129–136.
109. Guillaud, P. Process for Treatment of Vanadium Containing Fly Ash. U.S. Patent 3873669A, 3 March 1975.
110. Kudinova, A.; Kondrasheva, N.; Rudko, V. Influence of leaching parameters on the vanadium extraction from petroleum coke. *E3S Web Conf.* **2021**, *266*, 08002. [CrossRef]
111. Rezaei, H.; Shafaei, S.Z.; Abdollahi, H.; Shahidi, A.; Ghassa, S. A sustainable method for germanium, vanadium and lithium extraction from coal fly ash: Sodium salts roasting and organic acids leaching. *Fuel* **2022**, *312*, 122844. [CrossRef]
112. Ziyadanoğullari, R.; Aydın, I. Recovery of uranium, nickel, molybdenum, and vanadium from floated asphaltite ash. *Sep. Sci. Technol.* **2004**, *39*, 3113–3125. [CrossRef]
113. Jia, L.; Anthony, E.J.; Charland, J.P. Investigation of vanadium compounds in ashes from a CFBC firing 100% petroleum coke. *Energy Fuels* **2002**, *16*, 397–403. [CrossRef]
114. Jung, M.; Mishra, B. Vanadium Recovery from Oil Fly Ash by Carbon Removal and Roast-Leach Process. *Jom* **2018**, *70*, 168–172. [CrossRef]
115. Kadhim, M.J.; Hafiz, M.H.; Abbas, R.A. Optimization for leaching kinetics of vanadium pentoxide from thermal power plants fly ash using Taguchi approach. *Dijlah J.* **2019**, *2*.
116. Etsell, T.H.; Griffin, P.J. Vanadium Recovery from Ash from Oil Sands. CA Patent 1221243A, 5 May 1987.
117. Kologrieva, U.; Volkov, A.; Zinoveev, D.; Krasnyanskaya, I.; Stulov, P.; Wainstein, D. Investigation of Vanadium-Containing Sludge Oxidation Roasting Process for Vanadium Extraction. *Metals* **2021**, *11*, 100. [CrossRef]
118. Queneau, P.B.; Hogsett, R.F.; Beckstead, L.W.; Barchers, D.E. Processing of petroleum coke for recovery of vanadium and nickel. *Hydrometallurgy* **1989**, *22*, 3–24. [CrossRef]
119. Abdel-latif, M.A. Recovery of vanadium and nickel from petroleum flyash. *Miner. Eng.* **2002**, *15*, 953–961. [CrossRef]

120. Xiao, Y.; Mambote, C.R.; Jalkanen, H.; Yang, Y.; Boom, R. Vanadium recovery as FeV from petroleum fly ash. In Proceedings of the Twelfth International Ferroalloys Congress Sustainable Future, Helsinki, Finland, 6–10 June 2010; pp. 6–9.
121. Howard, R.; Richards, S.; Welch, B.; Moore, J. Pyrometallurgical processing of vanadiferous slag using plasma/induction heating. *INFACON* **1992**, *1*, 225–231.
122. Sun, S.; Yang, K.; Liu, C.; Tu, G.; Xiao, F. Recovery of nickel and preparation of ferronickel alloy from spent petroleum catalyst via cooperative smelting-vitrification process with coal fly ash. *Environ. Technol.* **2021**, *42*, 1–11. [CrossRef]
123. Johnson, T.P. Recovery of Nickel and Vanadium from Heavy Oil Residues Using DC Plasma Smelting. In *Extraction 2018*; Springer: Berlin/Heidelberg, Germany, 2018; pp. 1017–1028.
124. Seddiek, H.A.; Shetaia, Y.M.; Mahamound, K.F.; El-Aassy, I.E.; Hussien, S.S. Bioleaching of Egyptian Fly Ash Using Cladosporium cladosporioides. *Ann. Biol.* **2021**, *37*, 18–22.
125. Li, J.Z.; Wang, X.Y.; Wang, B.; Zhao, J.T.; Fang, Y.T. Investigation on the fates of vanadium and nickel during co-gasification of petroleum coke with biomass. *Bioresour. Technol.* **2018**, *257*, 47–53. [CrossRef] [PubMed]
126. Rasoulnia, P.; Mousavi, S.M. V and Ni recovery from a vanadium-rich power plant residual ash using acid producing fungi: Aspergillus niger and Penicillium simplicissimum. *RSC Adv.* **2016**, *6*, 9139–9151. [CrossRef]
127. Mirazimi, S.M.J.; Abbasalipour, Z.; Rashchi, F. Vanadium removal from LD converter slag using bacteria and fungi. *J. Environ. Manag.* **2015**, *153*, 144–151. [CrossRef] [PubMed]
128. Krebs, W.; Bachofen, R.; Brandl, H. Growth stimulation of sulfur oxidizing bacteria for optimization of metal leaching efficiency of fly ash from municipal solid waste incineration. *Hydrometallurgy* **2001**, *59*, 283–290. [CrossRef]
129. Wu, H.Y.; Ting, Y.P. Metal extraction from municipal solid waste (MSW) incinerator fly ash—Chemical leaching and fungal bioleaching. *Enzym. Microb. Technol.* **2006**, *38*, 839–847. [CrossRef]
130. Yang, J.; Wang, Q.H.; Wang, Q.; Wu, T.J. Heavy metals extraction from municipal solid waste incineration fly ash using adapted metal tolerant Aspergillus niger. *Bioresour. Technol.* **2009**, *100*, 254–260. [CrossRef]
131. Rasoulnia, P.; Barthen, R.; Lakaniemi, A.M. A critical review of bioleaching of rare earth elements: The mechanisms and effect of process parameters. *Crit. Rev. Environ. Sci. Technol.* **2020**, *51*, 378–427. [CrossRef]
132. Mirazimi, S.M.J.; Rashchi, F.; Saba, M. A new approach for direct leaching of vanadium from LD converter slag. *Chem. Eng. Res. Des.* **2015**, *94*, 131–140. [CrossRef]
133. Lee, J.-C.; Kurniawan; Kim, E.-Y.; Chung, K.W.; Kim, R.; Jeon, H.-S. A review on the metallurgical recycling of vanadium from slags: Towards a sustainable vanadium production. *J. Mater. Res. Technol.* **2021**, *12*, 343–364. [CrossRef]
134. An, Y.; Ma, B.; Li, X.; Chen, Y.; Wang, C.; Wang, B.; Gao, M.; Feng, G. A review on the roasting-assisted leaching and recovery of V from vanadium slag. *Process Saf. Environ. Prot.* **2023**, *173*, 263–276. [CrossRef]

Disclaimer/Publisher's Note: The statements, opinions and data contained in all publications are solely those of the individual author(s) and contributor(s) and not of MDPI and/or the editor(s). MDPI and/or the editor(s) disclaim responsibility for any injury to people or property resulting from any ideas, methods, instructions or products referred to in the content.

Article

Efficient Recovery of Cu from Wasted CPU Sockets by Slurry Electrolysis

Wenjing Chen [1], Manning Li [2] and Jiancheng Tang [1,3,*]

[1] School of Physics and Materials Science, Nanchang University, Nanchang 330031, China
[2] School of Advanced Manufacturing, Nanchang University, Nanchang 330031, China
[3] International Institute for Materials Innovation, Nanchang University, Nanchang 330031, China
[*] Correspondence: tangjiancheng@ncu.edu.cn; Tel.: +86-83969559

Abstract: In order to maximize the reuse of used electronic component resources, while reducing environmental pollution, Cu metal was recycled from wasted CPU sockets by the reformative slurry electrolysis method. However, the influences on the regulation of the Cu recovery rate and purity from waste CPU slots, by slurry electrolysis, has not been systematically elucidated in previous studies. In this work, the effects of H_2SO_4 concentration, slurry density, NH_4Cl concentration, current density, and reaction time, on the recovery rate and purity of Cu in slurry electrolysis, were researched by systematic experimental methods. The results showed that the recycled metal elements were mainly present as powders from the cathode, rather than in the electrolyte. Moreover, the metallic elements in the cathode powder consisted of mostly Cu and small amounts of Sn and Ni. The recovery rate and purity of Cu were up to 96.19% and 93.72%, respectively, with the optimum conditions being: an H_2SO_4 concentration of 2 mol/L, slurry density of 30 g/L, NH_4Cl concentration of 90 g/L, current density of 80 mA/cm^2, and reaction time of 7 h. Compared with previous studies, the Cu recovered in this experiment was present in the cathode powder, which was more convenient for the subsequent processing. Meanwhile, the recovery rate of Cu was effectively improved. This is an important guideline for the subsequent application of slurry electrolysis for Cu recovery.

Keywords: wasted CPU socket; slurry electrolysis; recovery rate of Cu; purity of Cu; current density; slurry density

Citation: Chen, W.; Li, M.; Tang, J. Efficient Recovery of Cu from Wasted CPU Sockets by Slurry Electrolysis. *Metals* **2023**, *13*, 643. https://doi.org/10.3390/met13040643

Academic Editor: Antoni Roca

Received: 13 February 2023
Revised: 5 March 2023
Accepted: 11 March 2023
Published: 23 March 2023

Copyright: © 2023 by the authors. Licensee MDPI, Basel, Switzerland. This article is an open access article distributed under the terms and conditions of the Creative Commons Attribution (CC BY) license (https://creativecommons.org/licenses/by/4.0/).

1. Introduction

Today, environmental protection is an urgent topic. With the rapid development of society, resources are rapidly being consumed and a large amount of garbage is generated, most of which is not fully used, or could be reused [1,2]. Among all waste, the growth rate of e-waste has accelerated with the frequency of technological updates [3–5]. In 2016, the global output of e-waste was 44.7 million tonnes, while in 2019, the figure reached 53.6 million tons, with an average annual growth rate of 6.64 percent. Asia alone generated 24.9 million tonnes of e-waste. The global production of e-waste is expected to exceed 70 million tonnes by 2028 [6].

The composition of e-waste is complex, and up to 69 elements have been found in it, including various precious metals (gold, copper, platinum, and palladium) [7,8]. Besides that, there are also various pollutants (e.g. plastics) found in e-waste [9]. If not handled properly, it can cause serious harm to the environment and human health. At the same time, e-waste, also known as "misplaced resources", generates a lot of economic losses [10,11]. The potential value of the e-waste generated in 2019, was estimated to be 57 billion US dollars [6]. Research has shown, that the Cu content in the printed circuit board (PCB) of an ordinary notebook computer is about 20 wt.%, which is much higher than the global average copper grade [12,13].

Currently, the main recycling technologies for e-waste are mechanical-physical treatment technology, biometallurgical recovery technology, and new recycling processes, such as supercritical fluid recycling treatment technology and slurry electrolysis recycling treatment technology [14–16]. Slurry electrolysis is a technology that simultaneously performs leaching and electrodeposition. The reaction of metals in slurry electrolysis is divided into two parts: anode leaching and cathode reduction.

CPU sockets need to be pre-processed before slurry electrolysis. Zhao et al. [17] separated the valuable metals from wasted mobile phone PCB particles using the liquid–solid fluidization technique. Hanafi et al. [18] reported that a ball-milling machine was preferable to a disc-milling machine, due to its uniform pulverization, but ball-milling required a longer time than disc-milling. Yi et al. [19] cut CPU sockets first, and then shredded them into pieces, with diameters less than 2 mm, using a cutting mill.

Electrolysis experiments are responsible for the recovery of Cu from e-waste. Veit et al. [20] recovered metallic copper from wasted printed wiring boards by acid leaching, and then electrolytic deposition of the acid leaching solution was carried out, which eventually obtained metallic copper powder, with a purity of up to 98%. Subsequent researchers have further explored methods for the efficient recovery of Cu by changing different variables. Chu et al. [21] used electrolysis to recover copper powder from wasted PCBs and studied the effect of different factors on current efficiency and copper powder particles. Min et al. [22] developed a multi-oxidation coupling with electrolysis strategy, for purifying PCB wastewater and recovering copper. Guimarães et al. [23] found that electrolyte stirring and temperature increase favor the cathodic recovery of copper from PCB powder concentrates, by direct electroleaching. Byung et al. [24] simultaneously extracted precious metals such as gold, palladium, and platinum, from scrap (PCBs) and cellular auto catalysts, by smelting. Zhao et al. [25] developed an efficient process based on a hammer mill, pneumatic column separator, and electrostatic separator, for recovering copper from scrap PCBs. Zhang et al. [3] used the slurry electrolysis method to study the ultrafine copper powder obtained by adding different additives. Wang et al. [26] performed electrolysis in a centrifuge, to recover high purity copper powder from a polymetallic solution. The purity, current efficiency, and recovery of copper on the centrifuged electrode, were significantly improved due to enhanced mass transfer compared to the non-rotating electrode. Liu et al. [27] investigated the recovery and purification of Pd from waste multilayer ceramic capacitors (MLCCs) using electrodeposition, and proposed an efficient and environmentally friendly process for the recovery of waste MLCCs. Cocchiara et al. [28] explored the electrochemical recovery of Cu by cyclic voltammetry, the results showed that H_2SO_4–$CuSO_4$–NaCl could efficiently leach Cu from wasted PCBs, so that electronic components can be more easily disassembled in their undamaged state, allowing for effective recycling and valorization of base materials.

At present, the object of metal recycling is mostly wasted printed circuit boards (PCBs). The composition of PCBs is affected by the manufacturer, age, and origin, but Cu is always one of the most abundant metals in these materials [29]. CPU sockets act as important components, utilized to connect CPUs and motherboards. To reduce resistance, the content of Cu in CPU sockets is relatively high [30]. Although a wide range of studies have focused on the recovery of precious metals from e-waste, researchers have mainly focused on the extraction of metals in the electrolyte. This experiment investigates the extraction of metals at the cathode, as well as the enhancement of the Cu recovery rate and purity.

2. Materials and Methods

2.1. Materials

The reagents used in the study included nitric acid (HNO_3, 65–68%), sulfuric acid (H_2SO_4, 98%), hydrochloric acid (HCl, 36%), ammonium chloride (NH_4Cl), and hydrogen peroxide (H_2O_2, 30 wt.%). All chemicals for the experiment used in the present study were analytically pure and purchased from Shanghai Aladdin Biochemical Technology Co., Ltd., Shanghai, China. All necessary leachate solutions used in the study were prepared from the materials mentioned above, in deionized water.

2.2. Waste Central Processing Unit Socket

CPU sockets were the same model (LGA1366) obtained from wasted computers, as shown in Figure 1a. The samples of CPU sockets were manually shredded into approximately 10 × 10 mm^2 pieces and stored for further preparation, as shown in Figure 1b. To obtain the extremely tiny powder, the pieces were broken by a cutting mill (QE-300), as displayed in Figure 1c. To separate the metallic powder from non-metallic powder, the CPU socket powder was put in a centrifuge, which caused the layering of the metallic and non-metallic powders, and then the non-metallic powder was removed by mechanical vibration. After the sampling, the metallic powder was washed with ethanol and acetone, and finally dried at 70 °C for 2 h.

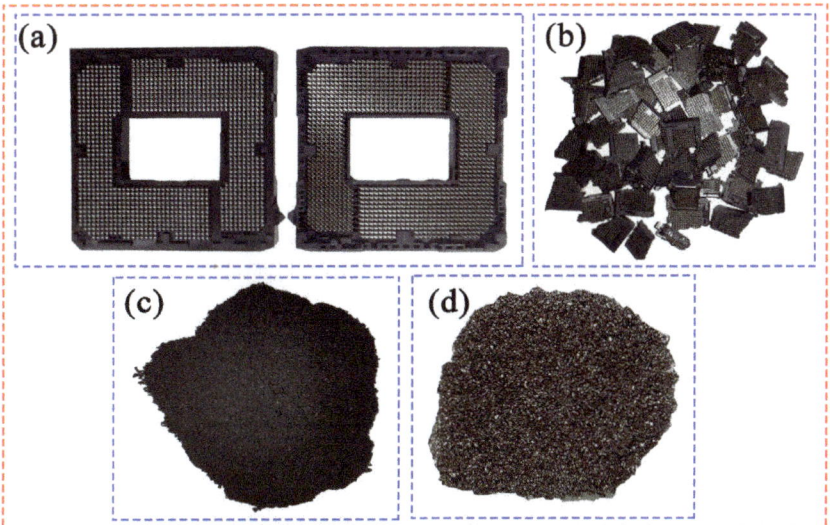

Figure 1. (**a**) CPU socket, (**b**) CPU socket pieces, (**c**) CPU socket powder, (**d**) metallic powder.

2.3. The Detection of Metallic Powder

X-ray diffraction (XRD) was used to detect the composition of metallic elements in the CPU sockets. To further probe the metal content in the CPU sockets, 0.1 g of collected CPU socket powder was put into a polytetrafluoroethylene (PTFE) crucible, and then 20 mL aqua regia (HNO$_3$:HCl = 1:3) was added. When the aqua regia was heated to 200 °C and then kept for 2 h, the lid of the crucible was opened to evaporate the solution to 1 mL. Finally, the remaining solution was fixed to 50 mL by adding deionized water, and stored. The leachate was analyzed by an inductively coupled plasma atomic emission spectrometer (ICP-AES, Optima 8000). Furthermore, the powder samples were also characterized via scanning electron microscopy (SEM, Quanta 200FEG), that was equipped with energy dispersive spectrometer mapping (EDS-Mapping).

2.4. Slurry Electrolysis Experiment

Figure 2a shows the instrument used in the slurry electrolysis experiment. The PTFE electrolysis cell was composed of a cathode part (7 × 6 × 4 cm^3) and anode part (7 × 6 × 6 cm^3). A graphite rod was utilized as the anode, while the cathode was titanium plate, and the electrodes were parallel to each other, with the distance kept constant. Electricity during the experiment was provided by a DC supplier (MS305DS, Dongguan Maihao Electronic Technology Co., Dongguan, China).

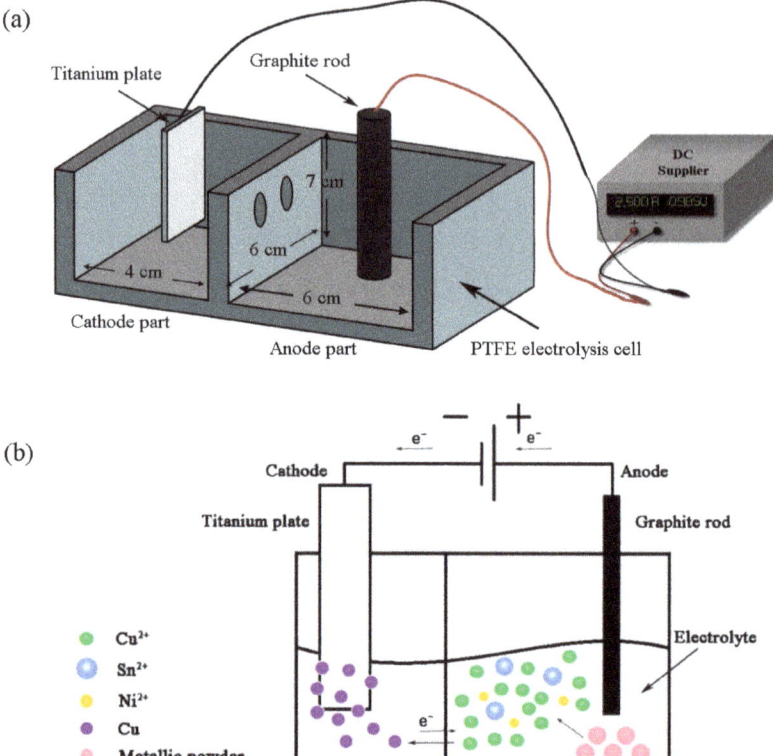

Figure 2. Schematic diagram of (**a**) slurry electrolysis experimental equipment, and (**b**) slurry electrolysis.

The CPU socket powder sample was put into the anode part, and 100 mL electrolyte, composed of 10 mL H_2O_2 (30 wt.%), 90 mL of different H_2SO_4 concentration solution, and varied weight of NH_4Cl (3–9 g) was added. Furthermore, the rotor of the magnetic stirrer was set at 300 rpm in the anode section. Finally, a DC supplier was connected to the graphite rod and titanium plate, and energized for the experiments. In the initial electrolysis experiments, without NH_4Cl, the current density was set to 80 mA/cm². This is the optimal current density for such electrolysis experiments, based on the study of Zhang et al. [3]. The metal powder in the anode was dissolved by the combined effect of the current and H_2SO_4.

To explore the influence of slurry density, the concentration of H_2SO_4 solution, NH_4Cl concentration, current density, and reaction time on the slurry electrolysis experiment, the slurry density was varied between 30, 50, and 70 g/L, the concentration of the H_2SO_4 solution was varied between 1, 2, 3, and 4 mol/L, the NH_4Cl concentration was varied between 30, 60, and 90 g/L, the current density was varied between 40, 80, and 120 mA/cm², and the reaction time was varied between 3, 5, and 7 h.

After each experiment, ICP-AES was applied, to analyze the metal concentration of the powders obtained from the cathode and anode, and the cathode powder was further characterized by SEM and EDS. The relative error of the EDS analysis varies with the elemental content; the smaller the content, the larger the relative error. The relative error is 2% when the content is >20 wt.%, 10% when the content is between 3 wt.%–20 wt.%, 30% when the content is between 1 wt.%–3 wt.%, and 50% when the content is < 1 wt.%. The whole experimental process is depicted in Figure 3.

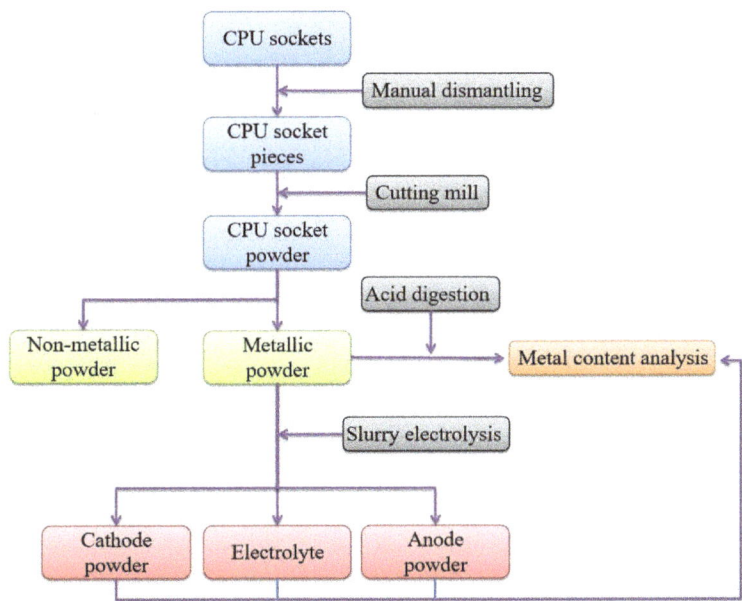

Figure 3. The procedure of the slurry electrolysis experiment.

2.5. Characterization

Recovery and separation of different metals from slurry electrolysis by electrodeposition are based on their different reduction potentials. The reactions in the cathode are shown by Equations (1)–(6) and the reactions in this experiment are shown in Figure 2b.

$$Cu^{2+} + 2e^- \rightarrow Cu \qquad E = 0.3419 \text{ V} \qquad (1)$$

$$Sn^{2+} + 2e^- \rightarrow Sn \qquad E = -0.1375 \text{ V} \qquad (2)$$

$$Ni^{2+} + 2e^- \rightarrow Ni \qquad E = -0.267 \text{ V} \qquad (3)$$

$$2H^+ + 2e^- \rightarrow H_2 \qquad E = 0 \text{ V} \qquad (4)$$

where E is the electrode potential, which is a value relative to the potential of the standard hydrogen electrode (SHE).

Metal recovery rate (r) and purity (P) are calculated by following Equations (5) and (6):

$$rij = \frac{mij}{M} \times 100\% \qquad (5)$$

where i is Cu, Sn, or Ni; j is the electrolyte and cathode; r_{ij} is the recovery rate of i in j (%); m_{ij} is the mass of i obtained in j (g); and M is the mass of metal contained in the waste CPU socket (g).

$$Pij = \frac{Mij}{Mj} \times 100\% \qquad (6)$$

where i is Cu, Sn, or Ni; j is the anode, electrolyte, and cathode; P_{ij} is the purity of metal of i in j (%); M_{ij} is the mass of i obtained in j (g); and M_j is the mass of metal obtained in j (g).

3. Results and Discussion

3.1. Characterization of the CPU Socket Powder Sample

As shown in Figure 4a, the main components of the CPU socket powder are pure metals, including Cu, Ni, and Sn, also, the peaks of Ag and Au were discovered.

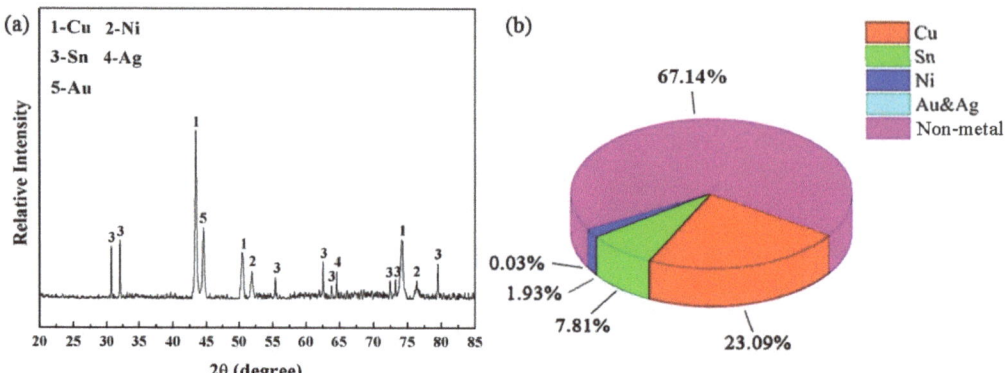

Figure 4. (**a**) XRD of the CPU socket metal powder, (**b**) compositions of the CPU socket powder.

The metal content of the metallic powder was detected by ICP-AES, which was consistent with the composition analysis by XRD. The amounts of different components in the CPU socket sample are summarized in Figure 4b.

Figure 5(a-1) shows an SEM image of the CPU socket powder fracture surface. It was detected, by EDS mapping, that the main elements of the CPU socket powder fracture surface were Cu and Sn. The results are displayed in Figure 5(a-2,a-3). Figure 5(b-1) shows an SEM image of the CPU socket powder surface, and Figure 5(b-2) shows the EDS mapping of the SEM image. It was indicated that only Ni was detected. In Figure 5(c-1), part 1 is the CPU socket powder surface and part 2 is the CPU socket powder fracture surface, Figure 5((c-2)–(c-4)) shows that the metal matrix is Cu and Sn, and Ni is plated on the surface as a coating.

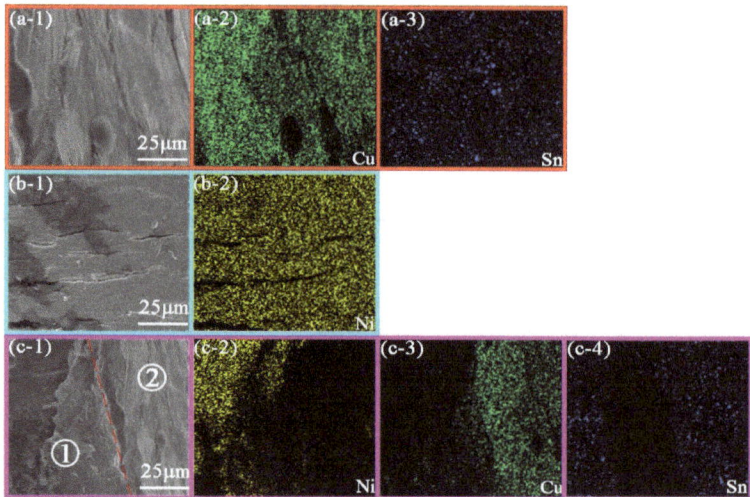

Figure 5. ((**a-1,b-1,c-1**)) SEM images of different parts of the CPU socket powder, (**a-2,a-3,b-2,c-2–c-4**) element mapping of Cu, Ni, and Sn for ((**a-1,b-1,c-1**)).

3.2. Characterization of Cathode Powder

A piece of metallic foil covering the titanium plate, and a large amount of powder adhering to the metallic foil, are shown in Figure 6a. They were obtained under the following experimental conditions: slurry concentration of 30 g/L, H_2SO_4 concentration

2 mol/L, current density 80 mA/cm^2, reaction time 7 h, and reaction temperature 15 °C. Based on the color of the metal, it can be deduced that the percentage of Cu in the metal was high. Metal foil was achieved after removing the powder from the titanium plate, and Figure 6b shows an SEM analysis of the side of the metal foil near the titanium plate. It is clear that a layer of dense metal, whose dominant component was detected to be 98.95% Cu (Figure 6c), is present. This is because the current efficiency of the hydrogen evolution reaction is not efficient enough to change the growth conditions of Cu. Figure 6d is the magnified image of area 1 in Figure 6b, it is clear that Cu grew in a dendritic form, which agrees with the data reported in [22].

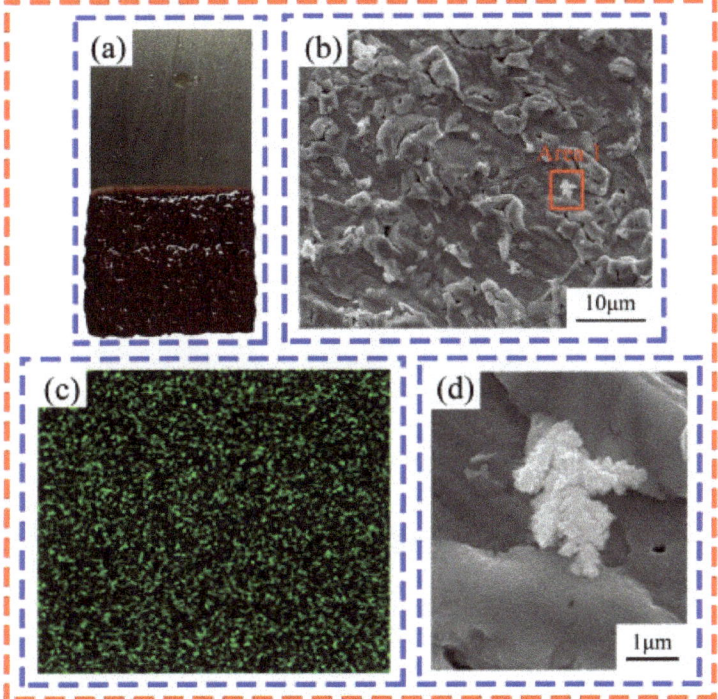

Figure 6. (**a**) Cathode powder, (**b**) SEM image of Cu foil, (**c**) element mapping of Cu of Figure 6b, and (**d**) the magnified image of area 1 in Figure 6b.

3.3. Effect of H_2SO_4 Concentration on Recovery Rate and Purity of Cu

Figure 7 demonstrates the effect of H_2SO_4 concentration on the purity and recovery rate of the metals. In this experiment, the increase in sulfuric acid concentration from 1 to 4 mol/L led to an increase in the recovery rate. The other conditions were: slurry concentration of 30 g/L, current density of 80 mA/cm^2, reaction time of 7 h, and reaction temperature of 15 °C. The recovery rates of Cu, Sn, and Ni in the cathode and electrolyte after the slurry electrolysis experiment are presented in Figure 7a,b. As shown in Figure 7a, the recovery rate of Cu in the cathode after the experiment increased from 65.10% to 77.77%, with the increase in H_2SO_4 from 1 to 4 mol/L. The recovery rates of Sn and Ni in the cathode increased with the H_2SO_4 concentration increasing from 3 mol/L to 4 mol/L, from 26.50% to 55.25% and 11.59% to 15.51%, respectively. Figure 7b indicates that the recovery rate of Cu in the electrolyte, after the experiment, remained at a fairly low level. At H_2SO_4 concentrations higher than 2 mol/L, the recovery rate of Sn in the electrolyte, after the experiment, rapidly increased, from 28.64% to 85.40%. The recovery rate of Ni in the electrolyte, after slurry electrolysis, remained in a certain range.

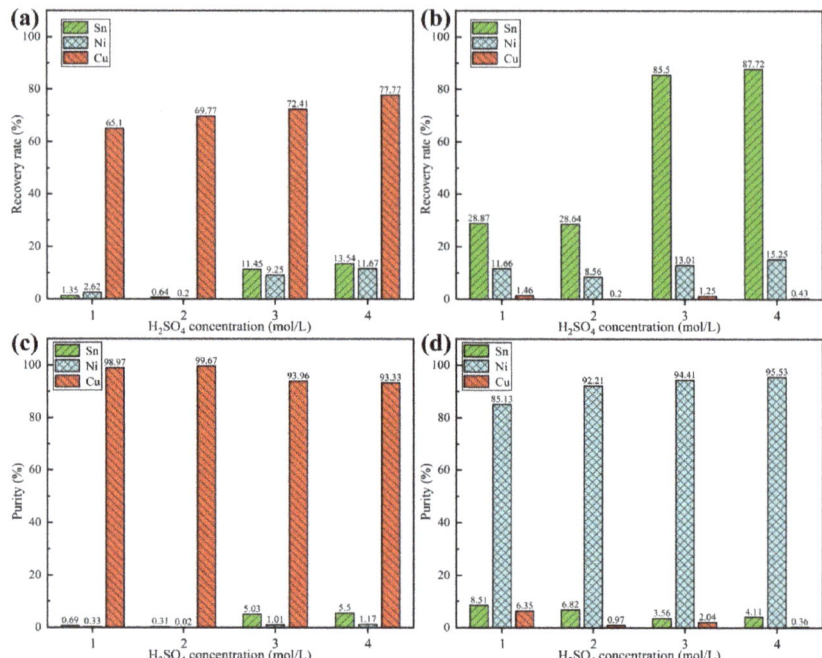

Figure 7. Effect of H_2SO_4 concentration on (**a**) recovery rate in the cathode, (**b**) recovery rate in the electrolyte, (**c**) purity in the cathode, and (**d**) purity in the electrolyte.

Figure 7c shows the purity of Cu in the cathode after the slurry electrolysis experiment, the purity of Cu first increased from 98.97% to 99.67% and then decreased to 93.33%, being highest at the H_2SO_4 concentration of 2 mol/L. The purity of the Sn and Ni in the cathode, after slurry electrolysis, remained at a low level as the H_2SO_4 concentration increased. The purity of Cu in the electrolyte after the experiment decreased, from 6.35% to 0.36%, when the H_2SO_4 concentration increased, as displayed in Figure 7d. The purity of Ni in the electrolyte decreased from 8.51% to 4.11% when the H_2SO_4 concentration increased, meanwhile, the purity of Sn in the electrolyte increased from 85.13% to 95.53%, as shown in Figure 7d.

As the H_2SO_4 concentration increased, not only did the recovery rate of Cu in the cathode remain stable, at near 70% when the H_2SO_4 concentrations were 2 and 4 mol/L, but the recovery rate of Ni and Sn increased. In the electrolyte, Cu could hardly be found, and the recovery rate of Ni remained around 10%, but the recovery rate of Sn increased. The slurry reacted more rapidly with the electrolyte when the H_2SO_4 concentration was higher, which caused a higher concentration of Cu^{2+}, Ni^{2+}, and Sn^{2+}, and more deposition of metal ions in the cathode. The increase in the concentration of H_2SO_4 leads to an increase in the concentration of H^+ in the electrolyte, which facilitates the leaching of metal from the anode, thus increasing the concentration of metal ions in the electrolyte, but the concentration of H^+ can lead to severe hydrogen precipitation reactions if it exceeds a certain limit. The recovery of Cu in the electrolyte is always smaller than that of Ni and Sn, because Cu has a more positive potential compared to Ni and Sn and can be deposited preferentially at the cathode. Compared to Ni and Sn, Sn has a more positive potential and can be deposited preferentially, so as the H_2SO_4 concentration increases, the recovery rate of Sn at the cathode grows faster than that of Ni, making the recovery rate of Sn greater than that of Ni. The recovery rate of Sn in both the cathode and electrolyte continued to increase with the increase in H_2SO_4 concentration, because the increase in H_2SO_4 concentration increased the Sn^{2+} concentration, thus promoting the deposition of Sn at the cathode. The fact that the recovery of Cu in the electrolyte remained close to 0, indicates that the rate of Cu deposition at the cathode was greater than the rate of Cu leaching at the anode, which resulted in a minimal amount of Cu in the electrolyte. Although the H_2SO_4 concentration had little effect on the cathodic Cu recovery, the increase in H_2SO_4 concentration increased the recovery of Sn and Ni from the cathode, resulting in a decrease in the cathodic Cu purity after the H_2SO_4 concentration reached 2 mol/L. In summary, the optimal H_2SO_4 concentration for cathodic recovery of Cu was 2 mol/L, when the cathodic recovery and purity of Cu reached the optimum.

3.4. Effect of Slurry Density on Recovery Rate and Purity of Cu

Figure 8 demonstrates the effect of slurry density on the purity and recovery rate of the metals. Experiments were conducted at different slurry densities, of 30, 50, and 70 g/L. The other conditions were: H_2SO_4 concentration 2 mol/L, current density 80 mA/cm^2, reaction time 7 h, and reaction temperature 15 °C. Figure 8a shows that the recovery rate of Cu in the cathode first increased, from 69.77% to 72.45%, and then decreased to 61.77% when the slurry density increased from 30 g/L to 70 g/L. Figure 8b shows that the recovery rate of Cu in the electrolyte was 0.10% at 30 g/L, 0.08% at 50 g/L, and 0.31% at 70 g/L, and obviously, the value is approximately 0%, thus Cu cannot be found in the electrolyte. The recovery rates of Sn and Ni in the cathode increased with the increase in slurry density. Sn increased from 0.64% to 24.66%, and Ni increased from 0.24% to 20.35%, when the slurry density increased from 30 to 70 g/L, as shown in Figure 8a. The recovery rate of Sn in the electrolyte decreased, from 28.64% to 18.56%, when the slurry density increased from 30 to 70 g/L. At the same time, the recovery rate of Ni in the electrolyte increased from 10.28% to 12.61%, as shown in Figure 8b.

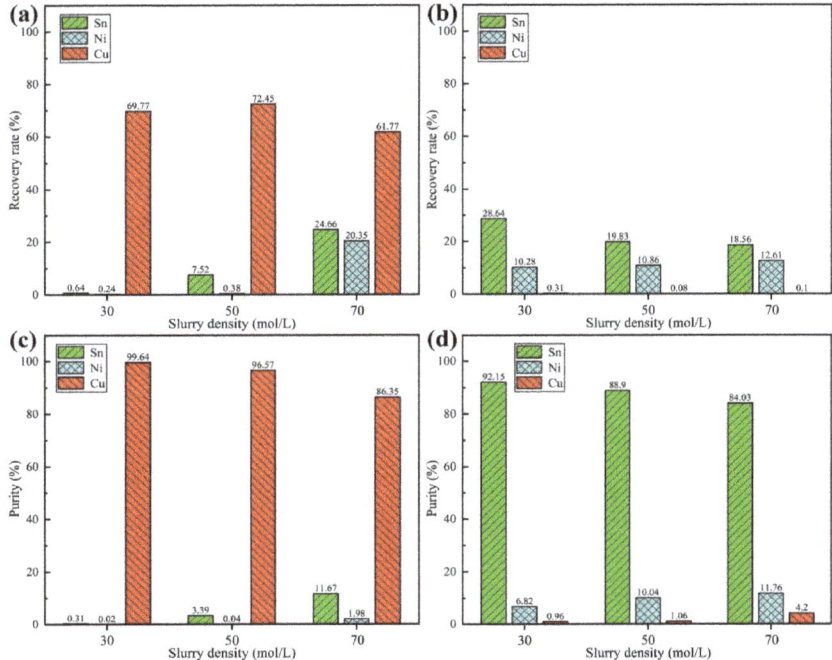

Figure 8. Effect of slurry density on (**a**) recovery rate in the cathode, (**b**) recovery rate in the electrolyte, (**c**) purity in the cathode, and (**d**) purity in the electrolyte.

Figure 8c illustrates that the purity of Cu in the cathode decreased from 99.64% to 86.35% when the slurry density increased from 30 to 70 g/L. Figure 8d shows that the purity of Cu in the electrolyte increased from 0.96% to 4.20%, when the slurry density increased from 30 g/L to 70 g/L. In Figure 8c, Sn and Ni increased from 0.31% and 0.02%, to 11.67% and 1.98%, when the slurry density increased from 30g/L to 70g/L, respectively. The purity of Sn in the electrolyte decreased from 92.15% to 84.03%, when the slurry density increased from 30 to 70 g/L. Meanwhile, the purity of Ni in the electrolyte increased from 6.82% to 11.76%, as shown in Figure 8d.

As the slurry density increased, the recovery rate of Cu in the cathode kept stable, at near 70%, when the slurry density was 30 g/L and 50 g/L, and decreased at 70 g/L, while the recovery rates of Ni and Sn kept increasing. In the electrolyte, Cu could hardly be found, and the recovery rate of Ni increased, while that of Sn decreased. This was because the slurry reacted more rapidly with the electrolyte when the slurry density became higher. It caused a higher concentration of metal ions (Cu^{2+}, Ni^{2+}, Sn^{2+}) and the rapid deposition of metal ions (Cu^{2+}, Ni^{2+}, Sn^{2+}) happened in the cathode, which led to a decrease in the recovery rate of Cu.

With the increase in slurry density, the purity of Cu in the cathode decreased, and the purity of Cu in the electrolyte remained near 0%. The purity of Sn in the cathode increased, and the purity of Sn in the electrolyte decreased when the slurry density increased. The purity of Ni in the cathode was kept at a low level and the purity of Ni in the electrolyte increased, with the increase in slurry density. Because slurry with a higher density reacted with the electrolyte, causing the densities of Cu^{2+}, Sn^{2+}, and Ni^{2+} to increase during electrolysis, thus speeding up ion deposition at the cathode part. The increase in slurry concentration, on the one hand, is beneficial to the leaching of scrap CPU slot metal powder in the anode, which increases the concentration of metal ions in the electrolyte and enhances the mass transfer, thus enhancing the deposition of metals in the cathode. Cu is not active enough, compared to Ni and Sn, so Ni and Sn are more easily leached in the anode. But on the other hand, the increase in slurry concentration means more metal powder of waste CPU slots in the anode chamber, which will reduce the specific surface area of the metal powder in contact with the electrolyte and thus affect the rate of metal leaching. This indicates that the recovery rate of Cu at the cathode does not increase or decrease monotonically; at the beginning of the slurry concentration increase, the metal ion concentration in the electrolyte increases, making the recovery rate of Cu increase. As the slurry concentration continues to increase, the specific surface area of the electrolyte in contact with the metal powder of the waste CPU slots becomes smaller and smaller, making the metal ion concentration within the electrolyte decrease, and no longer favorable for the leaching of metals. Since both Ni and Sn leached with higher priority than Cu, and were less affected by the decrease in specific surface area than Cu was, the amount of Ni and Sn leached in the electrolyte increased, and the purity of Cu^{2+} in the electrolyte decreased, ultimately leading to a decrease in the recovery of Cu. Although the recovery of Cu at the cathode was 70.25% when the slurry concentration was 50 g/L, which was slightly higher than the recovery of Cu at 30 g/L, of 66.96%. The purity of Cu at a slurry concentration of 50 g/L was lower than that at 30 g/L. Considering that this experiment aims at cathodic recovery of Cu with higher purity, it is in accordance with the experimental purpose to have the highest possible purity under the condition that the recovery rate is considerable.

3.5. Effect of NH_4Cl Concentration on Recovery Rate and Purity of Cu

Figure 9 indicates the effect of NH_4Cl concentration on the recovery rate and purity of Cu, Sn, and Ni. Experiments were conducted at different NH_4Cl concentrations of 30, 60, and 90 g/L. The other conditions were: H_2SO_4 concentration 2 mol/L, slurry concentration 30 g/L, current density 80 mA/cm^2, reaction time 7 h, and reaction temperature 15 °C. Figure 9a shows that when the NH_4Cl concentration was increased from 30 to 90 g/L, the recovery rates of the metals increased: Cu increased from 62.12% to 96.19%, Sn increased from 3.67% to 14.09%, and Ni increased from 3.78% to 17.17%. Figure 9c shows that the purity of Cu in the cathode was 97.47%, 96.53%, and 93.72%, respectively, when the NH_4Cl concentration was increased from 30 to 90 g/L. Meanwhile, the purity of the Sn and Ni remained at a low level (less than 5%). In Figure 9b, when the NH$_4$Cl concentration increased from 30 to 90 g/L, the recovery rates of Cu and Sn in electrolysis decreased from 3.71% and 1.03%, to 0.85% and 0.22%, respectively, but the recovery rate of Ni was kept at a high level in electrolysis. In Figure 9d, the purity of Cu and Sn in electrolysis decreased, while that of Ni in electrolysis increased, when the NH$_4$Cl concentration increased from 30 g/L to 90 g/L.

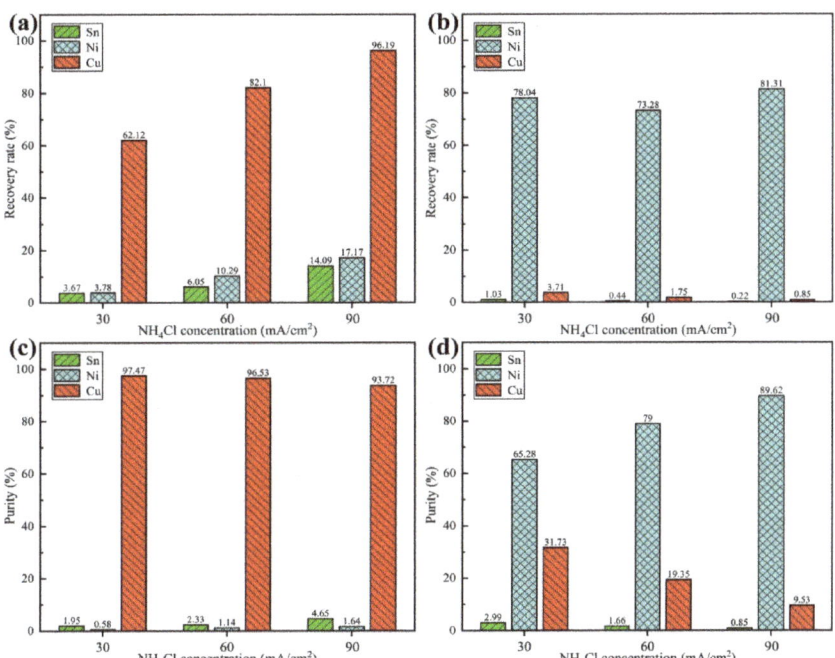

Figure 9. Effect of NH$_4$Cl concentration on (**a**) recovery rate in the cathode, (**b**) recovery rate in the electrolyte, (**c**) purity in the cathode, (**d**) purity in the electrolyte.

With the increase in NH$_4$Cl concentration, the recovery rate of Cu in the cathode increased to 96.19%, when the NH$_4$Cl concentration was 90 g/L. These values were rising, and the last one was close to 100%, so any increase after 90 g/L in the NH$_4$Cl concentration would not have a significant impact on the recovery rate of Cu. At the same time, the recovery rate of Sn and Ni both showed an increasing trend. The trend of the metal recovery rate shows that the increase in NH$_4$Cl concentration is beneficial to Cu recovery, reaching the highest recovery rate at 90 g/L. Because metal recovery rate increased with increasing NH$_4$Cl concentration, Cu recovery increased at a faster rate than Sn and Ni recovery. Although the purity of Cu decreased with increasing NH$_4$Cl concentration, the purity of Cu remains at a high level (over 90%). Because Cl$^-$ can form CuCl, CuCl$_2$, [CuCl$_2$]$^-$, [CuCl$_3$]$^{2-}$, and other compounds or ions with Cu, increasing the concentration of NH$_4$Cl both enhances the concentration of Cl$^-$ and facilitates the leaching of Cu from the anode. At the same time, increasing the concentration of NH$_4$Cl increases the electrical conductivity of the electrolyte and helps the dissolution of the metal, so that while Cu is dissolved, Sn and Ni are also leached, as impurities. The increase in metal concentration enhances the deposition ability of metals, so with the increase in NH$_4$Cl concentration, the recovery of all three metals at the cathode is increased, and the increase in Cu is the largest. Ni, as an active metal, can leach a lot in the electrolyte with the increase in electrolyte conductivity, but because Ni is not easily deposited in the cathode, so the electrolyte recovery of Ni is high. If the concentration of NH$_4$Cl exceeds 90 g/L, the room for improvement of Ni recovery and Sn recovery at the cathode far exceeds that of Cu recovery, thus leading to a decrease in the purity of Cu at the cathode, so there is no need to continue to increase the concentration of NH$_4$Cl. Overall, a NH$_4$Cl concentration of 90 g/L is the optimum concentration for Cu recovery.

3.6. Effect of Current Density on Recovery Rate and Purity of Cu

Figure 10 displays the effect of current density on the recovery rate and purity of Cu, Sn, and Ni. Experiments were conducted at different current densities, of 40, 80, and 120 mA/cm^2. The other conditions were: H$_2$SO$_4$ concentration 2 mol/L, slurry concentration 30 g/L, reaction time 7 h, and reaction temperature 15 °C. Figure 10a shows that the recovery rate of Cu increased from 61.94% to 98.73%, when the current density increased from 40 mA/cm^2 to 120 mA/cm^2. The increasing current also caused the recovery rate of Ni to increase from 2.70% to 87.92%, and the recovery rate of Sn from 1.09% to 15.28%. However, the increase in the value of the recovery rate of Cu was not obvious, while the increase in the recovery rate of Ni was nearly five times higher that of Cu, with the increase in current density from 80 mA/cm^2 to 120 mA/cm^2. Figure 10c shows that the purity of Cu in the cathode was 99.51%, 93.72%, and 86.36%, when the current density was increased from 40 mA/cm^2 to 120 mA/cm^2. Meanwhile, the purity of Sn and Ni increased at low current density levels. When the current density increased from 40 to 120 mA/cm^2, the recovery rate of Cu and Sn in electrolysis decreased from 27.55% and 1.18%, to 0.16% and 0.20%, respectively, but the recovery rate of Ni increased firstly and then decreased (Figure 10b). Additionally, the purity of Sn in electrolysis decreased when the current density increased. The purity of Cu in electrolysis decreased first and then increased when current density increased, while the purity of Ni in electrolysis increased first and then decreased (Figure 10d).

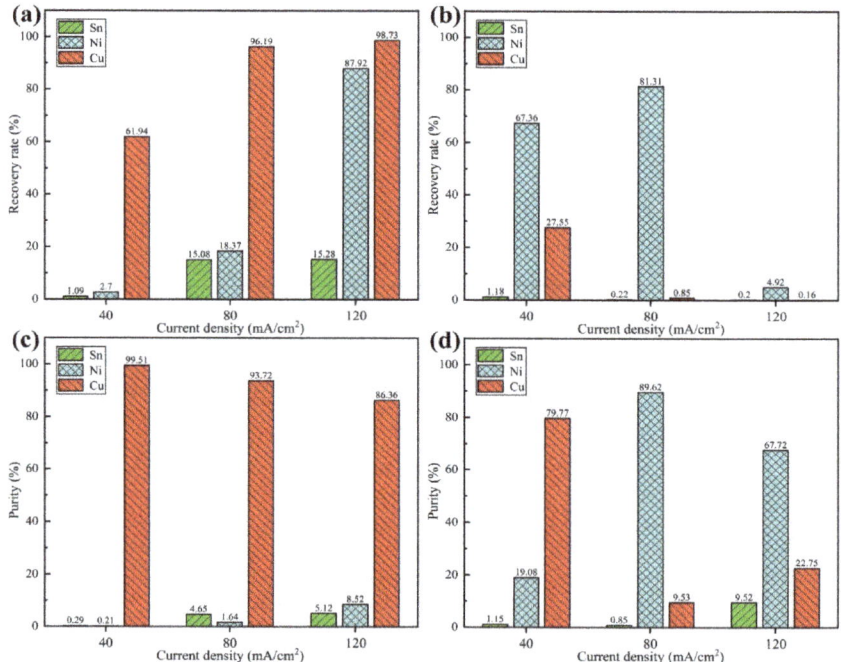

Figure 10. Effect of current density on (**a**) recovery rate in the cathode, (**b**) recovery rate in the electrolyte, (**c**) purity in the cathode, and (**d**) purity in the electrolyte.

With the increase in current density, the recovery rate of Cu in the cathode increased from 61.94% to 98.73%, when the current density was 120 mA/cm^2. These values were rising and the last one was close to 100%, so any increase after 120 mA/cm^2 in the current density would not have a significant impact on the recovery rate of Cu. Meanwhile, the recovery rates of Sn and Ni both showed an increasing trend, and that of Ni increased a lot when the current density increased from 80 mA/cm^2 to 120 mA/cm^2. The trend of the metal recovery rate shows that the increase in current density is beneficial to Cu recovery, reaching the highest recovery rate at 80 mA/cm^2. The recovery rate of Cu increased faster than that of Sn and Ni up to 80 mA/cm^2, but at 120 mA/cm^2, the recovery rate of Ni was very close to that of Cu. Moreover, the purity of Cu decreased from 40 to 120 mA/cm^2 of current density, the purity of Cu decreased to less than 90% when the current density was increased to 120 mA/cm^2. The increase in current density will enhance the mass transfer rate of metals in electrolysis, thus accelerating the cathodic deposition rate of the metals. When the current density is 40 mA/cm^2, a large number of metal ions are stored in the electrolyte and only some of the metal is deposited at the cathode, while the metal powder of the used CPU slot contains more Cu elements, so most of the metal ions in the electrolyte are Cu^{2+}. Because Cu has the advantage of depositing first at the cathode, the purity of the Cu recovered from the cathode is higher when the total metal deposition is relatively small. However, considering that there is still a large amount of Cu^{2+} present in the electrolyte, increasing the current density will improve the recovery of Cu at the cathode, so when the current density is increased to 80 mA/cm^2, the recovery of Cu at the cathode is greatly enhanced. The increase in current density can improve the recovery of cathodic Cu, but too high a current density makes other metal ions gain enough energy to accelerate the deposition at the cathode, so when the current density reaches 120 mA/cm^2, the recovery of Ni and Sn at the cathode increases significantly, which reduces the purity of the Cu recovered at the cathode. In general, 80 mA/cm^2 of current density is the best current density for Cu recovery.

3.7. Effect of Reaction Time on Recovery Rate and Purity of Cu

Figure 11 exhibits the effect of reaction time on the recovery rate and purity of Cu, Sn, and Ni. Experiments were conducted at different reaction times, of 3 h, 5 h, and 7 h. The other conditions were: H_2SO_4 concentration 2 mol/L, slurry concentration 30 g/L, current density 80 mA/cm^2, and reaction temperature 15 °C. Figure 11a shows that when reaction time increased from 3 to 7 h, the recovery rate of metals increased, for Cu this increased from 40.88% to 96.19%, for Sn from 0.45% to 14.09%, and for Ni from 0.36% to 17.17%. Figure 11c shows that the purity of Cu in the cathode was 99.54%, 97.46%, and 93.72%, respectively, when the reaction time was increased from 3, to 5, to 7 h. Meanwhile, the purity of Sn and Ni remained at a very low level (no more than 5%). In Figure 11b, when reaction time increased from 3 to 7 h, the recovery rate of Cu and Sn during electrolysis decreased, from 38.15% and 8.94%, to 0.85% and 0.22%, respectively, but the recovery rate of Ni stayed at a high level during electrolysis. In Figure 11d, the purity of Sn increased first and then decreased, the purity of Cu decreased, while the purity of Ni increased during electrolysis when the reaction time increased from 3 h to 7 h.

With the increase in reaction time, the recovery rate of Cu at the cathode increased to 96.19%, when the reaction time was 7 h. These values were rising and the last one was close to 100%, so any increase after 7 h in the reaction time would not have a significant impact on the recovery rate of Cu. Meanwhile, the recovery rates of Sn and Ni also showed an increasing trend with reaction time. The trend of metal recovery rate shows that the increase in reaction time was beneficial to Cu recovery, reaching the highest recovery rate at 7 h. Because the metal recovery rate went up with reaction time, the speed of Cu recovery was faster than that of Sn and Ni. Although the purity of Cu decreased when increasing the reaction time from 3 h to 7 h, the purity of Cu was kept at a high level close to 100%. Increasing the reaction time had little effect on the change in Cu purity. With the increase in reaction time, on the one hand, the metal powder of the used CPU slots fully reacted with the electrolyte and got more energy from the current, so that more metal was leached into the electrolyte, which increased the concentration of metal ions, on the other hand, the metal ions near the cathode received more electrons, which led to more deposition of metal ions in the electrolyte and reduced the concentration of metal ions. The Cu recovery of the cathode increased with the increase in the reaction time, and if the reaction time was increased past 7 h, it could further increase the Cu recovery of the cathode. However, it can be noted that the recovery of Cu in the electrolyte was close to 0 at 7 h. Based on the results of previous experiments, it can be speculated that the Ni recovery and Sn recovery of the cathode will increase after the Cu^{2+} concentration in the electrolyte is close to 0. This is because the potential of Cu^{2+} is more positive than that of the other metal ions, so Cu^{2+} is deposited preferentially, and when the Cu^{2+} concentration is low, the impurity metals will be deposited in large quantities, thus reducing the cathode recovery deposited, thus reducing the Cu purity of the cathode recovery. In short, a reaction time of 7 h is the optimum reaction time for recovering Cu.

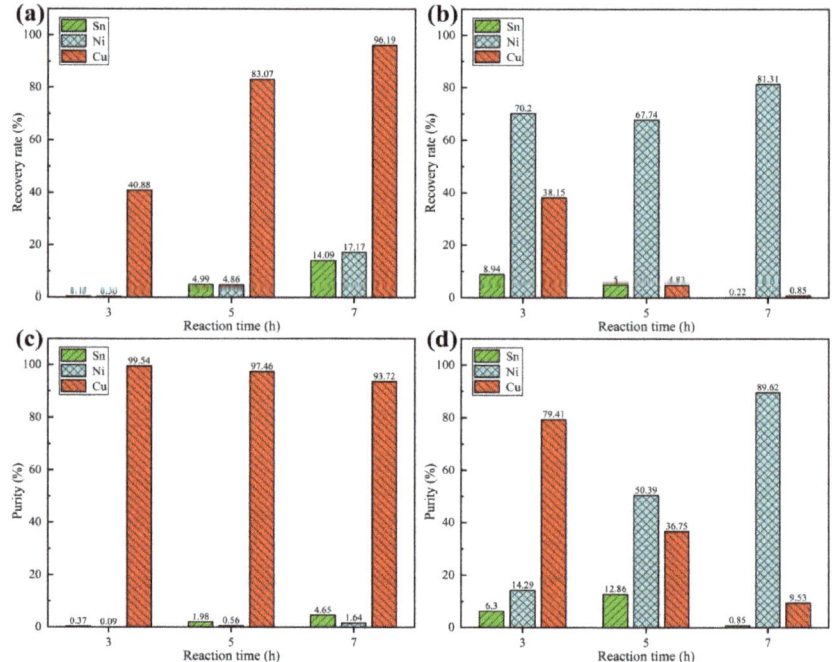

Figure 11. Effect of reaction time on (**a**) recovery rate in the cathode, (**b**) recovery rate in the electrolyte, (**c**) purity in the cathode, and (**d**) purity in the electrolyte.

4. Conclusions

In this paper, we used slurry electrolysis to recover Cu from used CPU slots, and studied the effects of different H_2SO_4 concentration, slurry concentration, current density, reaction time, and reaction temperature on the recovery and purity of Cu recovered from the cathode. In addition, the slurry electrolysis method was improved, to enhance the cathode Cu recovery rate, by adding NH_4Cl and changing the current mode, while ensuring the Cu purity of the cathode recovery. In general, the recovery rate and purity of the Cu in the cathode were up to 96.19% and 93.72%, respectively, with the optimum conditions being when the H_2SO_4 concentration was 2 mol/L, the slurry density was 30 g/L, the NH_4Cl concentration was 90 g/L, the current density was 80 mA/cm^2, and the reaction time was 7 h. Although some progress has been made in the research work on slurry electrolysis in this paper, there are still many aspects that deserve further study. For example, the effects of various experimental parameters on the powder particle size of recovered metals, current efficiency, etc., remain to be studied. Lastly, we hope to achieve higher recovery efficiency of Cu using the simplest device and method.

Author Contributions: W.C.: conceptualization, methodology, sample preparation, formal analysis, investigation, data curation, writing—original draft, and visualization. M.L.: investigation, validation, methodology, and writing—review and editing. J.T.: resources and supervision. All authors have read and agreed to the published version of the manuscript.

Funding: This work was supported by the National Natural Science Foundation of China (No. 51864034).

Institutional Review Board Statement: Not applicable.

Informed Consent Statement: Not applicable.

Data Availability Statement: All data generated or analysed during this study are included in this published article.

Conflicts of Interest: The authors declare that they have no known competing financial interests or personal relationships that could have appeared to influence the work reported in this paper.

References

1. Behnamfard, A.; Salarirad, M.M.; Veglio, F. Process development for recovery of copper and precious metals from waste printed circuit boards with emphasize on palladium and gold leaching and precipitation. *Waste Manag.* **2013**, *33*, 2354–2363. [CrossRef]
2. Li, J.; Zeng, X.; Chen, M.; Ogunseitan, O.A.; Stevels, A. "Control-alt-delete": Rebooting solutions for the e-waste problem. *Environ. Sci. Technol.* **2015**, *49*, 7095–7108. [CrossRef]
3. Zhang, S.; Li, Y.; Wang, R.; Xu, Z.; Wang, B.; Chen, S.; Chen, M. Superfine copper powders recycled from concentrated metal scraps of waste printed circuit boards by slurry electrolysis. *J. Clean Prod.* **2017**, *152*, 1–6. [CrossRef]
4. Liu, Y.; Song, Q.; Zhang, L.; Xu, Z. Separation of metals from Ni-Cu-Ag-Pd-Bi-Sn multi-metal system of e-waste by leaching and stepwise potential-controlled electrodeposition. *J. Hazard Mater.* **2020**, *408*, 124772. [CrossRef]
5. Zeng, X.; Mathews, J.A.; Li, J. Urban mining of e-waste is becoming more cost-effective than virgin mining. *Environ. Sci. Technol.* **2018**, *52*, 4835–4841. [CrossRef]
6. Forti, V.; Baldé, C.P.; Kuehr, R.; Bel, G. *The Global E-Waste Monitor 2020: Quantities, Flows and the Circular Economy Potential, Report*; United Nations University/United Nations Institute for Training and Research: Geneva, Switzerland, 2020. Available online: https://scholar.google.com/scholar_lookup?title=The%20global%20e-waste%20monitor%202020%3A%20quantities%2C%20flows%20and%20the%20circular%20economy%20potential.%20Report&author=V.%20Forti&publication_year=2020 (accessed on 12 February 2023).
7. Chen, M.; Huang, J.; Ogunseitan, O.A.; Zhu, N.; Wang, Y. Comparative study on copper leaching from waste printed circuit boards by typical ionic liquid acids. *Waste Manag.* **2015**, *41*, 142–147. [CrossRef] [PubMed]
8. Islam, A.; Ahmed, T.; Awual, M.R.; Rahman, A.; Sultana, M.; Aziz, A.A.; Monir, M.U.; Teo, S.H.; Hasan, M. Advances in sustainable approaches to recover metals from e-waste-a review. *J. Clean Prod.* **2020**, *244*, 118815. [CrossRef]
9. Panagiotis, E.; Samantha, A.; Henry, P.; Efthymios, K.; Yang, W. Reduction of brominated flame retardants (BFRs) in plastics from waste electrical and electronic equipment (WEEE) by solvent extraction and the influence on their thermal decomposition. *Waste Manag.* **2018**, *94*, 165–171.
10. Cucchiella, F.; D'Adamo, I.; Koh, S.C.L.; Rosa, P. Recycling of WEEEs: An economic assessment of present and future e-waste streams. *Renew. Sustain. Energ. Rev.* **2015**, *51*, 263–272. [CrossRef]
11. Menikpura, S.N.M.; Santo, A.; Hotta, Y. Assessing the climate co-benefits from waste electrical and electronic equipment (WEEE) recycling in Japan. *J. Clean Prod.* **2014**, *74*, 183–190. [CrossRef]
12. Yeh, C.; Xu, Y. Sustainable planning of e-waste recycling activities using fuzzy multicriteria decision making. *J. Clean Prod.* **2013**, *52*, 194–204. [CrossRef]
13. Calvo, G.; Mudd, G.; Valero, A.; Valero, A. Decreasing ore grades in global metallic mining: A theoretical issue or a global reality? *J. Res.* **2016**, *5*, 36. [CrossRef]
14. Zhang, S.; Forssberg, E. Mechanical separation-oriented characterization of electronic scrap. *Resour. Conserv. Recycl.* **1997**, *21*, 247–269. [CrossRef]
15. Ilyas, S.; Lee, J. Biometallurgical Recovery of Metals from Waste Electrical and Electronic Equipment: A Review. *Chembioeng. Rev.* **2014**, *1*, 148–169. [CrossRef]
16. Moosakazemi, F.; Ghassa, S.; Soltani, F.; Mohammadi, M.R.T. Regeneration of Sn-Pb solder from waste printed circuit boards: A hydrometallurgical approach to treating waste with waste. *J. Hazard Mater.* **2020**, *385*, 121589. [CrossRef]
17. Zhao, C.; Zhang, X.; Ding, J.; Zhu, Y. Study on recovery of valuable metals from waste mobile phone PCB particles using liquid-solid fluidization technique. *J. Chem Eng.* **2017**, *311*, 217–226. [CrossRef]

18. Hanafi, J.; Jobiliong, E.; Christiani, A.; Soenarta, D.C.; Kurniawan, J.; Irawan, J. Material Recovery and Characterization of PCB from Electronic Waste. *Procedia Soc. Behav. Sci.* **2012**, *14*, 169. [CrossRef]
19. Yi, X.; Qi, Y.; Li, F.; Shu, J.; Sun, Z.; Sun, S.; Chen, M.; Pu, S. Effect of electrolyte reuse on metal recovery from waste CPU slots by slurry electrolysis. *Waste Manag.* **2019**, *95*, 370–376. [CrossRef]
20. Veit, H.M.; Bernardes, A.M.; Ferreira, J.Z.; Tenório, J.A.S.; Malfatti, C.D.F. Recovery of copper from printed circuit boards scraps by mechanical processing and electrometallurgy. *J. Hazard Mater.* **2006**, *137*, 1704–1709. [CrossRef]
21. Chu, Y.; Chen, M.; Chen, S.; Wang, B.; Fu, K.; Chen, H. Micro-copper powders recovered from waste printed circuit boards by electrolysis. *Hydrometallurgy* **2015**, *156*, 152–157. [CrossRef]
22. Min, X.; Luo, X.; Deng, F.; Shao, P.; Wu, X.; Dionysiou, D.D. Combination of multi-oxidation process and electrolysis for pretreatment of PCB industry wastewater and recovery of highly-purified copper. *J. Chem. Eng.* **2018**, *354*, 228–236. [CrossRef]
23. Guimarães, Y.F.; Santos, I.D.; Dutra, A.J.B. Direct recovery of copper from printed circuit boards (PCBs) powder concentrate by a simultaneous electroleaching–electrodeposition process. *Hydrometallurgy* **2014**, *149*, 63–70. [CrossRef]
24. Kim, B.-S.; Lee, J.-C.; Seo, S.-P.; Park, Y.-K. A process for extracting precious metals from spent printed circuit boards and automobile catalysts. *JOM* **2004**, *56*, 55–58. [CrossRef]
25. Zhao, Y.; Wen, X.; Li, B.; Tao, D. Recovery of copper from waste printed circuit boards. *Min. Metall. Explor.* **2004**, *21*, 99–102. [CrossRef]
26. Wang, M.; Gong, X.; Wang, Z. Sustainable electrochemical recovery of high-purity Cu powders from multi-metal acid solution by a centrifuge electrode. *J. Clean Prod.* **2018**, *204*, 41–49. [CrossRef]
27. Liu, Y.; Zhang, L.; Song, Q.; Xu, Z. Recovery of palladium as nanoparticles from waste multilayer ceramic capacitors by potential-controlled electrodeposition. *J. Clean Prod.* **2020**, *257*, 120370. [CrossRef]
28. Cocchiara, C.; Dorneanu, S.; Inguanta, R.; Sunseri, C.; Ilea, P. Dismantling and electrochemical copper recovery from Waste Printed Circuit Boards in H_2SO_4–$CuSO_4$–NaCl solutions. *J. Clean Prod.* **2019**, *230*, 170–179. [CrossRef]
29. Huang, K.; Guo, J.; Xu, Z. Recycling of waste printed circuit boards: A review of current technologies and treatment status in China. *J. Hazard Mater.* **2009**, *164*, 399–408. [CrossRef]
30. Li, F.; Chen, M.; Shu, J.; Shirvani, M.; Li, Y.; Sun, Z.; Sun, S.; Xu, Z.; Fu, K.; Chen, S. Copper and gold recovery from CPU sockets by one-step slurry electrolysis. *J. Clean Prod.* **2019**, *213*, 673–679. [CrossRef]

Disclaimer/Publisher's Note: The statements, opinions and data contained in all publications are solely those of the individual author(s) and contributor(s) and not of MDPI and/or the editor(s). MDPI and/or the editor(s) disclaim responsibility for any injury to people or property resulting from any ideas, methods, instructions or products referred to in the content.

Article

Mechanism of Selective Chlorination of Fe from Fe$_2$SiO$_4$ and FeV$_2$O$_4$ Based on Density Functional Theory

Junyan Du [1], Yiyu Xiao [1], Shiyuan Liu [1,*], Lijun Wang [1,*] and Kuo-Chih Chou [2]

[1] Collaborative Innovation Center of Steel Technology, University of Science and Technology Beijing, Beijing 100083, China
[2] State Key Laboratory of Advanced Metallurgy, University of Science and Technology Beijing, Beijing 100083, China
* Correspondence: shiyuanliu@ustb.edu.cn (S.L.); lijunwang@ustb.edu.cn (L.W.)

Abstract: Vanadium slag is an important resource containing valuable elements such as Fe, V, Ti, and so on. A novel selective chlorination method for extracting these valuable elements from vanadium slag has been proposed recently. The proposed methods could recover valuable elements with a high recovery ratio and less of an environmental burden, while the study on the chlorination mechanism at the atom level was still insufficient. Fe$_2$SiO$_4$ and FeV$_2$O$_4$ are the two main phases of vanadium slag, and the iron element can be selectively extracted via the chlorination of NH$_4$Cl. The NH$_4$Cl decomposes into NH$_3$ gas and HCl gas, which was the true chlorination agent. As a result, the chlorination reactions of Fe$_2$SiO$_4$ and FeV$_2$O$_4$ with HCl were firstly calculated using FactSage 8.0. Then, this paper studied the characteristics of HCl adsorption on the Fe$_2$SiO$_4$(010) surface and the FeV$_2$O$_4$(001) Fe-terminated surface mechanism of the selective chlorination of Fe from Fe$_2$SiO$_4$ and FeV$_2$O$_4$ via DFT calculations. The processes of chlorination of Fe$_2$SiO$_4$ and FeV$_2$O$_4$ involved the processes of removing O atoms from them with HCl gas. The iron in Fe$_2$SiO$_4$ was selectively chlorinated because HCl could adsorb on the iron site but could not adsorb on the silicon site. The iron in FeV$_2$O$_4$ was selectively chlorinated because the electronegativity gap between V and O was more significant than that between the Fe and O elements.

Keywords: DFT; vanadium slag; selective chlorination

1. Introduction

Vanadium is an important element for alloying and is indispensable for manufacturing micro-alloyed steel [1,2]. Vanadium titanomagnetite ore is scattered in Australia, China, Russia, and South Africa as the main resource of vanadium element [3–5]. The vanadium slag produced after pyrometallurgical processes of vanadium titanomagnetite ore still contained many valuable elements with the composition of approximately 30–40 mass% total Fe, 13.52–19.03 mass% V$_2$O$_3$, 6.92–14.32 mass% TiO$_2$, 7.44–10.67 mass% MnO, and 0.93–4.59 mass% Cr$_2$O$_3$ [6]. Additionally, it produced about 1.4 million tons each year. The vanadium slag was mainly composed of encompassed structures, such as fayalite, manganoan ((Fe, Mn)$_2$SiO$_4$), titanomagnetite (Fe$_{2.5}$Ti$_{0.5}$O$_4$), and vuorelainenite ((Mn, Fe)(V, Cr)$_2$O$_4$) [7]. The main difficulty of extracting valuable elements from vanadium slag was the destruction of its encompassed structure [8].

The traditional method used to extract vanadium from the vanadium slag used the oxidation method, direct vanadium alloying, and ferrovanadium production with the vanadium slag [9–11]. The direct vanadium alloying with the vanadium slag demonstrated the shortcomings of low vanadium recovery, complex steelmaking operation, the introduction of impurities in steel, and high energy consumption which were disadvantages to the efficient utilization of resources [12,13]. Therefore, this made the oxidation technologies key to vanadium extraction from vanadium slag. Oxidation technologies included salt roasting, direct leaching, etc. [14]. The salt roasting of vanadium slag was widely used in the industry. The salt roasting of vanadium slag included the following steps in order: salt roasting of vanadium slag under an oxidation atmosphere, water leaching of the produced roasting product, separation of vanadium from the leachate and vanadium precipitation, calcination precipitation, and, finally, the reduction of V_2O_5 [15,16]. During this process, the Cr(III) was also oxidized to Cr(VI) which was one of the main carcinogen contaminants. The insoluble vanadium(III) in vanadium slag was converted into water-soluble vanadate(V) which was highly toxic [17]. The disposal of leaching lixivium and roasting tailings containing hazardous vanadium(V) and chromium(VI) became a great burden to environmental protection.

To alleviate the shortcomings of salt roasting technology and efficient resource utilization, our team proposed a new process to comprehensively extract the valuable elements with a high recovery ratio and less of an environmental burden from vanadium slag [8,12,18–21]. The vanadium slag was selectively chlorinated using a chlorination agent to accomplish the separation of valuable elements. For example, NH_4Cl was used as the chlorination agent to selectively extract iron and manganese from vanadium slag. The manganese and iron chlorination ratio can be reached at 95% and 72%, respectively [8]. During the chlorination process, the Cr and V were not oxidized to high valence toxic oxidation but remained at the original valence which was inspiring for tailing disposal and environmental protection. Our team also proposed other chlorination agents to selectively chlorinate V and Cr elements in vanadium slag and produced valuable products via extracted resources. In previous work, the optimal chlorination agent ratio and maximum chlorination rate were determined. Additionally, the thermodynamic mechanism of selective chlorination of the valuable element Fe in vanadium slag was clarified. However, the investigation of the selective chlorination mechanism of Fe in vanadium slag from the atomic level was necessary for a deep understanding of the mechanisms. The first principles calculation is applied more and more frequently to investigate the mechanism of chemical reactions and material properties due to the progress of computational algorithms and the rapid development of computer technologies.

2. Computational Methods

2.1. Computational Theory

All the calculations were performed using the CASTEP module in Material Studio 2017 (Accelrys, San Diego, CA, USA) software with the projector augmented wave (PAW) basis sets and periodic boundary conditions [22,23]. The electron–ion core interaction of Fe, Si, V, and O atoms was described using the ultrasoft potential in CASTEP. The generalized gradient approximation (GGA) defined by Perdew–Burke–Ernzerhof (PBE) was applied to describe the exchange–correlation energy [24]. The irreducible Brillouin zone (IBZ) was integrated using meshes generated via the Monkhorst–Pack method for self-consistent field calculation [25]. The total energy and atomic displacement convergence criteria were set to 1.5×10^{-5} eV/atom and 1×10^{-4} nm, respectively. The stress deviation criterion was set to be smaller than 0.05 GPa. The structure optimization was treated within the Fletcher–Goldfarb–Shanno algorithm (BFGS), and all the energy calculations were performed in reciprocal space [26].

2.2. Computational Details

In this paper, the geometry optimization of bulk FeV_2O_4, bulk Fe_2SiO_4, and their surfaces was applied to obtain the corresponding minimum energy structure that was the stable structure for further exploration of the chlorination mechanisms of FeV_2O_4 and Fe_2SiO_4. The bulk FeV_2O_4 was cubic while the bulk Fe_2SiO_4 was a face-centered orthodox lattice in terms of crystallography. For the geometry optimization of bulk FeV_2O_4 and bulk Fe_2SiO_4, the Monkhorst–Pack k-point mesh of $2 \times 2 \times 2$ and $1 \times 3 \times 4$ was adopted. The Kohn–Sham wave functions were expanded using an optimized kinetic energy cutoff of 700 eV for all calculations. The optimized lattice parameters of Fe_2SiO_4 and FeV_2O_4 exhibited in Table 1 are consistent with the reference data.

Table 1. Lattice parameters of Fe_2SiO_4 and FeV_2O_4.

	Calculated Results	Reference
Fe_2SiO_4	a = 10.647, b = 6.188, c = 5.014	a = 10.485, b = 6.093, c = 4.820 [27]
FeV_2O_4	a = b = c = 8.535	a = b = c = 8.439 [28]

The (001) surface of FeV_2O_4 was the most stable one. There were two classes of terminated surface for FeV_2O_4 defined as Fe-terminated surface and VO-terminated surface, which were similar to spinel $FeCr_2O_4$ [29]. However, the exposure of iron atoms was necessary to unravel the mechanism of the selective chlorination of iron in FeV_2O_4. As a result, in this paper, the Fe-terminated (001) surface of FeV_2O_4 was chosen to investigate the iron selective mechanism of FeV_2O_4, despite the existence of the Fe-terminated surface and VO-terminated surface. To eliminate the size effect, the calculated Fe-terminated (001) surface was a $2 \times 2 \times 1$ slab with 13 layers of atoms and 112 atoms as shown in Figure 1a. A vacuum layer of 15 Å was inserted between adjacent slabs. For the geometry optimization of the FeV_2O_4(001) Fe-terminated surface, the Monkhorst–Pack k-point mesh of $1 \times 1 \times 1$ was adopted. The three bottom layers of atoms were fixed during the DFT calculation of the FeV_2O_4(001) Fe-terminated surface. There were two classes of Fe site on the FeV_2O_4(001) Fe-terminated surface which were defined as Fe5 and Fe7 in this paper. Both the Fe-O bonds and V-O bonds on the FeV_2O_4(001) Fe-terminated surface were classified into two classes according to their Mulliken bond order. The corresponding Mulliken bond order from the DFT calculation is exhibited in Table 2. O7, O16, O17, O23, O44, O47, and O60 represent different oxygen atoms as denoted in Figure 1a.

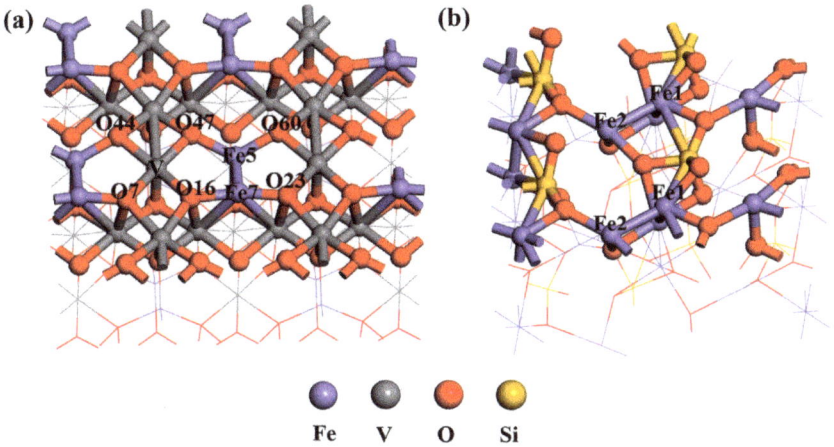

Figure 1. (a) FeV_2O_4(001) Fe-terminated surface, (b) Fe_2SiO_4(010) surface.

Table 2. Bond order of Fe-O bond and V-O bond in FeV_2O_4(001) Fe-terminated surface.

Bond	Bond Order
V-O44, V-O47	0.28
V-O7, V-O16	0.32
Fe5-O47, Fe5-O60	0.40
Fe7-O16, Fe7-O23	0.47

As for Fe_2SiO_4, the (010) index surface was chosen for its lowest surface energy according to the calculations. Additionally, the chosen Fe_2SiO_4(010) surface as shown in Figure 1b. was the lowest energy-terminated surface. A $2 \times 1 \times 1$ slab with 84 atoms of Fe_2SiO_4 was applied for further calculations to eliminate the size effect. For the geometry optimization of the Fe$_2$SiO$_4$(010) surface slab, the Monkhorst–Pack k-point mesh of 1 × 1 × 1 was adopted. The three bottom layers of the atom were fixed during the DFT calculations on the Fe$_2$SiO$_4$(010) surface. A vacuum layer of 15 Å was inserted between adjacent slabs. Figure 1b shows the optimized structure of the clean Fe$_2$SiO$_4$(010) surface. There were two crystallographically inequivalent Fe sites defined as the Fe1 site and the Fe2 site, as shown in Figure 1b. The Fe1 site atom was surrounded by two O atoms and one Fe atom, while the Fe2 site atom was surrounded by three O atoms and one Fe atom. The discrimination of the ambient chemical environment led to different atom charges of the Fe1 site and Fe2 site atom. The Mulliken charge of the iron atom in the Fe1 site and Fe2 site was 1.09 and 1.22, respectively.

3. Results and Discussion
3.1. Thermodynamic Calculations

The standard Gibbs free energies of Reactions (1) and (2) at different temperatures were calculated via the FactSage 8.0 program, using the pure substance database and the oxide database. The results are shown in Figure 2 and demonstrate that the reactions could occur in the temperature range of 0 to 600 °C and the standard Gibbs free energy of both reactions increased with an increase in temperature. Increasing the temperature was not conducive to Reactions (1) and (2). Meanwhile, it was found that Fe in Fe$_2$SiO$_4$ was easier to be chlorinated than Fe in FeV$_2$O$_4$. However, since the amount at each temperature in the chlorination process was not known, the following calculation was performed.

$$2HCl + FeV_2O_4 = FeCl_2 + V_2O_3 + H_2O \tag{1}$$

$$4HCl + Fe_2SiO_4 = 2FeCl_2 + SiO_2 + 2H_2O \tag{2}$$

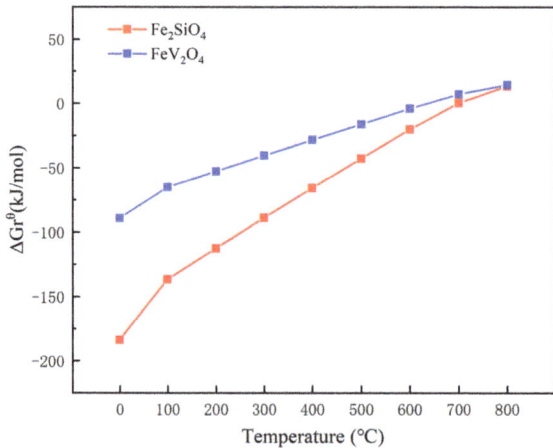

Figure 2. Gibbs free energy variation for Reactions (1) and (2) with different temperatures.

It can be seen in Figure 3 that as the temperature increased, Fe could be chlorinated in the form of FeCl$_2$. Si was present in the form of SiO$_2$. The amounts of Fe$_2$SiO$_4$ and HCl increased with increasing temperature, and the total amount of FeCl$_2$, SiO$_2$, and H$_2$O decreased with increasing temperature. This means that only iron could be selectively chlorinated to produce FeCl$_2$. It can be seen in Figure 4 that as the temperature increased, vanadium in FeV$_2$O$_4$ could not be chlorinated and the iron could be chlorinated to form FeCl$_2$. The amounts of V$_2$O$_3$ and H$_2$O increased with increasing temperature, and the total amount of FeCl$_2$ and FeV$_2$O$_4$ decreased with increasing temperature. In summary, the selective chlorination of Fe$_2$SiO$_4$ and FeV$_2$O$_4$ to form ferric chloride was found to be thermodynamically achievable using calculations with FactSage 8.0.

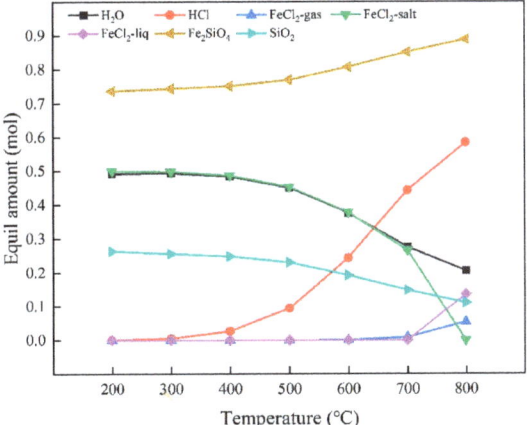

Figure 3. Changes in the content of each substance in the reaction of Fe_2SiO_4 with HCl at different temperatures.

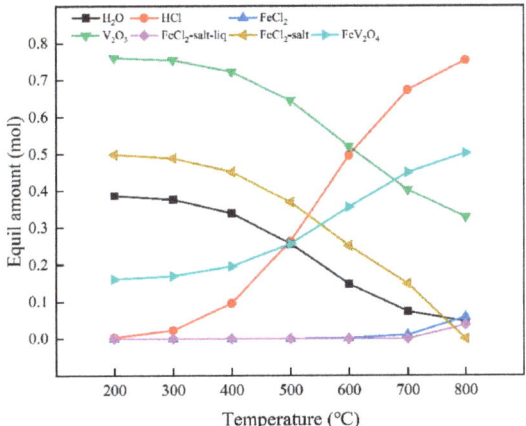

Figure 4. Changes in the content of each substance in the reaction of FeV_2O_4 with HCl with increasing temperature.

3.2. HCl Adsorption on Fe_2SiO_4

The adsorption of the HCl molecule on the Fe_2SiO_4 surface is no doubt an important step for the chlorination of Fe_2SiO_4. As a result, it is necessary to explore the characteristics of HCl adsorption on the Fe_2SiO_4 surface. As previously discussed, the (010) surface is the most stable surface, implied by it having the lowest surface energy. As a result, the Fe_2SiO_4(010) surface is believed to be the most likely exposed surface during chlorination. When we consider the chlorination of Fe_2SiO_4 via HCl gas, it is natural to think that the H atom bonds with the O atom while the Fe atom bonds with the Cl atom. However, it is unclear whether the HCl dissociated adsorption on the surface is either mainly led by the interaction between the adsorbed O and H atoms or between the adsorbed Fe atom and Cl atom. Moreover, it is also unclear in what manner the HCl would adsorb on the Fe_2SiO_4(010) surface. To clarify these questions, adsorption configurations were designed as shown in Figure 5. In Figure 5a, the HCl molecule was vertically put on the surface with a H atom near one O atom, while in Figure 5b, the HCl molecule was vertically put on the surface with a Cl atom near one Fe atom. The results showed that when the H atom was put near the O atom, the H atom and Cl atom in the HCl molecule were still bonded.

Moreover, the HCl molecule was pushed away from the surface. The most significant change was that the vertical HCl molecule rotated until it was horizontal to the surface. This suggested that it was preferable for the HCl molecule to horizontally adsorb on the Fe_2SiO_4(010) surface. When the Cl atom was put near the Fe atom, the HCl molecule was dissociated and adsorbed on the Fe_2SiO_4(010) surface. The distance between the adsorbed iron atom and the O atom was extended from 1.946 Å to 3.061 Å. This result indicates that the dominant force to break the H-Cl bond is the interaction between the Cl atom and the iron atom instead of the interaction between the H atom and O atom. In addition, it is preferable for the HCl molecule to adsorb on the Fe_2SiO_4(010) surface horizontally.

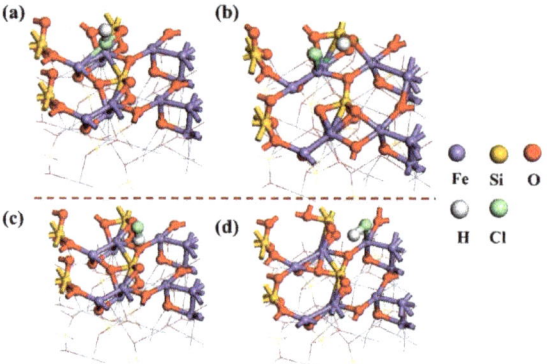

Figure 5. HCl adsorption on Fe_2SiO_4: (**a**) Configuration of HCl vertical adsorption with H atom near one O atom before optimization and (**b**) after optimization. (**c**) Configuration of HCl vertical adsorption with Cl atom near one Fe atom before optimization and (**d**) after optimization.

Therefore, the HCl molecule was put on the Fe_2SiO_4 surface horizontally to the Fe-O bond with the H atom above one O atom and the Cl atom above one Fe atom, as shown in Figure 6a,c. The two classes of iron sites were taken into consideration by putting the HCl molecule on the Fe1 site and Fe2 site, respectively. Therefore, two adsorption configurations were considered for the HCl adsorption on the iron site. Additionally, the adsorption results were interpreted for further understanding of the chlorination mechanism of Fe_2SiO_4 via HCl.

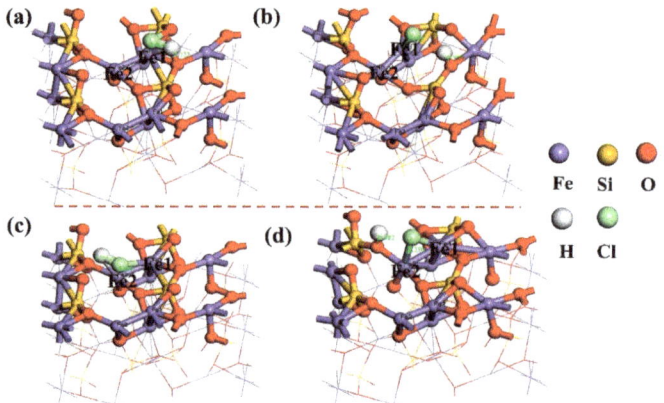

Figure 6. HCl adsorption on Fe_2SiO_4: (**a**) Configuration of HCl adsorption on Fe1 site before optimization. (**b**) Configuration of HCl adsorption on Fe1 site after optimization. (**c**) Configuration of HCl adsorption on Fe2 site before optimization. (**d**) Configuration of HCl adsorption on Fe1 site after optimization.

When the HCl molecule was put on the Fe1 site, the HCl molecule was dissociated and adsorbed on the Fe$_2$SiO$_4$(010) surface, as shown in Figure 6b. The H atom bonded with the O atom and the Cl atom bonded with the Fe atom in the Fe1 site after geometry optimization. The adsorption energy was −0.84 eV, as exhibited in Table 3. The adsorbed Fe atom and O atom were dragged away from each other by the attraction of the Cl atom and H atom, respectively. The distance of the adsorbed Fe1-O before and after adsorption was 1.922 Å and 3.174 Å, respectively. The distance of adsorbed Fe1-Cl changed from 2.289 Å to 1.985 Å. It can be concluded that the adsorbed Fe1 and O atoms were dragged away from the Fe$_2$SiO$_4$(010) surface by the Cl and O atoms. When the HCl molecule was put above the Fe2 site of the iron atom, similarly, the HCl molecule was dissociated and adsorbed on the Fe$_2$SiO$_4$ surface, as shown in Figure 6d. The dissociated H atom was adsorbed on the O atom and the Cl atom was adsorbed on the Fe2 atom, while the Cl atom was also attracted by the neighboring Fe1 site of the iron atom. The adsorption energy was −0.51 eV which indicates weaker adsorption than the previously discussed HCl adsorption on the Fe1 site. However, unlike the HCl adsorption on the Fe1 site of the iron atom, the adsorbed Fe atom and O atom were pulled closer to each other with a distance shortened from 2.047 Å to 1.985 Å. This may have resulted from the weaker attraction between the adsorbed Fe2 atom and the Cl atom than that between the Fe1 atom and Cl atom. The distances of Fe1-Cl and Fe2-Cl when the HCl molecule was initially put above the Fe2 site of the iron atom were 2.272 Å and 2.251 Å, respectively. In other words, the adsorbed Cl atom was in the bridge site of the Fe1 atom and Fe2 atom, as shown in Figure 5d. This phenomenon indicates that the attraction between the Fe1 site of the iron atom and Cl atom is stronger than that between the Fe2 site of the iron atom and Cl atom. From the previous analysis we gain the Mulliken charge of the Fe1 site of the iron atom and Fe2 site of the iron atom: 1.09 and 1.22, respectively. The lower charge of the Fe1 site of the iron atoms means that it is easier to lose electrons here than in the Fe2 site of the iron atoms. As a result, the Fe1 site is a preferable site to the Fe2 site for HCl adsorption. The Mulliken charge of the Fe1 site and Fe2 site atoms after adsorption was 0.35 and 0.52, respectively. It can be concluded that during the HCl adsorption process, the Cl captures an electron from the adsorbed Fe atoms and leads to a weakened adsorbed Fe-O bond. Additionally, the weakened Fe-O bond results in electron capture for the adsorbed Fe atom and electron loss for the adsorbed O atom. Therefore, both the adsorbed Fe atom and Cl atoms seize electrons while the adsorbed O atom loses electrons in total. The Fe1 site of the atom seizes more electrons from the O atom than the Fe2 site of the atom, indicating more serious destruction of the Fe-O bond. During the HCl adsorption process, the Cl captures electrons from adsorbed Fe atoms leading to a weakened adsorbed Fe-O bond.

Table 3. HCl adsorption energy on Fe$_2$SiO$_4$.

Adsorption Site	Adsorption Energy (eV)
Fe1	−0.84
Fe2	−0.51

The Si atom is also a possible adsorption site for the HCl molecule; as a result, we conducted the geometry optimization of structure where HCl is horizontally above the surface with a Cl atom right above one Si atom, while a H atom is right above one neighboring O atom. The structure before and after geometry optimization is shown in Figure 7a,b, respectively. The result shows that the HCl molecule adsorbed on the surface in a dissociated way after geometry adsorption. Unlike the HCl adsorption on the iron sites, the Cl atom was not bonded with the nearest Si atom but bonded with neighboring iron atoms. As a result, we can conclude that the Cl atom is inclined to interact with the iron atom instead of the Si atom. This is consistent with the experiment result of Liu et al. that found that Si element cannot be chlorinated by HCl gas which leads to the selective chlorination of Fe element [8].

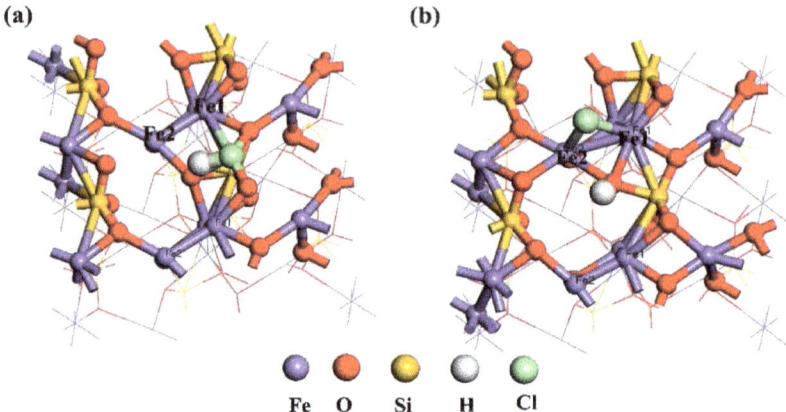

Figure 7. HCl adsorption on Fe_2SiO_4: (**a**) HCl adsorption on Si site before optimization. (**b**) HCl adsorption on Si site after optimization.

3.3. HCl Adsorption on FeV_2O_4

To unravel the HCl adsorption characteristics on the FeV_2O_4(010) surface and the mechanism of selective chlorination of Fe in FeV_2O_4, we conducted a series of DFT calculations as discussed henceforth. The chlorination of FeV_2O_4 via HCl means that the Cl atom bonds with the Fe atom or V atom while the H atom bonds with the O atom. As a result, we focus on the difference between the Fe atom, V atom, and their neighboring O atoms. We obtained the bond order of Fe-O and V-O bonds; we found that there are two classes of V-O bond and two classes of Fe-O bond, as well. The obtained bond order is exhibited in Table 2. As a result, we first conducted a calculation of HCl adsorption on these four sites. The HCl adsorption energy on these sites is exhibited in Table 4.

Table 4. HCl adsorption energy on FeV_2O_4.

Adsorption Site	Adsorption Energy (eV)
Fe7	−2.84
Fe5	−2.72
V-O47	−3.20
V-O16	−0.27

We first conducted HCl adsorption on the FeV_2O_4(001) Fe-terminated surface with the H atom above the O23 atom while the Cl atom was right above the Fe7 atom. The H atom bonds with the O23 atom while the Cl atom bonds with the Fe7 atom. The distance between the Fe7 atom and the O23 atom changed from 1.834 Å to 2.758 Å, while the distance between the Fe5 atom and the O60 atom changed from 1.862 Å to 2.994 Å. The Fe5 atom was attracted by the neighboring O47 atom. This means that the adsorption of HCl on Fe7-O23 results in a significant change in the neighboring Fe-O bond but not the adsorbed Fe7-O23 bond. Figure 8a,b show the structures before and after the geometric optimization of HCl adsorption at the Fe7 site, respectively. The more significant change related to Fe5-O60 over Fe7-O23 may have resulted from the smaller bond order of Fe5-O40, that is, the weaker bond strength of Fe5-O40. The geometry optimization of configuration was where HCl was parallel to the Fe5-O60 bond with the H atom being right above the O60 atom and the Cl atom being right above the Fe5 atom. It can be seen that HCl was dissociated and adsorbed on the FeV_2O_4(010) surface with the H atom bonding with the O60 atom and the Cl atom bonding with the Fe5 atom. The distance of Fe5-O60 changed from 1.862 Å to 2.254 Å, while the distance of Fe7-O23 remain unchanged. The adsorption energy of HCl adsorption on the Fe7-O23 and Fe5-O60 sites was −2.84 eV and −2.72 eV, respectively. The distance change for Fe-O was consistent with the adsorption energy. The minor restructure corresponds to smaller adsorption energy in the absent value. Figure 8c,d show the structures before and after the geometric optimization of HCl adsorption at the Fe5 site, respectively. From the previous result, it can be concluded that it is preferable for HCl to adsorb on the Fe7-O23 site which leads to a more significant configuration restructure. As a result, during the chlorination process of FeV_2O_4, the HCl molecule would adsorb on the Fe7-O23 site first, while the Fe5-O60 bond would break off first.

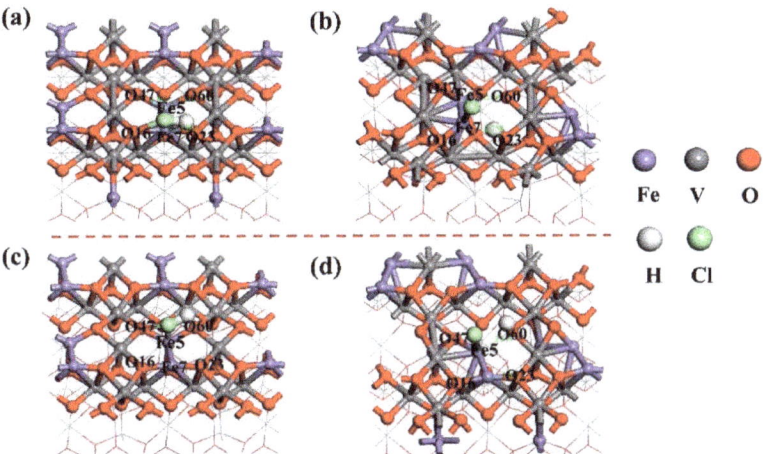

Figure 8. HCl adsorption on FeV_2O_4: (**a**) HCl adsorption on Fe7 site before optimization. (**b**) HCl adsorption on Fe7 site after optimization. (**c**) HCl adsorption on Fe5 site before optimization. (**d**) HCl adsorption on Fe5 site after optimization.

The HCl adsorption on the V-O site was conducted to explore the possibility of chlorination of the V element by HCl gas. There were also two classes of V-O bonds on the FeV_2O_4(010) surface. Therefore, the calculation of HCl adsorption on these two classes of sites was conducted. The HCl adsorption on the V-O47 site was first conducted. The result shows that the HCl molecule was broken off with the H atom adsorbed on the V atom and the Cl atom adsorbed on the O47 atom. The distance of V-O47 changed from 1.963 Å to 2.176 Å, while the distance of Fe5-O47 changed from 1.863 Å to 2.576 Å. In other words, the HCl adsorption on the V-O site leads to a more significant change in the neighboring Fe5-O47 bond, which roots from the weaker bond strength of Fe5-O47. Figure 9a,b show the structures before and after the geometric optimization of HCl adsorption at the V-O47 site, respectively. Similarly, the result of HCl adsorption on the V-O16 shows that the HCl molecule was also dissociated and adsorbed on the FeV_2O_4(010) surface. The most significant configuration restructure other than the dissociation of the HCl molecule is that the distance of V-O16 extended from 2.042 Å to 2.674 Å. Figure 9c,d show the structures before and after the geometric optimization of HCl adsorption at the V-O16 site, respectively. The adsorption energy of HCl adsorption on V-O47 and V-O16 was −3.20 eV and −0.27 eV, respectively. The difference in adsorption energy results from the different configuration restructures. When HCl was adsorbed on the V-O47 site, the most significant change in surface was the breakoff of the Fe5-O47 bond; when the HCl was adsorbed on the V-O16 site, the most significant change in surface was the elongation of the V-O16 bond. This phenomenon implies that the V-O bond's breakoff needs more energy than the Fe-O bond's breakoff, with HCl as the chlorination agent. This is consistent with the experiment result that the Fe element is selectively chlorinated from FeV_2O_4 via HCl gas.

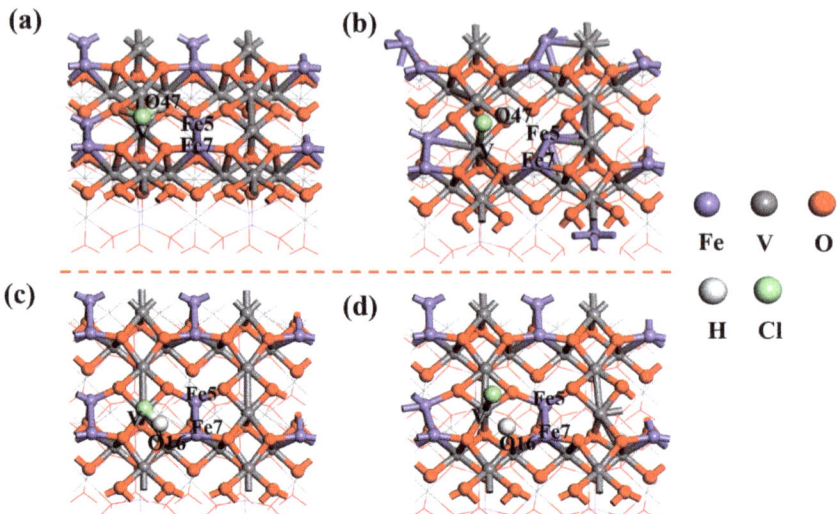

Figure 9. HCl adsorption on FeV$_2$O$_4$: (**a**) HCl adsorption on V-O47 site before optimization. (**b**) HCl adsorption on V-O47 site after optimization. (**c**) HCl adsorption on V-O16 site before optimization. (**d**) HCl adsorption on V-O16 site after optimization.

3.4. Selective Chlorination Mechanism of Fe$_2$SiO$_4$ and FeV$_2$O$_4$

The essence of chlorination of Fe$_2$SiO$_4$ and FeV$_2$O$_4$ is, as a matter of fact, the process of removing the O atom in the form of water while the remaining metal atom bonds with the Cl atom to form chloride. The formation of one water molecule needs two H atoms, that is, two HCl molecules. From the previous discussion, we can conclude that the adsorption of one HCl molecule leads to the formation of one OH$^-$ on the surface. There are two possible pathways for water formation during chlorination, as follows: (1) one HCl molecule is adsorbed on the surface while the H atom bonds with one previous OH$^-$ to form a water molecule; (2) the H atom in the formed OH$^-$ moves to its formed neighbor OH$^-$ to form a water molecule. However, when we optimized the structure using the first pathway, it was found that the H atom in the HCl could not bond with the formed OH$^-$ but was attracted by the neighboring O atom to form another OH$^-$. Therefore, the formation of a water molecule may be a process whereby one adsorbed H atom moves into its neighboring OH$^-$ forming one water molecule, and then the water molecule departs from the surface. The optimized structure with H drifting to neighboring OH$^-$ formed a cluster of H$_2$O$^-$ on the surface, as shown in Figure 10. Figure 10a,b show the optimized structure with H$_2$O$^-$ on the Fe$_2$SiO$_4$(010) surface and the FeV$_2$O$_4$(001) Fe-terminated surface, respectively. The angle of H-O-H and length of the H-O bond in the H$_2$O$^-$ cluster are exhibited in Table 5. The angle of the H-O-H angle and the H-O bond length on both the Fe$_2$SiO$_4$(010) surface and the FeV$_2$O$_4$(001) Fe-terminated surface are close to those of the water molecule which are about 104.45° and 0.96 Å [30]. Therefore, the oxides in vanadium can be chlorinated by two HCl molecules with one O atom removed from the oxide. In other words, the metal atom is the active site for Cl atom adsorption while the O atom is the active site for H atom adsorption. Although the chlorination process can be considered a reduction process, the O atom in the olivine or spinel FeV$_2$O$_4$ acts as a catalyzer, just as Xiaolu Xiong et al. proposed that O adsorption is conducive to the chlorination of Cu$_2$S by NH$_4$Cl [31]. As a result, it is reasonable to conjecture that the pre-oxidation of vanadium before chlorination is a good way to enhance the chlorination rate.

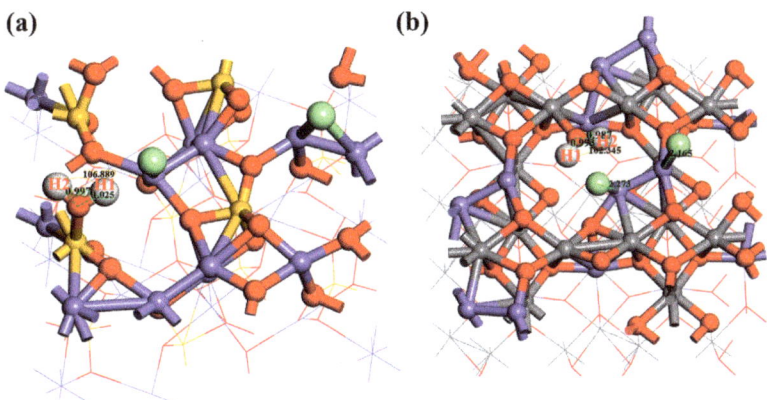

Figure 10. Optimized (**a**) Fe$_2$SiO$_4$(010) and (**b**) FeV$_2$O$_4$(001) Fe-terminated surface with H$_2$O$^-$.

Table 5. Structure parameter of formed H$_2$O$^-$.

	Angle of H-O-H (Degree)	Bond Length of H1-O, H2-O (Å)
Fe$_2$SiO$_4$	106.889	1.025, 0.997
FeV$_2$O$_4$	102.345	0.993, 0.987

We conducted a transition state search of the process of a H atom moving to combine with neighboring OH$^-$ on the Fe$_2$SiO$_4$(010) surface and the FeV$_2$O$_4$ Fe-terminated surface. The results show that the energy barrier of the two processes is 1.80 eV and 0.76 eV, respectively, as shown in Figure 10a,b. Figure 11a,b show the energy path of the chlorination process of Fe$_2$SiO$_4$ and FeV$_2$O$_4$, respectively. It can be concluded that the HCl adsorption on the Fe$_2$SiO$_4$(010) surface and the FeV$_2$O$_4$(001) Fe-terminated surface is thermodynamically favorable while the formation of water is an endothermic process. The energy barrier of H moving on the Fe$_2$SiO$_4$(010) surface is greater than that on the FeV$_2$O$_4$(001) Fe-terminated surface which results from the stronger H-O bond on the Fe$_2$SiO$_4$(010) surface. However, it is preferable for the chlorination of Fe$_2$SiO$_4$ to happen according to the thermodynamic calculations [8]. The HCl adsorption on the Fe site of the FeV$_2$O$_4$(001) Fe-terminated surface is thermodynamically preferable according to previous calculations. During the adsorption process, the breaking bond is the Fe-O bond; while during the water departure process, the breaking bond is the Si-O bond or the V-O bond on the Fe$_2$SiO$_4$(010) and FeV$_2$O$_4$(001) Fe-terminated surface, respectively. Additionally, the electronegativity difference between Si and O was smaller than that between the V and O bond, implying a stronger Si-O bond than the V-O bond. As a result, the conclusion can be made that the rate-determining step in the chlorination of vanadium can be water departure from the surface instead of H moving to neighboring OH$^-$ to form water or HCl adsorption.

It can be concluded that the Cl atom cannot adsorb on the Si atom of Fe$_2$SiO$_4$ according to the previous calculation. Additionally, the adsorption of HCl on the Fe site results in the breakoff of the Fe-O bond and the weakening of the corresponding Si-O bond. Additionally, the H$_2$O$^-$ formation on the surface would weaken the Si-O bond further, eventually resulting in water departure under high-temperature conditions. The difference in the chlorination of FeV$_2$O$_4$ with Fe$_2$SiO$_4$ is that both the Fe atom and V atom are possible adsorption sites for HCl, while only the Fe atom is the adsorption site on the Fe$_2$SiO$_4$(010) surface. Additionally, the HCl adsorption on V-O47 is preferable with a relatively large adsorption energy of 3.20 eV. However, the neighboring Fe-O bond breaks off instead of the adsorbed V-O bond, as described in Section 3.2. The chlorination process is the process of the H atom in the HCl molecule fighting for the O atom with the Fe atom and V atom on the FeV$_2$O$_4$(001) surface. As a result, the selective chlorination of the Fe element or selective chlorination of the V element depends on the difference in attraction between V-O and Fe-O. Additionally, the electronegativity gap between V and O is more significant than that between Fe and O atoms [32,33]. The more significant the electronegativity gap between elements, the stronger the interaction between them. In other words, after the O atom was removed as a water molecule, it was preferable for the Cl atom to combine with the Fe atom instead of the V atom. Therefore, the Fe element can be selectively chlorinated by HCl gas from FeV_2O_4.

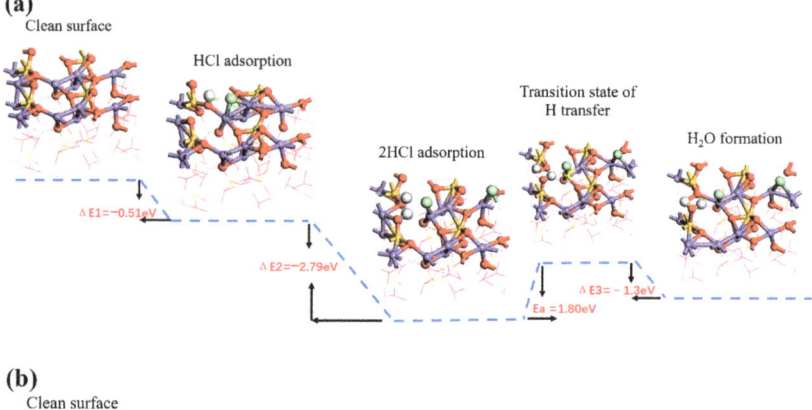

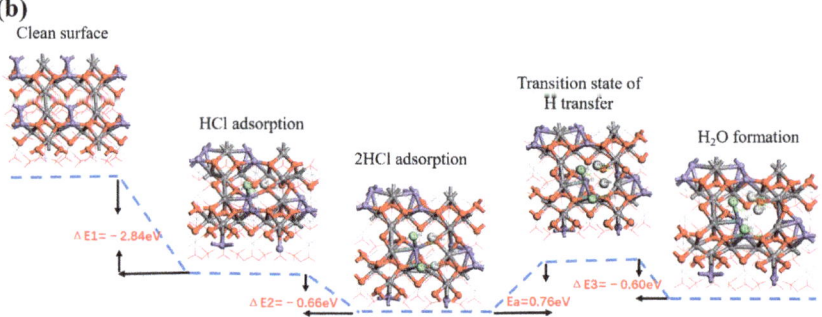

Figure 11. Energy path of chlorination of (**a**) Fe_2SiO_4 and (**b**) FeV_2O_4 by HCl.

4. Conclusions

In summary, the feasibility was calculated using FactSage 8.0 and it was found to be thermodynamically possible to achieve selective chlorination of Fe_2SiO_4 and FeV_2O_4 to form iron chloride. However, FactSage 8.0 could not further explain the microscopic chlorination process, so a first principles calculation was performed. The adsorption and chlorination mechanisms of Fe_2SiO_4 and FeV_2O_4 via HCl gas were studied using DFT calculations. The HCl molecule could be dissociated and adsorbed on both the Fe_2SiO_4(010) surface and the FeV_2O_4(001) Fe-terminated surface. However, the HCl could not adsorb on the silicon site of the Fe_2SiO_4(010) surface which implies that the silicon element cannot be chlorinated by HCl gas. As a result, the iron element in Fe_2SiO_4 could be selectively chlorinated by HCl gas. It was preferable for the HCl molecule to adsorb on the Fe1 site compared to the Fe2 site, which indicated that the Fe1 iron atom may be chlorinated first. Moreover, the DFT calculation showed that the dissociation adsorption of HCl was mainly caused by the attraction between the Fe atom and the Cl atom. The HCl could be adsorbed on both the Fe site and the V site on the FeV_2O_4(001) Fe-terminated surface. However, the electronegativity gap between V and O was more significant than that between Fe and O elements, which implied a stronger attraction between V and O elements. Therefore, the iron element in FeV_2O_4 could be selectively chlorinated by HCl gas. The HCl adsorption in Fe_2SiO_4 and FeV_2O_4 led to the formation of OH^- on their surface. The neighboring adsorbed H atom could be attracted by OH^- to form a water molecule. Thus, Fe_2SiO_4 and FeV_2O_4 could be chlorinated by removing the O atom through water formation. The chlorination mechanism of Fe_2SiO_4 and FeV_2O_4 investigated may provide possible ideas for the chlorination of other oxides.

Author Contributions: Conceptualization, J.D. and Y.X.; methodology, J.D. and Y.X.; validation, J.D.; formal analysis, Y.X.; investigation, Y.X., S.L. and L.W.; resources, S.L. and L.W.; writing—original draft preparation, J.D. and Y.X.; writing—review and editing, S.L. and L.W.; visualization, S.L. and L.W.; supervision, S.L., L.W. and K.-C.C.; funding acquisition, S.L., K.-C.C. and L.W. All authors have read and agreed to the published version of the manuscript.

Funding: This research was funded by the National Natural Science Foundation of China (Nos. 52274406, 51922003, 51904286) and Interdisciplinary Research Project for Young Teachers of USTB (Fundamental Research Funds for the Central Universities) (FRF-IDRY-21-015).

Data Availability Statement: The data presented in this study are available from the corresponding author, upon reasonable request.

Acknowledgments: The authors are grateful for the financial support of this work from the National Natural Science Foundation of China (Nos. 52274406, 51922003, 51904286) and Interdisciplinary Research Project for Young Teachers of USTB (Fundamental Research Funds for the Central Universities) (FRF-IDRY-21-015).

Conflicts of Interest: The authors declare no conflict of interest.

References

1. Liu, L.; Du, T.; Tan, W.; Zhang, X.; Yang, F. A novel process for comprehensive utilization of vanadium slag. *Int. J. Miner. Metall. Mater.* **2016**, *23*, 156–160. [CrossRef]
2. Fang, H.; Li, H.; Xie, B. Effective chromium extraction from chromium-containing vanadium slag by sodium roasting and water leaching. *ISIJ Int.* **2012**, *52*, 1958–1965. [CrossRef]
3. Li, X.; Xie, B. Extraction of vanadium from high calcium vanadium slag using direct roasting and soda leaching. *Int. J. Miner. Metall. Mater.* **2012**, *19*, 595–601. [CrossRef]
4. Moskalyk, R.; Alfantazi, A. Processing of vanadium: A review. *Miner. Eng.* **2003**, *16*, 793–805. [CrossRef]
5. Li, X.; Bing, X.; Wang, G.E.; Li, X.-J. Oxidation process of low-grade vanadium slag in presence of Na_2CO_3. *Trans. Nonferrous Met. Soc. China* **2011**, *21*, 1860–1867. [CrossRef]
6. Zhao, Y.H. Melting Properties and Melt Structure of Vanadium Slag Containing Chromium Oxide. Master's Thesis, Northeastern University, Shenyang, China, 2015.
7. Liu, S. *Fundametal Studies on Selective Chlorination of Valuable Elements (Fe, Mn, V, Cr and Ti) from Vanadium Slag and Utilizations towards High Valuable Added*; Beijing University of Science and Technology: Beijing, China, 2019.
8. Liu, S.; Wang, L.; Chou, K. Selective Chlorinated Extraction of Iron and Manganese from Vanadium Slag and Their Application to Hydrothermal Synthesis of $MnFe_2O_4$. *ACS Sustain. Chem. Eng.* **2017**, *5*, 10588–10596. [CrossRef]
9. Zhang, J.; Zhang, W.; Zhang, L.; Gu, S. Mechanism of vanadium slag roasting with calcium oxide. *Int. J. Miner. Process.* **2015**, *138*, 20–29. [CrossRef]
10. Li, H.; Fang, H.; Wang, K.; Zhou, W.; Yang, Z.; Yan, X.; Ge, W.-S.; Li, Q.-W.; Xie, B. Asynchronous extraction of vanadium and chromium from vanadium slag by stepwise sodium roasting–water leaching. *Hydrometallurgy* **2015**, *156*, 124–135. [CrossRef]
11. Chen, D.; Zhao, L.; Liu, Y.; Qi, T.; Wang, J.; Wang, L. A novel process for recovery of iron, titanium, and vanadium from titanomagnetite concentrates: NaOH molten salt roasting and water leaching processes. *J. Hazard. Mater.* **2013**, *244*, 588–595. [CrossRef]
12. Liu, S.; Wang, L.; Chou, K. A Novel Process for Simultaneous Extraction of Iron, Vanadium, Manganese, Chromium, and Titanium from Vanadium Slag by Molten Salt Electrolysis. *Ind. Eng. Chem. Res.* **2016**, *55*, 12962–12969. [CrossRef]
13. Aas, H.; Nordheim, R. Process for the Production of Ferro-Vanadium Directly from Slag Obtained from Vanadium-Containing Pig Iron. U.S. Patent 3,579,328, 18 May 1971.
14. De Aguiar, E.M.M.M.; Botelho Junior, A.B.; Duarte, H.A.; Espinosa, D.C.R.; Tenório, J.A.S.; Baltazar, M.P.G. Leaching of Ti and V from the non-magnetic fraction of ilmenite-based ore: Kinetic and thermodynamic modelling. *Can. J. Chem. Eng.* **2022**, *100*, 3408–3418. [CrossRef]
15. Lee, J.-C.; Kurniawan; Kim, E.-Y.; Chung, K.W.; Kim, R.; Jeon, H.-S. A review on the metallurgical recycling of vanadium from slags: Towards a sustainable vanadium production. *J. Mater. Res. Technol.* **2021**, *12*, 343–364. [CrossRef]
16. Ning, P.; Lin, X.; Wang, X.; Cao, H. High-efficient extraction of vanadium and its application in the utilization of the chromium-bearing vanadium slag. *Chem. Eng. J.* **2016**, *301*, 132–138. [CrossRef]

17. Li, M.; Wei, C.; Fan, G.; Wu, H.; Li, C.; Li, X. Acid leaching of black shale for the extraction of vanadium. *Int. J. Miner. Process.* **2010**, *95*, 62–67. [CrossRef]
18. Liu, S.; Wang, L.; Chou, K. Synthesis of metal-doped Mn-Zn ferrite from the leaching solutions of vanadium slag using hydrothermal method. *J. Magn. Magn. Mater.* **2018**, *449*, 49–54. [CrossRef]
19. Liu, S.; He, X.; Wang, Y.; Wang, L. Cleaner and effective extraction and separation of iron from vanadium slag by carbothermic reduction-chlorination-molten salt electrolysis. *J. Clean. Prod.* **2021**, *284*, 124674. [CrossRef]
20. Liu, S.; Wang, L.; Chou, K. Innovative method for minimization of waste containing Fe, Mn and Ti during comprehensive utilization of vanadium slag. *Waste Manag.* **2021**, *127*, 179–188. [CrossRef]
21. Liu, S.; Xue, W.; Wang, L. Extraction of the Rare Element Vanadium from Vanadium-Containing Materials by Chlorination Method: A Critical Review. *Metals* **2021**, *11*, 1301. [CrossRef]
22. Clark, S.J.; Segall, M.D.; Pickard, C.J.; Hasnip, P.J.; Probert, M.I.J.; Refson, K.; Payne, M.C. First principles methods using CASTEP. *Z. Kristallogr.* **2005**, *220*, 567–570. [CrossRef]
23. Corso, A.D. Pseudopotentials periodic table: From H to Pu. *Comput. Mater. Sci.* **2014**, *95*, 337–350. [CrossRef]
24. Perdew, J.; Zunger, A. Self-interaction correction to density-functional approximations for many-electron systems. *Phys. Rev. B* **1981**, *23*, 5048. [CrossRef]
25. Monkhorst, H.; Pack, J. Special points for Brillouin-zone integrations. *Phys. Rev. B* **1976**, *13*, 5188. [CrossRef]
26. Fischer, T.; Almlof, J. General methods for geometry and wave function optimization. *J. Phys. Chem.* **1992**, *96*, 9768–9774. [CrossRef]
27. Santoro, R.; Newnham, R.; Nomura, S. Magnetic properties of Mn_2SiO_4 and Fe_2SiO_4. *J. Phys. Chem. Solids* **1966**, *27*, 655–666. [CrossRef]
28. Nii, Y.; Sagayama, H.; Arima, T.; Aoyagi, S.; Sakai, R.; Maki, S.; Nishibori, E.; Sawa, H.; Sugimoto, K.; Ohsumi, H. Orbital structures in spinel vanadates A V_2O_4 (A = Fe, Mn). *Phys. Rev. B* **2012**, *86*, 125142. [CrossRef]
29. Sun, L. Adhesion and electric structure at $Fe_3O_4/FeCr_2O_4$ interface: A first principles study. *J. Alloys Compd.* **2021**, *875*, 160065. [CrossRef]
30. Liu, W.; Jiang, P.; Xiao, Y.; Liu, J. A study of the hydrogen adsorption mechanism of $W_{18}O_{49}$ using first-principles calculations. *Comput. Mater. Sci.* **2018**, *154*, 53–59. [CrossRef]
31. Xiong, X.; Sun, C.; Li, G.; Yu, C.; Xu, Q.; Zou, X.; Cheng, H.; Zhu, K.; Li, S.; Lu, X. A novel approach for metal extraction from metal sulfide ores with NH_4Cl: A combined DFT and experimental studies. *Sep. Purif. Technol.* **2021**, *267*, 118626. [CrossRef]
32. Sahoo, P.; Debroy, T.; McNallan, M. Surface tension of binary metal—Surface active solute systems under conditions relevant to welding metallurgy. *Metall. Trans. B* **1988**, *19*, 483–491. [CrossRef]
33. Gao, X.; Wachs, I. Molecular engineering of supported vanadium oxide catalysts through support modification. *Top. Catal.* **2002**, *18*, 243–250. [CrossRef]

Disclaimer/Publisher's Note: The statements, opinions and data contained in all publications are solely those of the individual author(s) and contributor(s) and not of MDPI and/or the editor(s). MDPI and/or the editor(s) disclaim responsibility for any injury to people or property resulting from any ideas, methods, instructions or products referred to in the content.

Article

Improved Process for Separating TiO₂ from an Oxalic-Acid Hydrothermal Leachate of Vanadium Slag

Qingdong Miao [1], Ming Li [1], Guanjin Gao [1], Wenbo Zhang [2,3], Jie Zhang [2,3] and Baijun Yan [2,3,*]

[1] State Key Laboratory of Vanadium and Titanium Resources Comprehensive Utilization, Panzhihua 617000, China
[2] State Key Laboratory of Advanced Metallurgy, Beijing 100083, China
[3] School of Metallurgical and Ecological Engineering, University of Science and Technology Beijing, Beijing 100083, China
* Correspondence: baijunyan@ustb.edu.cn

Abstract: In the present study, a process of separating high-quality TiO_2 from an oxalic-acid leachate of vanadium slag was proposed. It consists of two steps; oxalic acid was firstly recovered from the leachate by the cooling-crystallization method, and subsequently TiO_2 was separated from the oxalic-acid recovered leachate by the hydrothermal precipitation method. The experimental results indicate that oxalic acid can be recovered from the leachate by cooling crystallization at 5 °C, and after the recovery of oxalic acid, the purity of final TiO_2 product can also be improved. For example, when the leachate was cooled directly at 5 °C for 5 h, about 7% of oxalic acid was recovered, and the purity of final TiO_2 product improved from 95.7% to 96.6%. Furthermore, it was found that when some HCl solution was added to the leachate, both the recovery percentage of oxalic acid and the purity of TiO_2 product increased. For instance, when 15 vol% of HCl solution relative to pregnant leachate was added, about 35% oxalic acid was recovered by cooling crystallization at 5 °C for 3 h, and the anatase TiO_2 product with a purity of 99.2% was obtained by hydrothermal precipitation at 140 °C for 2.5 h.

Keywords: TiO_2; hydrothermal separation; oxalic-acid leachate; cooling crystallization

1. Introduction

Vanadium titanomagnetite (VTM) ore is an important deposit, not only for its plenty reserves of vanadium, titanium, chromium, and iron but also for its globally abundant distribution in countries such as Russia, China, South Africa, New Zealand, Canada, and Australia [1,2]. In the VTM deposit, the principal oxide minerals are titanomagnetite ($Fe_{3-x}Ti_xO_4$) and ilmenite ($FeTiO_3$). In titanomagnetite, a small amount of vanadium and chromium exist simultaneously by substituting titanium [3]. After the beneficiation process of VTM, the obtained titanomagnetite concentrate is used as feedstock for iron-making, and ilmenite concentrate is used for producing titanium dioxide [4,5].

The titanomagnetite concentrate is generally smelted through the traditional blast furnace process, and liquid iron containing vanadium, titanium, and chromium is produced [4,6,7]. Then, to recover the vanadium that is reduced and dissolved in the liquid iron during the blast furnace smelting process, the liquid iron is blown in an oxygen-converter, which produces the vanadium, titanium, chromium, and some impurities, and some iron was oxidized to form a by-product called vanadium slag [8–10]. Although this by-product is generally called vanadium slag, a considerable amount of titanium and chromium is also contained in this it beside vanadium [11]. Typically, apart from 10~19% V_2O_3, 7~14% TiO_2 and 0.9~5% Cr_2O_3 exist in vanadium slag simultaneously. However, traditionally, vanadium slag was just adopted as a feedstock for vanadium production, and no more attention was paid to the recovery of titanium and chromium from vanadium slag.

Sodium salt roasting-water leaching and calcification roasting-acid leaching are the most common processes for vanadium production from vanadium slag [12–14]. In the sodium-salt-roasting–water-leaching process, the vanadium slag is roasted in an air atmosphere with the addition of sodium salt (one or a combination of NaCl, Na_2CO_3, or Na_2SO_4) to transform trivalent vanadium that occurs in vanadium slag in water-soluble sodium vanadate, and vanadium pentoxide is prepared by the following water-leaching, purification, precipitation, and calcination steps [12,15]. In the calcification roasting-acid leaching process, other than sodium salt, lime or limestone is used as an additive, and the trivalent vanadium that occurs in vanadium slag is transformed into acid-soluble calcium vanadate through roasting step and then leached by sulfuric acid [16,17]. No matter which of these two processes is adopted, only the vanadium that occurs in vanadium slag can be extracted. The valuable components of titanium and chromium that occur in vanadium slag are left in the residues, which causes not only an enormous waste of resources but also potential risk to the environment [18–20].

To solve the problems mentioned above, a novel approach to the co-extraction of vanadium, titanium, and chromium from vanadium slag was proposed in our previous publications [21,22]. The processes of co-extraction are as follows: firstly, the vanadium slag without roasting is directly hydrothermally leached by an oxalic-acid solution, by which the trivalent vanadium ion, tetravalent titanium ion, and trivalent chromium ion that occur in vanadium slag are simultaneously leached out; next, the separations of titanium oxide, vanadium oxide, and chromium oxide from the pregnant leachate are achieved through hydrothermal precipitation at temperatures higher than leaching. Based on this novel approach, the direct leaching coefficients of vanadium, titanium, and chromium in vanadium slag can reach 97.9%, 98.6%, and 93.3%, respectively, under the conditions of a temperature of 125 °C, an oxalic acid concentration of 25%, a liquid-to-solid mass ratio of 8:1, a reaction time of 90 min, the iron powder addition of 3.2%, and a stirring speed of 500 r/min. Additionally, a product of spherical anatase TiO_2 with a purity of 95.7% can be separated hydrothermally from the leachate after precipitation at 150 °C for 2.5 h [23].

This novel hydrothermal leaching and separation process is efficient and clean for the co-extraction of vanadium, titanium, and chromium from vanadium slag and for separating titanium oxide. However, two problems still need to be solved. One is that the consumption of oxalic acid needs to be reduced due to the relatively high price of oxalic acid. The other is that the quality or purity of the TiO_2 product need be improved. To these ends, in this study, a route of the cooling crystallization of oxalic acid from the leachate followed by the hydrothermal precipitation of TiO_2 was proposed and attempted. Unexpectedly, it was found that the cooling crystallization method was not only effective in recovering oxalic acid from the leachate but favorable for improving the purity of the TiO_2 product. So, after the effectiveness of the route was confirmed, the factors, such as acidity, temperature, etc., on the recovery extent of oxalic acid and quality of TiO_2 product were investigated systematically. Finally, an improved process of recovering oxalic acid and producing high-quality TiO_2 was presented.

2. Experimental

2.1. Co-Extraction of Vanadium, Titanium, and Chromium from Vanadium Slag

The detailed procedure for co-extraction can be found in previous publications [21,22]. For the sake of brevity, only the primary steps are narrated here. The used vanadium slag is analyzed by X-ray fluorescence spectroscopy (XRF-1800, Shimadzu Co., Ltd, Tokyo, Japan) and characterized by X-ray diffraction (XRD, MAC Science Co. Ltd., Kanagawa, Japan). The results indicated that it is composed of 10.61% V_2O_5, 8.65% TiO_2, 4.38% Cr_2O_3, 43.51% Fe_2O_3, 10.61% SiO_2, 7.98% MnO, 3.32% CaO, 3.08% MgO, and 1.64% Al_2O_3, and it is mainly constituted of spinel, olivine, and pyroxene mineral phases. About 1.25 g of vanadium slag, 0.04 g of iron powder, 10 g of distilled water, and 3.33 g of oxalic acid were placed into an autoclave of 50 mL; then, the leaching reaction was carried out at 125 °C for 90 min with a stirring rate of 500 r/min. After hydrothermal leaching reaction, the reacted mixture was filtered and a pregnant leachate was obtained. This obtained pregnant leachate was used as the raw solution for recovery of oxalic acid and the separation of TiO_2. Its composition was determined by ICP optical emission spectrometer (ICP-OES, OPTIMA 7000DV, Waltham, MA, USA), and the results are presented in Table 1.

Table 1. Chemical composition of the pregnant leachate (g/L).

Elements	V	Ti	Cr	Fe	Mn	Si	Ca	Mg	Al
Concentration (g/L)	2.42	2.30	1.14	0.31	0.79	0.47	0.42	0.29	0.35

2.2. Recovery of Oxalic Acid from Leachate through Cooling Crystallization

About 20 mL of pregnant leachate was placed into a glass beaker; then, the beaker was placed into a thermoelectric cooling cup, which can hold the temperature constant at 5 °C. When the leachate was cooled for 20 min, about 0.02 g of oxalic acid was added as seed crystal. Thereafter, the leachate was stirred once every 10 min to make the temperature homogeneous.

In order to evaluate the recovery extent of oxalic acid, a tiny amount of leachate was taken out using a pipette during the cooling process at setting time intervals to track the change of oxalate concentration. The concentration of oxalate was measured by ion chromatography. Then, the percentage of recovery of oxalic acid was calculated according to Equation (1).

$$\beta = \left(1 - \frac{C_t}{C_0}\right) \times 100\% \qquad (1)$$

where β denotes the recovery fraction of oxalic acid, and C_t and C_0 signify the concentration of oxalate in leachate at the sampling time and the initial time, respectively.

2.3. Separation of TiO_2 from Oxalic-Acid Recovered Leachate

After a cooling crystallization reaction, the crystalized solid was filtered out. The filtered solid was identified by XRD, and its composition was measured by ICP-OES. The obtained filtrate, which is referred to as oxalic-acid recovered leachate in the following text, was further used for the separation of TiO_2.

About 15 mL of oxalic-acid recovered leachate was placed into a 50 mL autoclave to perform the hydrothermal precipitation reaction for 2.5 h. Additionally, the hydrothermal temperature, as well as the property of the oxalic-acid recovered leachate, were changed to investigate their effects on the precipitation extent and purity of TiO_2.

After precipitation reaction, the precipitate was filtered out. Then, XRD measurement was performed to identify its phase structure, ICP-OES measurements were carried out to determine the impurity and the purity of the product, and SEM measurements were adopted to characterize the morphology of the product.

To determine the hydrothermal precipitation fraction of titanium from the oxalic-acid recovered leachate, the concentration of titanium in oxalic-acid recovered leachate, $C_{Ti,i}$, and the concentration of titanium in the filtrate after hydrothermal precipitation reaction, $C_{Ti,f}$, were measured by ICP-OES. Then, the precipitation fraction of titanium, η, was calculated according to Equation (2).

$$\eta = \left(1 - \frac{C_{Ti,f}}{C_{Ti,i}}\right) \times 100\% \qquad (2)$$

3. Results and Discussion

3.1. Practicability of the Design for Recovery of Oxalic Acid and Separation of TiO_2

The design of this study consists of two steps; one is to recover oxalic acid from the pregnant leachate through the cooling crystallization method, the following is to separate TiO_2 from the oxalic-acid recovered leachate by the hydrothermal precipitation method. So, first of all, the feasibility of the whole design was confirmed.

To perform the confirmation, the pregnant leachate with a composition shown in Table 1 was directly cooled to 5 °C and held for 5 h; then, the crystalized precipitate was filtered out and identified by XRD measurement. The measured XRD pattern is shown in Figure 1, from which it can be seen that the obtained precipitate is the well crystalized oxalic acid phase.

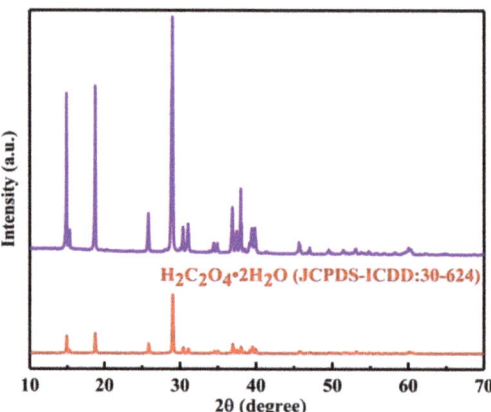

Figure 1. XRD pattern of the precipitate by cooling crystallization.

To further estimate the purity of the obtained oxalic acid phase, it was analyzed by ICP-OES. The contents of the main impurities are shown in Table 2. From the contents of impurities, the purity of the crystalized oxalic acid can be estimated to be higher than 99%.

Table 2. Contents of main impurities in the crystalized oxalic acid phase.

Elements	Fe	Ca	Si	Others
Content (wt. %)	0.03	0.02	0.06	trace

In the following step, the oxalic-acid recovered leachate was directly used for separating TiO_2 by hydrothermal precipitation method. The sample was heated to 140 °C in autoclave and held for 2.5 h; then, the sample was filtered. The solid phase filtered out was identified by XRD, and its composition was determined by ICP-OES.

Figure 2 shows the measured XRD pattern. By comparison with the standard pattern of anatase TiO_2 shown together in Figure 2, it can be seen that the obtained solid product is single-phase TiO_2 with an anatase structure. The composition of the product measured by ICP-OES is shown in Table 3, which indicates that the purity of the product is acceptable.

To evaluate the effect of recovering oxalic acid on the quality of the obtained TiO_2 product, our previous reported results were adopted as a reference for comparison. In our previous study [23], the pregnant leachate without the recovery of oxalic acid was used as a raw material of hydrothermal precipitation. The composition of the obtained TiO_2 product, as shown in Table 4, is cited.

By comparing the composition in Table 3 with that in Table 4, it can be seen that the purity of TiO_2 is improved slightly. Especially, the contents of V_2O_5 and Cr_2O_3 impurities can be reduced through recovering oxalic acid from the pregnant leachate.

The above results indicate that not only is the design of the present study feasible, it is also favorable for improving quality of the TiO_2 product.

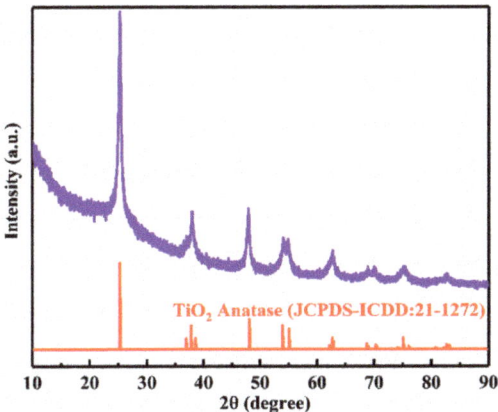

Figure 2. XRD pattern of the precipitate by hydrothermal precipitation.

Table 3. Composition of the product of hydrothermal precipitation.

Components	TiO_2	V_2O_5	Cr_2O_3	Fe_2O_3	MnO	SiO_2	CaO	MgO	Al_2O_3
Content (wt. %)	96.6	1.08	0.01	0.29	0.22	1.19	0.24	0.05	0.26

The content of TiO_2 was obtained by subtracting the contents of impurities.

Table 4. Composition of the TiO_2 product in previous study [23].

Components	TiO_2	V_2O_5	Cr_2O_3	Fe_2O_3	MnO	SiO_2	CaO	MgO	Al_2O_3
Content (wt. %)	95.7	2.44	0.10	0.24	0.15	1.27	0.04	0.03	<0.01

The content of TiO_2 was obtained by subtracting the contents of impurities.

3.2. Optimization of the Oxalic-Acid Recovering Process

Now that oxalic acid can be recovered from the pregnant leachate by the cooling crystallization method, the more effective conditions for recovering oxalic acid need to be further investigated. We know that oxalic acid is a weak acid; four species (H^+, $H_2C_2O_4$, $HC_2O_4^-$ and $C_2O_4^{2-}$) exist in its aqueous solution. Inspecting the chemical reactions among these species, it can be concluded easily that increasing the concentration of H^+ will definitely cause an increase in $H_2C_2O_4$ content. The effect was calculated according to their chemical equilibrium reactions, and the results are shown graphically in Figure 3.

From Figure 3, it can be seen clearly that the content of the $H_2C_2O_4$ molecule increases significantly when the pH value is less than 2. This increase in content for the $H_2C_2O_4$ molecule will promote its crystallization and precipitation. Hence, by adding HCl solution into pregnant leachate to change its pH value, the effect of acidity on the percentage of recovery of oxalic acid was studied.

A HCl solution with a concentration of 36.64% was used as the additive, and three different adding amounts 5 vol%, 10 vol%, and 15 vol% relative to the pregnant leachate were investigated for comparison. After adding HCl solution, the pregnant leachate was placed into a cooling cup of 5 °C. At various time intervals, sampling was carried out to monitor the change of oxalate concentration in pregnant leachate, and the percentage of recovery of oxalic acid by cooling crystallization was calculated. The obtained results at various conditions are exhibited in Figure 4.

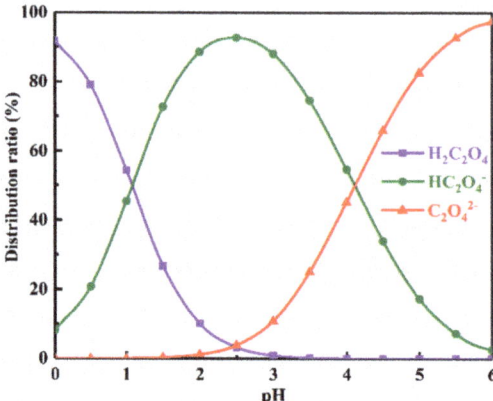

Figure 3. Effect of pH value on the distribution of various species in oxalic acid solution.

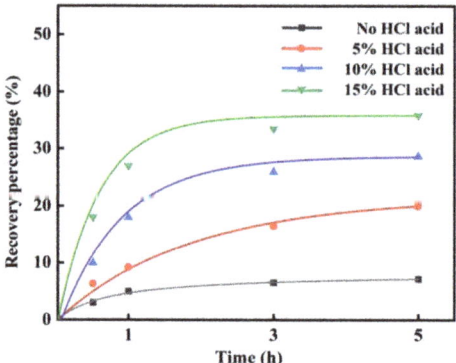

Figure 4. Recovery extent of oxalic acid at various HCl adding conditions.

From the results shown in Figure 4, it can be found that, compared with the case without adding HCl, both the extent and the rate of recovery of oxalic acid increased sharply. When 15 vol% HCl was added, the percentage of recovery reached about 35% after cooling crystallization for 5 h, which is about seven times higher than no adding of HCl and about two times higher than the case of adding 5 vol% HCl. Concerning the rate of recovery, the recovery fraction reached nearly to the maximum after 1 h when 15 vol% HCl was added, whereas 3 h was need to reach the maximum when 10 vol% HCl was added. Hence, from the perspective of recovering oxalic acid, adding 15 vol% is favorable for improving the recovery efficiency.

3.3. Optimization of the TiO_2 Separating Process

By adding HCl solution, the recovery fraction of oxalic acid can be improved significantly. Then, the effect of the added HCl on the following step, i.e., the hydrothermal precipitation of TiO_2, needs to be evaluated. So, the effects on hydrothermal temperature, precipitation extent, and the quality of produced TiO_2 were investigated.

About 15 mL leachate of oxalic acid recovered was placed into a 50 mL autoclave and then heated and held at different temperatures of 120 °C, 130 °C, 140 °C, and 150 °C, respectively, for 2.5 h. After hydrothermal precipitation, the filtered solid was identified by XRD and analyzed by ICP-OES to determine the contents of impurities, while the filtrate was also analyzed by ICP-OES to measure the content of Ti to determine the precipitation extent.

In Figure 5, the effects of the added amount of HCl solution, as well as the hydrothermal temperature, on the precipitation fractions of TiO$_2$ are presented. By comparing the results shown in Figure 5 with the case of no HCl added, it can be found that by adding HCl, the precipitating temperature can be lowered significantly. For example, without adding HCl, almost no TiO$_2$ was precipitated at 120 °C, whereas when 5 vol% HCl solution was added, the precipitation fraction of TiO$_2$ was already higher than 90% at 120 °C, reaching 96.5% at 130 °C, and was close to the complete extent at 140 °C. However, with the further addition of HCl, this positive effect on the precipitating temperature will deteriorate firstly and then resume gradually. As 10 vol% HCl solution was added, the precipitation fractions of TiO$_2$ at all temperatures were less than the 5 vol% HCl solution added case; then, when 15 vol% HCl solution was added, the degraded precipitation fraction showed a recovering trend. This change of precipitation extent of TiO$_2$ implies that the adding of HCl causes the change of occurrence state of species in solution, which needs to be elucidated deeply in the future work.

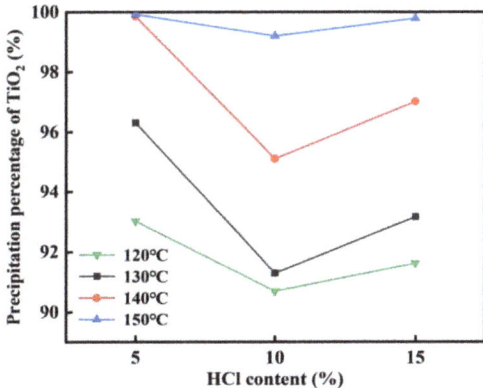

Figure 5. Effects of added HCl on the hydrothermal precipitation extent of TiO$_2$.

Since the hydrothermal precipitation extent at 140 °C for each case is relatively high and acceptable, the precipitates obtained at this temperature were used to characterize the quality of the product. Firstly, their phase structures were identified by XRD; the measured patterns are shown in Figure 6. It can be seen that the adding of HCl has almost no effect on the structure and crystallinity of the product. Although the crystallinity of the products is not perfect, they can be easily identified as anatase TiO$_2$ through comparison with standard pattern.

The contents of impurities in the products measured by ICP-OES are summarized in Table 5, where the contents for TiO$_2$ were calculated by subtracting all impurities. As can be seen, the purity of the product increases gradually when more HCl is added. When 15 vol% HCl solution was added, a product with TiO$_2$ content higher than 99% was obtained.

To provide an overview of the effect of added HCl on the contents of impurities, their changing trend with an increasing amount of added HCl is summarized in Figure 7. It can be seen that, apart from MgO and Cr$_2$O$_3$, the content of the other impurities can be suppressed to some extent through adding HCl. Especially, the impurities V$_2$O$_5$, Al$_2$O$_3$, and MnO can be eliminated effectively.

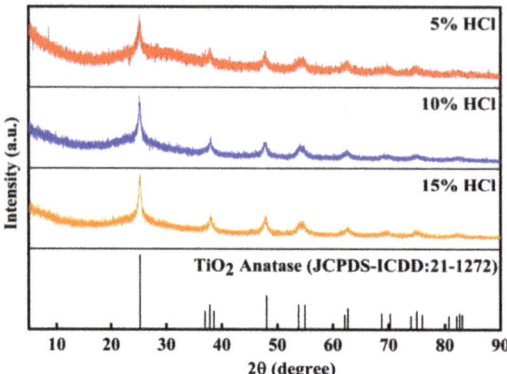

Figure 6. XRD patterns of the hydrothermal product obtained at 140 °C.

Table 5. Compositions of hydrothermal precipitation product at 140 °C (wt. %).

Sample	TiO$_2$	V$_2$O$_5$	Cr$_2$O$_3$	Fe$_2$O$_3$	MnO	SiO$_2$	CaO	MgO	Al$_2$O$_3$
No HCl added	96.6	1.08	0.01	0.29	0.22	1.19	0.24	0.05	0.26
5 vol% HCl added	96.4	0.09	0.06	0.1	0.04	2.81	0.29	0.03	0.11
10 vol% HCl added	98.1	0.08	0.03	0.1	0.02	1.37	0.15	0.03	0.04
15 vol% HCl added	99.2	0.07	0.03	0.1	0.01	0.36	0.11	0.04	0.03

The content of TiO$_2$ was obtained by subtracting the contents of impurities.

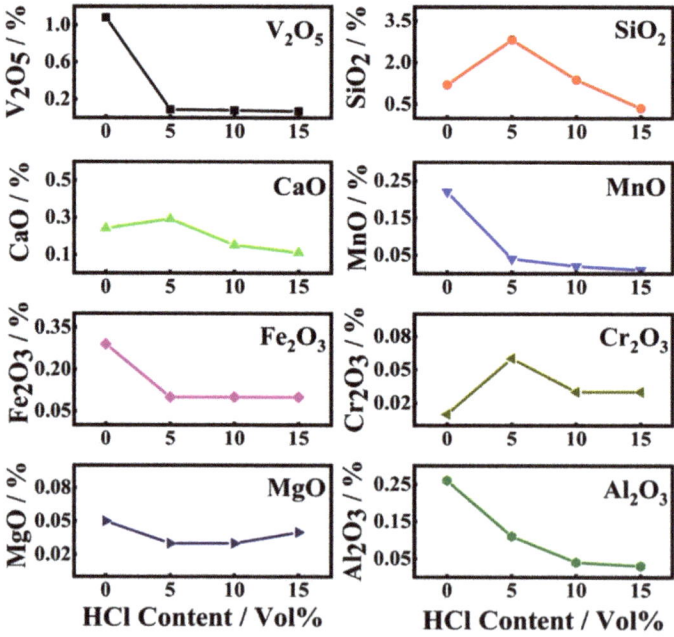

Figure 7. Effect of added HCl on the changing trend of impurities in TiO$_2$ product.

To further inspect the effect of added HCl on the micromorphology of hydrothermal product, the products of various HCl adding amounts were observed through SEM, and the typical micrographics of the products are shown in Figure 8. It can be seen that both the particle size and the morphology of the product were altered significantly after HCl was added to the pregnant leachate. The possible reason may be adding HCl can promote the recovery of oxalic acid and subsequently cause a remarkable decrease in oxalate concentration in solution. With the decrease in the oxalate concentration, the decomposing temperature for the complex of Ti ions and oxalate may be lowered, which will ultimately influence the hydrothermal precipitating extent as well as the morphology of the product.

Figure 8. Effect of added HCl on the changing trend of impurities in TiO_2 product.

4. Conclusions

As one of a series of studies about the clean and efficient utilization of vanadium slag, this study is concerned with the processing of a Ti-bearing oxalic acid hydrothermal leachate of vanadium slag. From an economic point of view, a process of first recovering part of the oxalic acid from the pregnant leachate by the cooling crystallization method and then separating TiO_2 by the hydrothermal precipitation method was designed. The experimental results confirmed that this process is not only feasible but also favorable for improving the quality of the separated TiO_2 product. Then, an optimization of the process was carried out, and the obtained results are as follows:

(1) Adding some HCl to the pregnant leachate can further improve the recovery extent and rate of oxalic acid. When 15 vol% HCl was added, the recovery percent of oxalic acid reached about 35% after cooling crystallization at 5 °C for 3 h, which is about seven times higher than the recovery percent when not adding HCl and about two times higher than the recovery percent when adding 5 vol% HCl.

(2) The adding of HCl in the pregnant leachate does not result in negative effects on the hydrothermal precipitation of TiO_2 from the leachate. Additionally, it helps to ameliorate the quality of the precipitated TiO_2. When 15 vol% HCl was added, the TiO_2 product with a purity higher than 99.2% was obtained.

Author Contributions: Conceptualization Q.M. and M.L.; methodology, G.G.; formal analysis, W.Z.; draft preparation J.Z.; funding acquisition, project administration, B.Y. All authors have read and agreed to the published version of the manuscript.

Funding: This research was funded by National Natural Science Foundation of China grant number 52174274 and the State Key Laboratory of Vanadium and Titanium Resources Comprehensive Utilization grant number 2021P4FZG01A.

Data Availability Statement: Not applicable.

Conflicts of Interest: The authors declare that they have no conflict of interest.

References

1. Li, W.; Fu, G.; Chu, M.S. An effective and cleaner process to recovery iron, titanium, vanadium, and chromium from Hongge vanadium titanomagnetite with hydrogen-rich gases. *Ironmak. Steelmak.* **2021**, *48*, 33–39. [CrossRef]
2. Li, W.; Fu, G.; Chu, M.S. Reduction kinetics of hongge vanadium titanomagnetite-oxidized pellet with simulated shaft furnace gases. *Steel Res. Int.* **2017**, *88*, 1600228. [CrossRef]
3. Rorie, G.; Aleksandar, N. The extraction of vanadium from titanomagnetites and other sources. *Miner. Eng.* **2020**, *146*, 106106.
4. Sun, H.Y.; Wang, J.S.; Han, Y.H. Reduction mechanism of titanomagnetite concentrate by hydrogen. *Int. J. Miner. Process.* **2013**, *125*, 122–128. [CrossRef]
5. Kushnarev, A.V.; Mironov, K.V.; Zagainov, S.A. Improvement in vanadium-containing titanomagnetite processing technology. *Mater. Sci. Eng.* **2020**, *996*, 012062. [CrossRef]
6. Tang, W.D.; Yang, S.T.; Xue, X.X. Effect of Titanium on the Smelting Process of Chromium-Bearing Vanadium Titanomagnetite Pellets. *JOM* **2021**, *73*, 1362–1370. [CrossRef]
7. Wang, S.; Guo, Y.F.; Zheng, F.Q. Behavior of vanadium during reduction and smelting of vanadium titanomagnetite metallized pellets. *Trans. Nonferrous Met. Soc. China* **2020**, *30*, 1687–1696. [CrossRef]
8. Smirnov, L.A.; Kushnarev, A.V.; Fomichev, M.S. Oxygen-converter processing of vanadium-bearing hot metal. *Steel Transl.* **2013**, *43*, 587–592. [CrossRef]
9. Goncharov, K.V.; Agamirova, A.S.; Olyunina, T.V. Reduction Roasting of a Titanomagnetite Concentrate with the Formation of a Titanium–Vanadium Slag Suitable for the Subsequent Recovery of Vanadium and Titanium. *Russ. Metall.* **2022**, *2022*, 707–713. [CrossRef]
10. Jiao, K.X.; Chen, C.L.; Zhang, J.L. Analysis of titanium distribution behaviour in vanadium-containing titanomagnetite smelting blast furnace. *Can. Metall. Q.* **2018**, *57*, 274–282. [CrossRef]
11. Smirnov, L.A.; Rovnushkin, V.A.; Smirnov, A.L. Conversion of vanadium hot metal in a flux-free converter process. *Steel Transl.* **2015**, *45*, 356–360. [CrossRef]
12. Li, M.; Xiao, L.; Liu, J.J. Effective Extraction of Vanadium and Chromium from High Chromium Content Vanadium Slag by Sodium Roasting and Water Leaching. *Mater. Sci. Forum* **2016**, *863*, 144–148. [CrossRef]
13. Gao, F.; Du, H.; Wang, S.N. A Comparative Study of Extracting Vanadium from Vanadium Titano-Magnetite Ores: Calcium Salt Roasting Vs Sodium Salt Roasting. *Miner. Process. Extr. Metall. Rev.* **2022**, *43*, 1–13. [CrossRef]
14. Zhang, J.H.; Zhang, W.; Zhang, L. Mechanism of vanadium slag roasting with calcium oxide. *Int. J. Min-Eral Process.* **2015**, *138*, 20–29. [CrossRef]
15. Wen, J.; Jiang, T.; Liu, Y. Extraction behavior of vanadium and chromium by calcification roasting-acid leaching from high chromium vanadium slag: Optimization using response surface methodology. *Miner. Process. Extr. Metall. Rev.* **2019**, *40*, 56–66. [CrossRef]
16. Wen, J.; Jiang, T.; Wang, J.P. Cleaner extraction of vanadium from vanadium-chromium slag based on MnO_2 roasting and manganese recycle. *J. Clean. Prod.* **2020**, *261*, 121205. [CrossRef]
17. Zhang, Y.; Zhang, T.A.; Dreisinger, D.; Lv, C.; Lv, G.; Zhang, W. Recovery of vanadium from calcification roasted-acid leaching tailing by enhanced acid leaching. *J. Hazard. Mater.* **2019**, *369*, 632–641. [CrossRef]
18. Li, H.Y.; Wang, C.J.; Yuan, Y.H. Magnesiation roasting-acid leaching: A zero-discharge method for vanadium extraction from vanadium slag. *J. Clean. Prod.* **2020**, *260*, 121091. [CrossRef]
19. Chen, D.S.; Zhao, L.S.; Liu, Y.H. A novel process for recovery of iron, titanium, and vanadium from titanomagnetite concentrates: NaOH molten salt roasting and water leaching processes. *J. Hazard. Mater.* **2013**, *244*, 588–595. [CrossRef]
20. Lee, J.C.; Kurniawan; Kim, E.Y. A review on the metallurgical recycling of vanadium from slags: Towards a sustainable vanadium production. *J. Mater. Res. Technol.* **2021**, *12*, 343–364. [CrossRef]
21. Dong, Z.H.; Zhang, J.; Yan, B.J. Co-extraction of vanadium titanium and chromium from vanadium slag by oxalic acid hydrothermal leaching with synergy of Fe powder. *Metall. Mater. Trans. B* **2021**, *52*, 3961–3969. [CrossRef]
22. Dong, Z.H.; Zhang, J.; Yan, B.J. A new approach for the comprehensive utilization of vanadium slag. *Metall. Mater. Trans. B* **2022**, *53*, 2198–2208. [CrossRef]
23. Dong, Z.H.; Zhang, J.; Yan, B.J. Hydrothermal separation of titanium vanadium and chromium from a pregnant oxalic acid leachate. *Materials* **2022**, *15*, 1538. [CrossRef] [PubMed]

Disclaimer/Publisher's Note: The statements, opinions and data contained in all publications are solely those of the individual author(s) and contributor(s) and not of MDPI and/or the editor(s). MDPI and/or the editor(s) disclaim responsibility for any injury to people or property resulting from any ideas, methods, instructions or products referred to in the content.

Article

Technology for Complex Processing of Electric Smelting Dusts of Ilmenite Concentrates to Produce Titanium Dioxide and Amorphous Silica

Zaure Karshyga [1], Almagul Ultarakova [1], Nina Lokhova [1], Azamat Yessengaziyev [1,*], Kaisar Kassymzhanov [1] and Maxat Myrzakulov [2]

1. The Institute of Metallurgy and Ore Beneficiation, Satbayev University, Almaty 050013, Kazakhstan
2. Institute of Energy and Mechanical Engineering, Satbayev University, Almaty 050013, Kazakhstan
* Correspondence: a.yessengaziyev@satbayev.university; Tel.: +7-707-7229946

Abstract: This paper presents the results of research on the development of a technology intended to process electric smelting dusts of ilmenite concentrate with the extraction of silicon and titanium and the production of products in the form of their dioxides. Dusts were processed for silicon separation using the ammonium fluoride method. The optimum conditions for the fluorination and sublimation process of silicon compounds from the electric smelting dust of the ilmenite concentrate were determined: a temperature of 260 °C, a 6 h duration, and mass ratio of dust to ammonium bifluoride of 1:0.5 ÷ 0.9. The sublimation degree of silicon compounds was ~84–91%. The sublimation of titanium fluorides from the remaining sinter was carried out at a temperature of 600 ± 10 °C for 2 h, the mass ratio titanium-containing residue: ammonium bifluoride of 1:0.5, and the degree of sublimation of titanium fluorides was 99%. Iron, manganese, and chromium impurities in the sublimation of titanium fluorides sublimate to a rather low degree. Pyrohydrolysis of titanium fluoride sublimes at 600 °C and allows for the conversion of fluorides into titanium dioxide by 99.5% in 4–5 h. Titanium dioxide of rutile modification with 99.8% TiO_2 was obtained after hydrochloric acid purification and calcination. A technological scheme for the complex processing of dust from the electric smelting of ilmenite concentrates with the production of silica and titanium dioxide is proposed.

Keywords: dust; sublimate; fluoroammonium processing; silicon dioxide; titanium dioxide

1. Introduction

Ilmenite concentrate is used in the production of titanium metal as a feedstock. Electric melting of ilmenite concentrates to produce titanium slag and pig iron is accompanied by high dust emissions since the charge is fed in a loose state. Silicon contained in the charge is sublimed in the process of melting and falls into the thin sleeve filters together with gases entrained into the gas duct system, condensing in the form of amorphous silica SiO_2. High silica content in the dust makes it impossible to return it back to the process, so it is stored in designated storage fields for production wastes.

Considerable amounts of titanium are lost together with the dust generated in the electric smelting process of ilmenite concentrate. Its content in the dusts reaches 50%. Additional extraction of titanium from the dusty waste will not only reduce losses but also allow the acquisition of additional commercial products.

Some of the most demanded products in the market of titanium raw materials are titanium dioxide and amorphous silica. Titanium dioxide is used as a white pigment in the paint, optical coatings, pharmaceuticals, ceramics, food, and paper industries. Titanium dioxide has been used in the photoelectrochemical decomposition of water to produce hydrogen [1,2], to purify water from organics [3–9], to clean the environment from toxic substances [10,11], in high-performance solar cells [12], as a material with antibacterial and antiviral effect for medical applications [13,14], and to create self-cleaning surfaces [15].

Due to the lack of effective technology, most of the waste from titanium production is not currently recycled. Processing is focused on traditional raw materials, the main of which are ilmenite ores and concentrates.

There are two industrial methods of titanium raw material processing for the production of titanium dioxide: sulfate or sulfuric acid and chlorine. In the sulfate method, the titanium-containing raw material (usually ilmenite concentrate) is treated with concentrated sulfuric acid, and the sulfate solution containing sulfuric salt is decomposed to produce titanium dioxide [16–20]. Sulfate technology allows the use of poorer and cheaper raw materials; however, it has several drawbacks, the main of which is the formation of large amounts of waste solutions. According to the chlorine technology [21], rutile or ilmenite is firstly subjected to the action of chlorine gas in the presence of carbon (coke, etc.) at high temperature, and titanium tetrachloride is formed, which is then oxidized by oxygen to its dioxide at 1300–1800 °C. Compared with the sulfate method, the chlorine method is more environmentally friendly. However, it is selective towards raw materials and requires the processing of high-quality rutile.

Fluoroammonium refining is becoming one of the promising methods of rare metal extraction. Ammonium fluoride, ammonium bifluoride, or their mixtures are used as fluorinating agents.

Ammonium hydrofluoride, unlike fluorine, hydrogen fluoride, and hydrofluoric acid, under normal conditions does not represent a significant environmental hazard and becomes a strong fluorinating agent when heated. The physicochemical basis for the process of fluorination with ammonium bifluoride is that oxygen-containing compounds of transition and many non-transition elements in interaction with NH_4HF_2 form very convenient for processing fluoro- or oxofluorometallates of ammonium, whose physicochemical properties ensure product solubility and the possibility to separate mixtures by sublimation [22]. A great advantage of these complex salts is their selective tendency to sublimation or thermal dissociation to non-volatile fluorides which guarantees a deep separation of the components, and the stepwise separation of NH_4F vapors allows for the collection of the desublimate of the latter and use in a closed cycle.

In [23], during the fluorination of titanium slag with ammonium hydrofluoride, at 380 °C, the degree of sublimation of ammonium hexafluorosilicate was 99%, after sublimation of silicon hexafluoride, titanium dioxide with impurities of other oxides remained in the solid product. The separation of titanium dioxide from other components was carried out using a solution of ammonium hydrofluoride. The precipitation of the titanium compound from the ammonium fluoride solution was carried out by adding a 25% solution of ammonia water. However, titanium compounds $(NH_4)_2TiOF_4$ or $(NH_4)_3TiOF_5$ precipitated from the solution. A further shift of the equilibrium towards the formation of $Ti(OH)_4$ required multiple washing of the precipitate with ammonia water. The content of TiO_2 in the resulting product was more than 90%. A method using another fluorinating agent, ammonium fluoride NH_4F, is known [24]. The method consists in treating the initial flotation quartz-leucoxene concentrate with ammonium fluoride at a mass ratio of 0.6–1.25:1 and 195–205 °C. The compounds of silicon and titanium were separated by heat treatment of the resulting product at 295–305 °C and the sublimation of ammonium fumoride and obtaining the residue of artificial rutile containing 90–95% titanium dioxide. The method for processing titanium-containing ilmenite concentrate raw material [25] includes fluorination of raw materials, thermal treatment of the pro-fluorinated mass, separation of fluorination products through sublimation, and pyrohydrolysis of the residue after sublimation to produce iron oxide. Ammonium fluoride, ammonium bifluoride, or their mixtures was used in the fluorination process as a fluoride reagent in a stream of inert gas. Subliminal products were collected with water to produce a solution of ammonium fluorotitanate, and hydrated titanium dioxide was precipitated by water ammonia solution, followed by heat treatment of the precipitate to obtain anhydrous titanium dioxide. In the method [26], the fluorination of an ilmenite or quartz-leucoxene concentrate was performed at a temperature of 110–195 °C or without heating, and the subsequent separation of silicon from titanium was carried out by sublimation of ammonium silicofluoride at a temperature of 305–450 °C or by aqueous leaching. The content of silicon dioxide in the titanium-containing residue was 0.3 wt%. The titanium content in silicon sublimes was less than 1 wt%. In another method [27], the ilmenite concentrate was treated with a solution of ammonium fluoride or hydrodifluoride with the separation of titanium from insoluble fluoroammonium salts of iron. Titanium fluoroammonium salts precipitated from the solution were mixed with finely dispersed silicon dioxide, then the mixture was pyrohydrolyzed at a stepwise increase in temperature to 850–900 °C with exposure at each stage for 20–60 min. Titanium dioxide with an anatase structure containing 99.5% TiO_2 and 0.5% SiO_2 was obtained.

As the review of fluoride methods for processing titanium-containing raw materials shows, the separation of silicon from titanium after treatment with a fluorine-containing agent was carried out both by sublimation of silicon fluoride and by leaching with the transfer of silicon into solution in the form of a silicofluoride compound.

The ammonium fluoride processing method makes it possible to regenerate the used fluoride reagents rather well. This has significant advantages over the sulfate method, which produces a large amount of dilute waste sulfuric acid contaminated with various impurities. This makes it difficult to return sulfuric acid back to the process. Also, the method requires a high content of titanium (not less than 46 wt% TiO_2) in the ore material. Moreover, the decomposition of titanium-containing raw materials is carried out with concentrated sulfuric acid, which poses a certain danger, since gas and reaction mass are released in this case. In the chlorine method during the processing of ilmenite, difficulties arise at the stage of separation of titanium, silicon, aluminum, and iron chlorides due to the proximity of their physical and chemical properties, and it is also necessary to strictly observe the technological regulations and safety measures due to the existing danger of phosgene formation during chlorination in the presence of carbon-containing reducing agents. Despite the fact that the use of the ammonium fluoride method requires the use of corrosion-resistant equipment and high sealing of technological stages, this method reduces the number of technological operations. The number of reagents, with the possibility of their regeneration, improves the quality of the products obtained and creates the possibility of using a safer and more environmentally friendly method.

Production waste is a complex multi-component raw material that is formed in technological processes and accumulates in its composition components similar in properties. The processing of such raw materials is already a problem. The ammonium fluoride method makes it possible to separate the target components with high selectivity and obtain end products of the appropriate quality from them.

Information about the use of fluoroammonium treatment in the available patent and scientific literature refers to natural titanium-containing raw materials. There are sporadic studies on the application of the fluoroammonium-processing method to titanium slurries with the production of calcium nitrate and titanium dioxide [28,29]. Silicon with alkali forms a water-soluble sodium silicate; therefore, in our previous studies [30], to separate silicon from titanium, electric smelting dust of ilmenite concentrate was leached with a sodium hydroxide solution. The influence of sodium hydroxide solution concentration, duration, leaching temperature, and S:L ratio on the leaching process was studied. The optimum parameters of sodium hydroxide leaching of electric smelting dust of ilmenite concentrate were determined: NaOH concentration of 110–115 g/dm^3; S:L ratio of 1:5; a temperature of 80–90 °C; and a duration of 90–120 min. The silicon extraction in the alkaline solution was 77.7%. Physicochemical studies of electrical melting dust of ilmenite concentrate showed that the silicon is in the form of a magnesium silicate phase, which, as a result of alkaline leaching, is not completely decomposed, partially remaining in the cake.

Studies were performed using high-temperature fluoroammonium processing to ensure the most complete decomposition of silicon-containing phases and to separate the silicon impurity from titanium. Taking into account the differences in the physicochemical properties of the dust constituents, it was of interest to determine the optimal conditions for ammonium fluoride processing with the separation of silicon and titanium and the production of products in the form of their oxides with a high content of the main component.

2. Materials and Methods

Materials: all of the reagents used were ammonium bifluoride, aqueous ammonia, and hydrochloric acid were of a grade not lower than "chemically pure".

The fine dust of electric smelting of ilmenite concentrates were provided by the "Ust-Kamenogorsk Titanium-Magnesium Plant" JSC, the content of the main components of the dust is shown in Table 1.

Table 1. Composition of electric smelting dust of ilmenite concentrate (wt%).

Ti	Si	Fe	Cr	Mn	Zn	Al	Mg	O	Others
26.30	12.15	18.14	0.47	3.12	0.52	0.45	0.80	37.03	1.02

X-ray diffraction analysis (XRD) of the dust (Figure 1) showed that the substance of the dust sample is in an X-ray amorphous state and the diffractogram background is high, iron in the dust is mainly in the trivalent state, and the harmful impurity silicon is connected with titanium, magnesium, and iron.

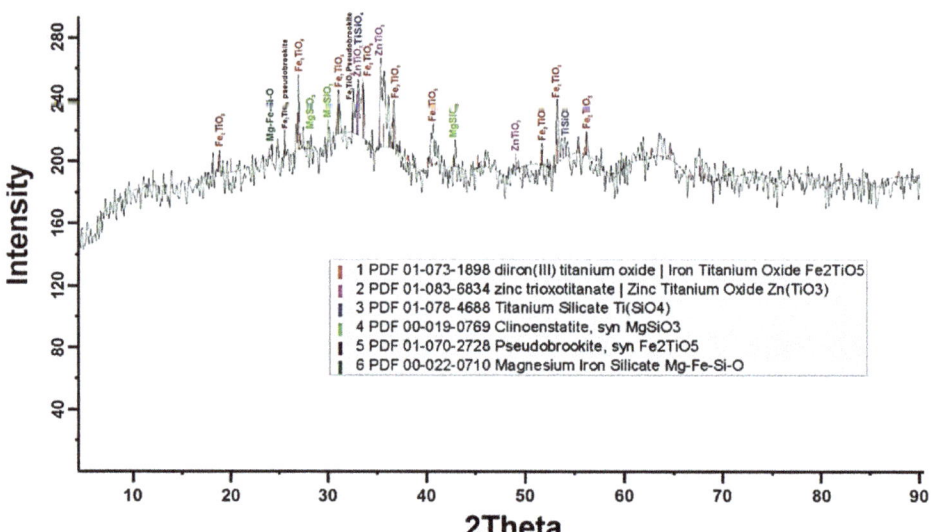

Figure 1. Diffraction pattern of electric smelting dust of ilmenite concentrate.

Analysis methods: X-ray diffraction analysis was performed on a D8 ADVANCE "BRUKER AXS GmbH" diffractometer (Karsruhe, Germany), Cu-Kα emission. The database PDF-2 International Center for Diffraction Data ICDD (Swarthmore, PA, USA) was used.

X-ray fluorescence analysis was performed using an Axios PANalytical spectrometer with wave dispersion (Almelo, The Netherlands).

The chemical analysis of the samples was performed using an Optima 8300 DV inductively coupled plasma optical emission spectrometer (Perkin Elmer Inc., Waltham, MA, USA).

Experimental procedure: to carry out the processes of sublimation of silicon or titanium fluorides, the dust or residue from the sublimation of silicon, respectively, was thoroughly mixed with ammonium bifluoride in the required ratio. The charge sample was placed in an alundum boat and installed in a LOP LT-50/500–1200 tubular electric furnace. Argon was supplied through a horizontal pipe, and the furnace was heated to a predetermined temperature and maintained at this temperature for a certain period of time. At the end of the experiment, sublimates of ammonium hexafluorosilicate or titanium fluorides were captured at the end of the tube, and the gas-air mixture was captured in a flask with ammonia water. Preliminary experiments on the fluorination of dust from electric smelting of ilmenite concentrate determined the rate of argon supply, which makes it possible to remove fluoride fumes from the reaction zone. The argon feed rate for the used installation was 1.0–1.5 dm^3/h. The degree of fluorination and sublimation of silicon was estimated from the change in the content of the controlled component in the solid residue according to the formula:

$$E_{Si} = \frac{(C_0 - C_i)}{C_0} \cdot 100\ \% \tag{1}$$

where C_0 is the amount of silicon in the initial dust, g; C_i is the amount of silicon in the residue after fluorination and sublimation, g.

The fluorination and sublimation plants are shown in Figure 2.

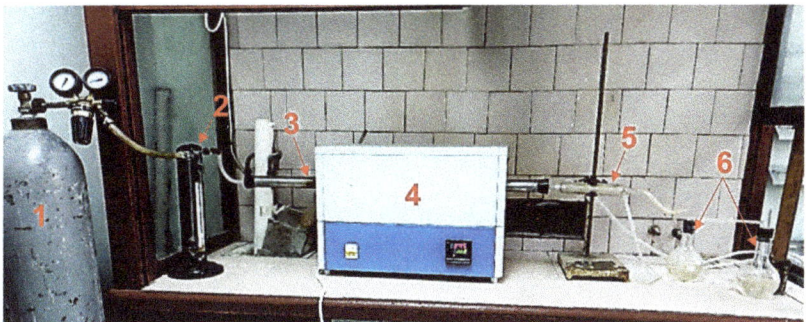

Figure 2. Assembly for laboratory research on fluoroammonium processing of electric melting ilmenite concentrate dust: 1—cylinder with argon, 2—flow meter, 3—pipe with alundum boat and sample, 4—horizontal furnace, 5—water-cooled refrigerator-condenser, and 6—gas trap system (10% NH$_4$OH solution).

Silicon fluoride sublimates were dissolved in water and treated with water in a ratio of solid to liquid equal to 1:10. After dissolution in water, silicon sublimes were subjected to ammonia hydrolysis. The hydrolysis of a solution of ammonium and oxonium hexafluorosilicates was conducted as follows: in a solution heated to 40 °C containing hexafluorosilicate ion, 10% or 25% ammonia solution was added in portions with active stirring to pH 7.5–8, which upon reaching it was necessary to keep the suspension for 80–90 min by stirring to form and precipitate silica flakes.

During the process of pyrohydrolysis of sublimes of titanium fluorides, a weighed sample of titanium fluorides was placed in an alundum boat and loaded into an electric furnace. After heating to 100 °C, steam was supplied to the furnace, where a boat with a sample of sublimes of titanium fluorides was previously installed. The steam rate was 1.5–2.0 dm^3/h.

Titanium dioxide was purified from impurities by hydrochloric acid solution in thermostatic reactors with a volume of 0.5 dm^3. Purification was made under established conditions and constant stirring of the pulp. Pulp stirring was performed with a glass stirrer. The stirring speed was 450 rpm.

3. Results and Discussion

3.1. Study of Silica Separation from Electric Smelting Dust of Ilmenite Concentrates

Dusts of electric smelting of ilmenite concentrates during their formation, as noted above, can concentrate silicon, the presence of which does not allow their return to the smelting process. Therefore, initially, it is advisable to remove silicon from the dust for its further processing to produce titanium dioxide. It is of interest to separate silicon and obtaining an additional product of amorphous silica.

The authors [23] suggest that the conversion of silica and other silicon compounds into fluorominium salts during the reaction with ammonium hydrofluoride produces $(NH_4)_3SiF_7$ and $(NH_4)_2SiF_6$.

However, in [31], it is argued that the formation of $(NH_4)_2SiF_6$ is thermodynamically advantageous because of the large negative enthalpy of formation and free energy compared with other silicon fluorides.

The interaction of ammonium hydrofluoride with the components of electric smelting dust of ilmenite concentrate at elevated temperatures can proceed by the reactions:

$$TiO_2 + 3NH_4HF_2 = (NH_4)_2TiF_6 + NH_3\uparrow + 2H_2O\uparrow \quad (2)$$

$$SiO_2 + 3NH_4HF_2 = (NH_4)_2SiF_6 + NH_3\uparrow + 2H_2O\uparrow \quad (3)$$

$$4FeO + 12NH_4HF_2 + O_2 = 4(NH_4)_3FeF_6 + 6H_2O\uparrow \quad (4)$$

$$Al_2O_3 + 6NH_4HF_2 = 2(NH_4)_3AlF_6 + 3H_2O\uparrow \quad (5)$$

$$2MnO + 3NH_4HF_2 = 2NH_4MnF_3 + NH_3\uparrow + 2H_2O\uparrow \quad (6)$$

The influence of various factors on the sublimation of silicon from electric melting dusts was studied.

3.1.1. Influence of the Temperature of the Fluorination Process

The investigations were performed in the temperature range of 200–280 °C. The duration of the experiments was 6 h, and the ratio of masses of electrofusion dust of ilmenite concentrate to ammonium hydrodifluoride was 1:0.9.

It is shown in Table 2 that with an increase in the process temperature from 200 to 280 °C, the yield of the residue from fluorination decreases, which is associated with an increase in the consumption of the fluorinating agent for the reaction with dust components and the removal of silicon fluoride from the reaction zone by the argon flow.

Table 2. Influence of fluorination temperature on the controlled elements content in the residue.

Temperature, °C	Residue Yield from Fluorination,%	Content of Components in the Residue, wt%							
		Si	Ti	Fe	Cr	Mn	O	F	Others
200	79.6	8.4	17.3	12.9	0.40	3.2	9.1	30.1	18.60
230	68.4	6.4	17.6	14.6	0.51	2.9	-	31.1	26.89
250	54.4	2.7	25.7	19.6	0.65	4.5	10.1	24.3	12.45
260	47.8	1.7	28.7	20.7	0.76	4.7	13.9	26.6	2.94
280	44.9	1.5	29.6	22.0	0.81	5.0	14.8	25.9	0.39

The analysis of the data presented in Table 2 and Figure 3 shows that the process temperature of 260 °C results in the sublimation degree of silicon fluoride up to 84.2% and its content in the titanium-containing intermediate product decreases by eight times compared with the initial dust. Further temperature increase has an insignificant influence on the sublimation degree of silicon fluoride and its content in the residue from fluorination.

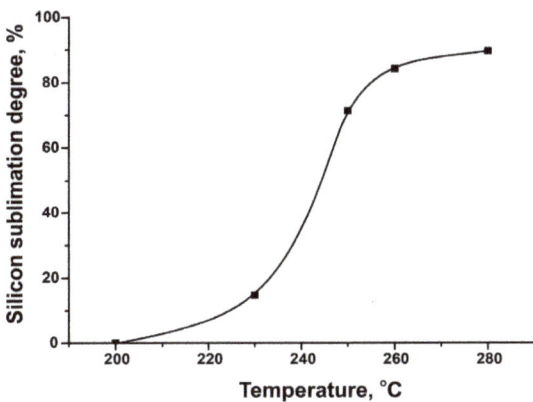

Figure 3. Influence of temperature on silicon fluoride sublimation.

It should be considered that the optimal temperature for fluorination of electric smelting of ilmenite concentrate to separate the silicon should be 260 °C.

3.1.2. Effect of the Fluoridation Duration

The study of the influence of fluorination process duration was performed in a series of experiments with different time intervals of 2–8 h at 260 °C and mass ratio of dust to ammonium bifluoride of 1:0.9. The results of studies are shown in Table 3 and Figure 4.

Table 3. Results of fluorination of electrosmelting dust of ilmenite concentrate. Influence of the process duration.

Duration of Experiment, h	Residue Yield from Fluorination,%	Content in the Residue, wt%							
		Si	Ti	Fe	Cr	Mn	O	F	Others
2	65.3	5.4	21.7	17.2	0.57	3.8	8.1	25.4	17.83
4	58.2	4.8	25.4	17.0	0.72	4.3	10.7	26.9	10.18
6	47.8	1.7	28.7	20.7	0.76	4.7	13.9	26.6	2.94
8	43.7	1.5	30.3	22.2	0.80	5.1	13.0	27.0	0.10

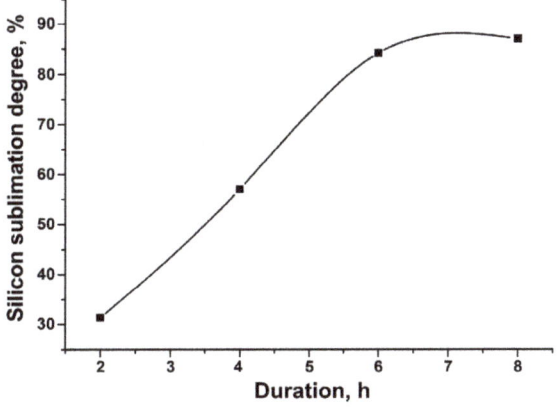

Figure 4. The dependence of the degree of sublimation of silicon from silicon dust of electrofluorination of ilmenite concentrate during fluorination.

The curve in Figure 4 on the segment of 2–6 h describes a rectilinear dependence of silicon fluoride sublimation degree on the process duration. The sublimation rate of silicon decreases and its content in the residue insignificantly changes in 6 h of fluorination (Table 3).

XRD analysis showed that formed ammonium hexafluoride does not sublimate completely in 2–4 h. The reaction of ilmenite fluorination is not completed (Figure 5).

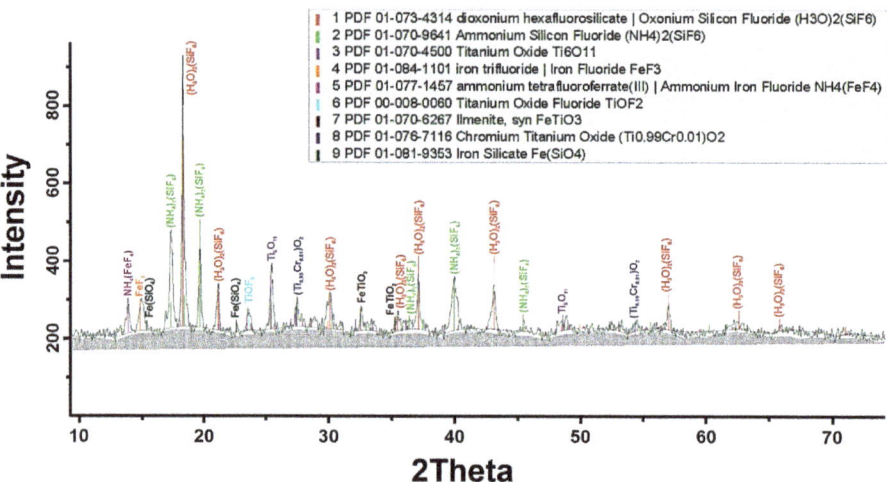

Figure 5. A diffractogram of the residue from fluorination of electrosmelting dust of ilmenite concentrate (2 h, 260 °C, dust:NH$_4$HF$_2$ = 1:0.9).

An increase in the fluorination duration up to 6 h allowed almost complete extraction of silicon in the sublimations (Figure 6) [32].

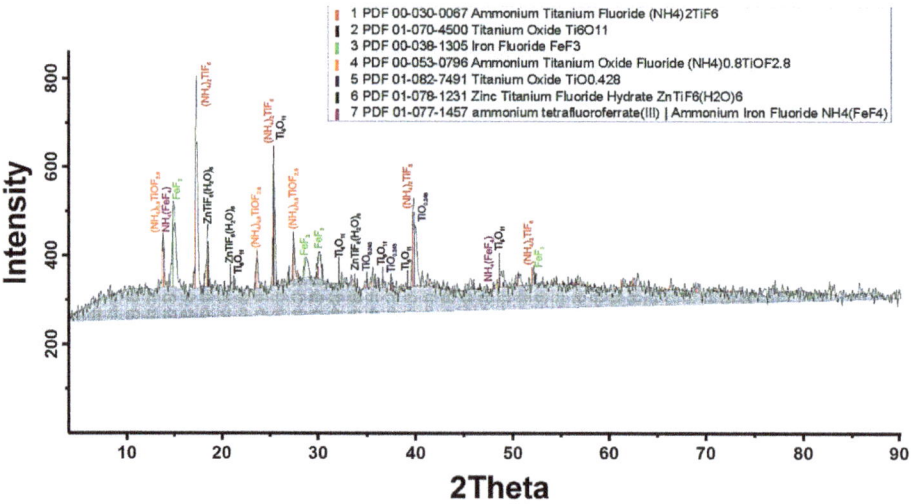

Figure 6. A diffractogram of the residue from fluorination of electrofusion dust of ilmenite concentrate (6 h, 260 °C, dust:NH$_4$HF$_2$ = 1:0.9).

3.1.3. Influence of Mass Ratio of Dust from Electric Smelting of Ilmenite Concentrate to Ammonium Bifluoride

The study of the influence of specific ammonium bifluoride consumption on fluorination of electrical melting components of ilmenite concentrate dust with formation and sublimation of silicon fluoride was performed at a temperature of 260 °C with a 6 h duration.

Table 4 represents the data on the change of the content of controlled elements in the residue from fluorination, which shows that the silicon fluoride sublimation degree was 90.6% at a mass ratio of dust to ammonium bifluoride of 1:0.5.

Table 4. Results of fluorination of electric melting dust of ilmenite concentrate. Influence of dust:NH_4HF_2 mass ratio.

Dust:NH_4HF_2 Mass Ratio	Residue Yield from Fluorination,%	Content in the Residue, wt%							
		Si	Ti	Fe	Cr	Mn	O	F	Others
1:0.5	32.1	1.5	29.7	20.6	0.73	4.6	30.2	8.7	3.97
1:0.9	47.8	1.7	28.7	20.7	0.76	4.7	13.9	26.6	2.94
1:1.5	70.8	3.05	19.7	15.4	0.55	3.4	11.0	27.8	19.1

With an increase in the consumption of ammonium hydrodifluoride (mass ratio 1:0.9), the silicon fluoride sublimation degree remains satisfactory. At the same time, ilmenite, iron oxide, and a part of titanium oxides are completely fluorinated.

Further increase of the specific flow rate of ammonium hydrofluoride decreases the degree of silicon fluoride sublimation (Figure 7). It is connected with the formation of a big mass of the melted dust mixture with ammonium hydrofluoride that made it difficult to free silicon fluoride in the gas phase.

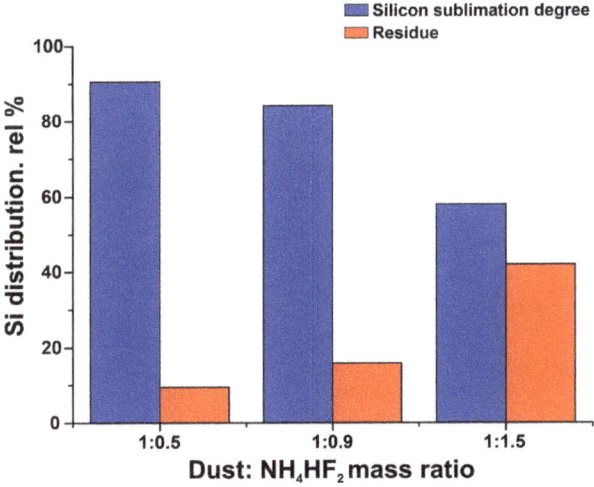

Figure 7. Dependence of silicon sublimation degree from smelting dust of ilmenite concentrate during fluorination on specific flow rate of ammonium bifluoride.

Thus, the optimal conditions for fluorination of electric melting dust of ilmenite concentrate were experimentally established: 260 °C temperature, 6 h duration, and a mass ratio of dust to ammonium bifluoride of 1:0.5–0.9. The degree of sublimation of silicon fluoride was ~84–91% under these conditions.

3.2. Silicon Dioxide Production

The obtained silicon-containing sublime is represented by oxonium hexofluorosilicate $(H_3O)_2SiF_6$ (~98%) and ammonium hexofluorosilicate $(NH_4)_2SiF_6$ (~2%) [32]. Ammonium and oxonium hexafluorosilicates are well soluble in water at room temperature. Precipitation of silicon oxide was performed with ammonia for 30–90 min, and the suspension was held for the formation and precipitation of silicon oxide flakes. The obtained amorphous product had the following composition (wt%): 81.6 SiO_2; 12.9 NH_4F; 0.045 Fe; 0.005 Cu; 0.025 Zn; 0.014 As; 0.003 Sr; and 0.017 Pb. The product contains insignificant amounts of heavy metals and arsenic. The presence of ammonium fluoride is due to its absorption by amorphous particles, and it cannot be removed by washing the sediment with water. Ammonium fluoride is known to decompose when it is heated. The sample was heated and incubated at 530–560 °C for 60–80 min to ensure the most complete decomposition of ammonium fluoride and its removal from the sample composition [33]. The resulting silica had the following composition (wt%): 96.3 SiO_2; n/d F; 0.14 Fe_2O_3; 0.16 Al_2O_3; 0.02 ZnO; 0.03 CaO; and 0.15 TiO_2. According to the content of silicon dioxide and accompanying impurities of iron, calcium, and magnesium, the product meets the requirements of the state standard GOST 18307-78 for the "White soot" brand BS-100 [34].

3.3. Titanium Fluoride Sublimation and Study of the Behavior of Impurity Components during Fluorination

The residue from dust fluorination with sublimation of silicon fluorides mainly consists of titanium compounds (Figure 6). The studies were performed to determine the optimal conditions of the process intended to provide the most complete sublimation of titanium fluorides from the residue. The influence of temperature (450 to 650 °C) and duration (0.5 to 4 h) on titanium fluorides sublimation degree were studied [35]. The optimal conditions of the sublimation process of titanium fluorides were a temperature of 600 ± 10 °C, a duration of 2 h, and a mass ratio of titanium-containing residue (ammonium bifluoride) of 1:0.5. Sublimation degree of titanium fluorides under these conditions reached 99%. Fluoride sublimations were composed of the following phases: $(NH_4)_2TiF_6$, $(NH_4)_3TiF_7$, $(NH_4)FeF_6$, and NH_4HF_2.

Strict requirements for the content of chromophoric impurities are applied to the pigment titanium dioxide which imparts a different color to a white pigment even at a very small content.

The sublimation of iron, manganese, chromium, and residual silicon during the sublimation of titanium fluorides was studied in this connection. Studies on the effect of the process duration on the sublimation degree of iron, manganese, chromium, and silicon were performed at 610 °C in the range of 30–240 min. The results of the experiments on the sublimation of impurities are shown in Figure 8.

The results showed that silicon in the form of ammonium fluorosilicate is almost completely extracted into sublimations in the first 30 min of the process.

The curves of dependence of the sublimation degree on the duration of the experiment for iron, manganese, and chromium are similar and only differ in the value of the sublimation degree (Figure 8). Sublimation of iron within 2 h of the process occurs at 15.1%, of manganese at 9.3%, and of chromium at 13.6%. This behavior is due to the conditions of the experiments. Heating of the furnace to 610 °C occurs at a rate of 20 °C/min which contributes to the sublimation of impurity components.

It should be noted that in spite of a certain sublimation of iron, manganese, and chromium, their content in the residue increases. The content of iron increased by 1.8, manganese by 1.9, and chromium by 2.0 times during 2 h (Table 5).

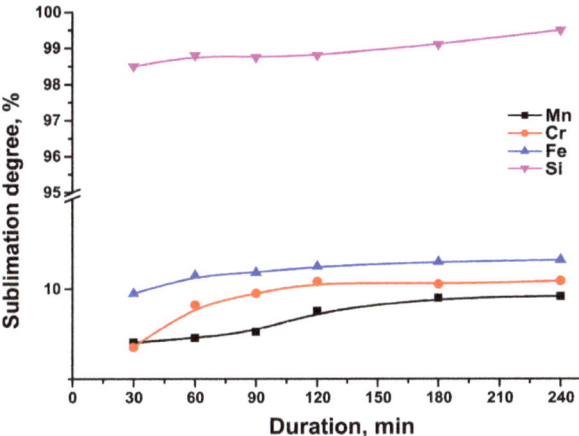

Figure 8. Influence of process duration on sublimation of impurities from fluorinated dust of electric melting of ilmenite concentrate.

Table 5. Effect of the duration of the experiment on the content of impurity components in the cinder.

Duration, min	Content * in the Cinder, wt%			
	FeO	SiO$_2$	MnO	Cr$_2$O$_3$
init.	23.0	1.28	2.35	0.39
30	40.6	0.071	4.32	0.64
60	40.4	0.058	4.36	0.69
90	41.6	0.057	4.39	0.74
120	42.1	0.051	4.44	0.76
180	43.7	0.043	4.50	0.77
240	45.4	0.036	4.58	0.79

* component content is given in terms of the oxide form.

Studies of the process of sublimation of impurity components during the fluorination of titanium from sinter showed that, under experimental conditions, such impurities as iron, manganese, and chromium sublimate with a fairly low degree together with titanium.

3.4. Titanium Dioxide Production and Purification

Pyrohydrolysis of titanium fluorides. Fluoride technology involving pyrohydrolysis of titanium fluorides to produce titanium dioxide is one of the ways to produce high-quality titanium oxide products.

Obtained fluoride sublimations of titanium have the following composition (wt%): 59.0 Ti; 0.6 Fe; 0.005 Si; 0.09 Cr; 0.14 Al; 20.4 F; and 19.8 O. Studies of titanium fluoride pyrohydrolysis with the investigation of the influence of temperature (in the range from 300 to 700 °C) and duration of the process (in the range from 1 to 5 h) have shown that optimal parameters are 600 °C and a duration of 4–5 h [36].

Obtained titanium dioxide was contaminated with impurities and had a grayish hue, therefore *hydrochloric acid treatment* of the product was performed.

Studies of the hydrochloric acid leaching process of pyrohydrolysis products were performed to provide the most complete purification of titanium dioxide. The influence of hydrochloric acid concentration (in the range from 5 to 15% HCl), S:L ratio (from 1:4 to 1:8), and the process duration (in the range from 5 to 60 min) on the conversion degree of manganese, iron, and chromium impurities into the solution was studied. As a result of this research, the following optimum conditions of the purification process were determined: 12.5–15% HCl; a temperature of 25–30 °C; ratio S:L of 1:6 ÷ 8; and a duration of 20–30 min.

Titanium dioxide after acid processing consisted of two modifications: 94% anatase and 6% rutile. High quality of pigmental titanium dioxide is provided by the rutile phase; therefore, titanium dioxide was calcined at a temperature of 900 °C for 2 h to obtain the rutile modification. The TiO_2 content was 99.8 wt%, and the content of impurities of silicon, chromium, manganese, and iron in recalculation on their oxides was 0.0005, 0.032, 0.005, and 0.039 wt%, respectively, in the obtained rutile product [36]. According to the content of the main component and accompanying impurities, the product complies with the requirements of the state standard GOST 9808-84 for pigment titanium dioxide [37].

3.5. Process Flow of Complex Processing of Electric Smelting Dust of Ilmenite Concentrates

The conducted studies formed the basis for the development of a technology for the integrated processing of fine dusts from the electric smelting of ilmenite concentrates to extract silicon and titanium and the production of oxide products (Figure 9).

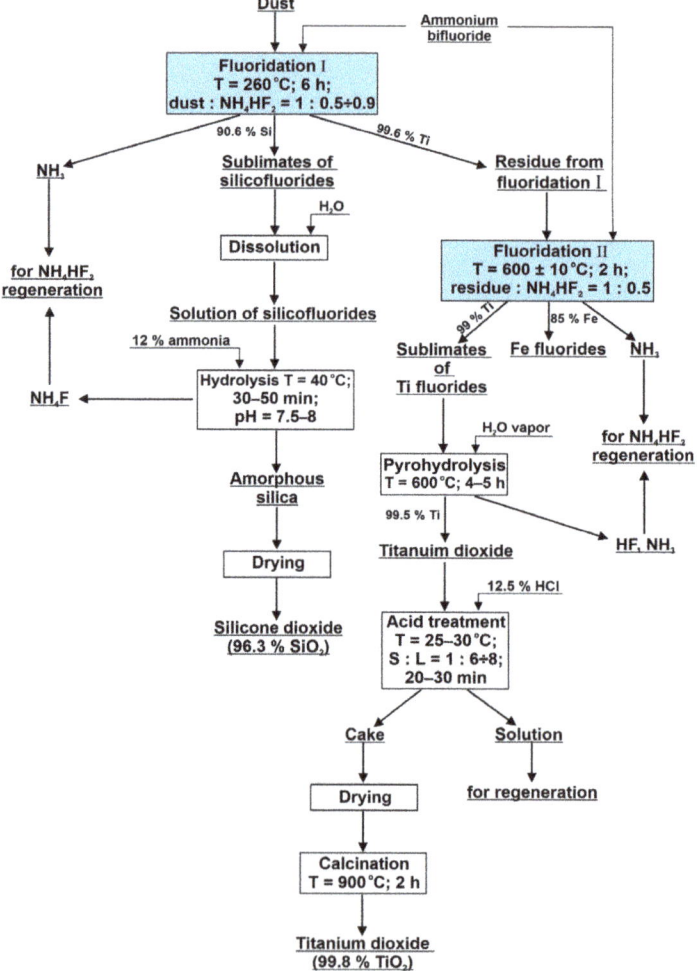

Figure 9. Basic process flow of fine dust processing of electric smelting of ilmenite concentrates.

The dust from the smelting of ilmenite concentrates is hydrofluorinated in a molten ammonium bifluoride at 260 °C, producing silicon fluorides in the form of sublimations and separating them from iron and titanium.

During fluorination of electric smelting dust of ilmenite concentrate, a mixture of gases (water vapor and ammonium fluoride) is formed and sent for condensation (regeneration of ammonium fluoride).

Aspiration containing silicon fluoride is dissolved in water, and amorphous silicon dioxide is precipitated by a 12% ammonia solution. Under these conditions, amorphous silica contains an impurity of fluorine ions. When amorphous silica is dried at 200–300 °C, the fluorine ion is removed along with water vapor, and the remainder is pure amorphous silica.

The residue from fluorination containing iron and titanium is subject to fluorination at 590–610 °C. The titanium component is completely separated from iron and impurities.

Sublimated titanium fluoride goes to the oxidative pyrohydrolysis operation (interaction with superheated water vapor and air oxygen).

Heat-treatment conditions of the final titanium product are maintained depending on the desired structure of the resulting titanium dioxide.

Thus, the complex processing of electric smelting dusts of ilmenite concentrate with the production of silica and titanium dioxide products is possible according to the given regimes.

4. Conclusions

Based on the research results, a technology intended to produce amorphous silica and titanium dioxide from fine-dispersed dust of electrical smelting of ilmenite concentrate was proposed.

A study of the effect of different parameters on the fluorination and sublimation of silicon from the electric smelting dust of ilmenite concentrate showed that the optimum conditions of the process are a temperature 260 °C, duration of 6 h, and a mass ratio of dust to ammonium hydrodifluoride of 1:0.5 ÷ 0.9. As a result, the sublimation degree of silicon fluoride compounds was equal to ~84–91%. After ammonia hydrolysis of silicon fluoride sublimates dissolved in water and further drying of the precipitate, amorphous silica was obtained with a content of 96.3% SiO_2.

The sublimation of titanium fluorides from the remaining sinter was carried out at a temperature of 600 ± 10 °C for 2 h, the mass ratio of titanium-containing residue to ammonium bifluoride was 1:0.5, and degree of sublimation of titanium fluorides was 99%.

A study of impurities behavior during the sublimation of titanium fluorides from the sinter showed that iron, manganese, and chromium were sublimated with a sufficiently low degree. The content of these impurities in the residue after sublimation of titanium fluorides increases by approximately two times.

After pyrohydrolysis of titanium fluoride sublimes, hydrochloric acid purification, and subsequent calcination of titanium dioxide, the product of rutile modification with a content of 99.8% TiO_2 was obtained.

The process flow for complex processing of electric smelting dusts of ilmenite concentrates with the production of titanium dioxide and silicon dioxide was proposed according to the results of the research.

Author Contributions: Conceptualization, Z.K., A.U. and N.L.; methodology, N.L. and K.K.; software, A.Y. and M.M.; validation, Z.K., A.U. and N.L.; formal analysis, A.U., Z.K. and N.L.; investigation, N.L., A.Y., K.K. and M.M.; resources, A.Y., K.K. and M.M.; data curation, Z.K., A.U., N.L. and K.K.; writing—original draft preparation, Z.K. and N.L.; writing—review and editing, Z.K. and A.Y.; supervision, Z.K.; project administration, Z.K. and A.U.; funding acquisition, Z.K. and A.U. All authors have read and agreed to the published version of the manuscript.

Funding: This research is funded by the Science Committee of the Ministry of Science and High Education of the Republic of Kazakhstan, Grant Project No. AP08855505 and Grant Project No. AP09258788.

Institutional Review Board Statement: Not applicable.

Informed Consent Statement: Not applicable.

Data Availability Statement: The data and results presented in this study are available in the article.

Conflicts of Interest: The authors declare that there is no conflict of interest regarding the publication of this manuscript.

Symbols and Abbreviations

IMOB JSC	Institute of Metallurgy and Ore Beneficiation Joint Stock Company
UK TMC JSC	"Ust-Kamenogorsk Titanium-Magnesium Combine" Joint Stock Company.
XRD	X-ray diffraction analysis
S:L	ratio of solid phase weight (in grams) to liquid phase volume (in ml)

References

1. Fujishima, A.; Honda, K. Electrochemical Photolysis of Water at a Semiconductor Electrode. *Nature* **1972**, *238*, 37–38. [CrossRef]
2. Ngo, T.Q.; Posadas, A.; Seo, H.; Hoang, S.; McDaniel, M.D.; Utess, D.; Triyoso, D.H.; Mullins, C.B.; Demkov, A.A.; Ekerdt, J.G. Atomic layer deposition of photoactive $CoO/SrTiO_3$ and CoO/TiO_2 on Si(001) for visible light driven photoelectrochemical water oxidation. *J. Appl. Phys.* **2013**, *114*, 084901. [CrossRef]
3. Frank, S.N.; Bard, A.J. Heterogeneous photocatalytic oxidation of cyanide ion in aqueous solutions at titanium dioxide powder. *J. Am. Chem. Soc.* **1977**, *99*, 303–304. [CrossRef]
4. Frank, S.N.; Bard, A.J. Heterogeneous photocatalytic oxidation of cyanide and sulfite in aqueous solutions at semiconductor powders. *J. Phys. Chem.* **1977**, *81*, 1484–1488. [CrossRef]
5. Liu, H.; Cheng, S.; Wu, M.; Wu, H.; Zhang, J.; Li, W.; Cao, C. Photoelectrocatalytic Degradation of Sulfosalicylic Acid and Its Electrochemical Impedance Spectroscopy Investigation. *J. Phys. Chem. A* **2000**, *104*, 7016–7020. [CrossRef]
6. Yu, J.; Yu, H.; Ao, C.H.; Lee, S.C.; Yu, J.C.; Ho, W. Preparation, characterization and photocatalytic activity of in situ Fe-doped TiO_2 thin films. *Thin Solid Film.* **2006**, *496*, 273–280. [CrossRef]
7. Janus, M.; Choina, J.; Morawski, A.W. Azo dyes decomposition on new nitrogen-modified anatase TiO_2 with high adsorptivity. *J. Hazard. Mater.* **2009**, *166*, 1–5. [CrossRef]
8. Korina, E.; Stoilova, O.; Manolova, N.; Rashkov, I. Polymer fibers with magnetic core decorated with titanium dioxide prospective for photocatalytic water treatment. *J. Environ. Chem. Eng.* **2018**, *6*, 2075–2084. [CrossRef]
9. Sraw, A.; Kaur, T.; Pandey, Y.; Sobti, A.; Wanchoo, R.K.; Toor, A.P. Fixed bed recirculation type photocatalytic reactor with TiO_2 immobilized clay beads for the degradation of pesticide polluted water. *J. Environ. Chem. Eng.* **2018**, *6*, 7035–7043. [CrossRef]
10. Hoffmann, M.R.; Martin, S.T.; Choi, W.; Bahnemann, D.W. Environmental Applications of Semiconductor Photocatalysis. *Chem. Rev.* **1995**, *95*, 69–96. [CrossRef]
11. MiarAlipour, S.; Friedmann, D.; Scott, J.; Amal, R. TiO_2/porous adsorbents: Recent advances and novel applications. *J. Hazard. Mater.* **2018**, *341*, 404–423. [CrossRef]
12. O'Regan, B.; Grätzel, M. A low-cost, high-efficiency solar cell based on dye-sensitized colloidal TiO_2 films. *Nature* **1991**, *353*, 737–740. [CrossRef]
13. Pan, J.; Leygraf, C.; Thierry, D.; Ektessabi, A.M. Corrosion resistance for biomaterial applications of TiO_2 films deposited on titanium and stainless steel by ion-beam-assisted sputtering. *J. Biomed. Mater. Res.* **1997**, *35*, 309–318. [CrossRef]
14. Heidenau, F.; Mittelmeier, W.; Detsch, R.; Haenle, M.; Stenzel, F.; Ziegler, G.; Gollwitzer, H. A novel antibacterial titania coating: Metal ion toxicity and in vitro surface colonization. *J. Mater. Sci. Mater. Med.* **2005**, *16*, 883–888. [CrossRef]
15. Wang, R.; Hashimoto, K.; Fujishima, A.; Chikuni, M.; Kojima, E.; Kitamura, A.; Shimohigoshi, M.; Watanabe, T. Light-induced amphiphilic surfaces. *Nature* **1997**, *388*, 431–432. [CrossRef]
16. Weintraub, G. Process of Obtaining Titanic Oxid. U.S. Patent 1014793A; IPC C22B34/125 (EP, US); Y10S423/02 (EP), 16 January 1912.
17. Joseph, B. Titanium Compound. U.S. Patent 1504669A; IPC C22B34/12, 12 August 1924.
18. Weizmann, C.; Blumenfeld, J. Improvements Relating to the Treatment of Solutions for the Separation of Suspended Matter. UK Patent 228814A; IPC C01F15/00 (EP); C01G19/00 (EP); C01G23/001 (EP), 3 February 1925.
19. Mecklenburg, W. Production of Titanium Dioxide. U.S. Patent 1758528A; IPC C01G23/053 (EP), 13 May 1930.
20. Belenky, E.F.; Riskin, I.V. *Chemistry and Technology of Pigments*; Goskhimizdat: Leningrad, Russia, 1960; p. 756.
21. Jelks, B. *Titanium: Its Occurrence, Chemistry and Technology*, 2nd ed.; Ronald Press: New York, NY, USA, 1966; p. 691.
22. Rakov, E.G. *Ammonium Fluorides*; Results of Science and Technology; Inorganic Chemistry; All-Union Institute of Scientific and Technical Information: Moscow, Russia, 1988; Volume 15, p. 154.
23. Dmitriev, A.N.; Smorokov, A.A.; Kantaev, A.C.; Nikitin, D.S.; Vit'kina, G.Y. Fluorammonium-processing method of titanium slag. Izvestiya vysshee uchebnykh obrazovatel'nykh uchebov [Proceedings of Higher Educational Institutions]. *Ferr. Metall.* **2021**, *64*, 178–183.
24. Fedun, M.P.; Bakanov, V.K.; Pastikhin, V.V. Method of Processing of Titanium-Silicon-Containing Concentrates. R.F. Patent 2264478, 20 November 2005.

25. Andreev, A.A.; D'jachenko, A.N. Method of Processing of Raw Materials Containing Titanium. R.F. Patent 2365647, 27 August 2009.
26. Andreev, A.A.; D'jachenko, A.N. Method to Process Titanium-Silicon-Containing Stock. Patent RF 2377332, 27 August 2009.
27. Gordienko, P.S.; Pashnina, E.V.; Shabalin, I.A.; Dostovalov, D.V. Method of Processing Titanium-Containing Mineral Raw Materials. Patent RF 2717418, 23 March 2020.
28. Ultarakova, A.A.; Yessengaziyev, A.M.; Kuldeyev, E.I.; Kassymzhanov, K.K.; Uldakhanov, O. Kh. Processing of titanium production sludge with the extraction of titanium dioxide. *Metalurgija* **2021**, *60*, 411–414.
29. Yessengaziyev, A.M.; Ultarakova, A.A.; Burns, P.C. Fluoroammonium method for processing of cake from leaching of titanium-magnesium production sludge. *Complex Use Miner. Resour.* **2022**, *320*, 67–74. [CrossRef]
30. Yessengaziyev, A.; Ultarakova, A.; Lokhova, N.; Karshigina, Z.; Kasymzhanov, K. Study of the Alkaline Treatment Effect on Separation of Silica from the Electric Melting Dust of Ilmenite Concentrates. In Proceedings of the XXIth International Multidisciplinary Scientific Geo Conference, Science and Technologies in Geology, Exploration and Mining—SGEM 2021, Albena, Bulgaria, 16–22 August 2021; pp. 601–609. [CrossRef]
31. Niwano, M.; Kurita, K.; Takeda, Y. Formation of hexafluorosilicate on Si surface treated in NH_4F investigated by photoemission and surface infrared spectroscopy. *Appl. Phys. Lett.* **1993**, *62*, 1003–1005. [CrossRef]
32. Karshyga, Z.; Ultarakova, A.; Lokhova, N.; Yessengaziyev, A.; Kassymzhanov, K. Processing of Titanium-Magnesium Production Waste. *J. Ecol. Eng.* **2022**, *23*, 215–225. [CrossRef]
33. Ultarakova, A.A.; Karshyga, Z.B.; Lokhova, N.G.; Naimanbaev, M.A.; Yessengaziyev, A.M.; Burns, P. Methods of silica removal from pyrometallurgical processing wastes of ilmenite concentrate. *Complex Use Miner. Resour.* **2022**, *322*, 79–88. [CrossRef]
34. GOST 18307-7; White Soot. Specifications. Revised Edition; PPC Standards Publishing House: Moscow, Russia, 1998; 18p.
35. Karshyga, Z.B.; Ultarakova, A.A.; Lokhova, N.G.; Yessengaziyev, A.M.; Kuldeyev, E.I.; Kassymzhanov, K.K. Study of fluoroammonium processing of reduction smelting dusts from ilmenite concentrate. *Metalurgija* **2023**, *62*, 145–148.
36. Ultarakova, A.; Karshyga, Z.; Lokhova, N.; Yessengaziyev, A.; Kassymzhanov, K.; Mukangaliyeva, A. Studies on the Processing of Fine Dusts from the Electric Smelting of Ilmenite Concentrates to Obtain Titanium Dioxide. *Materials* **2022**, *15*, 8314. [CrossRef]
37. GOST 9808-84; Pigment Titanium Dioxide. Specifications. 2nd Revised Edition; PPC Standards Publishing House: Moscow, Russia, 2004; 18p.

Article

Vacuum Carbon Reducing Iron Oxide Scale to Prepare Porous 316 Stainless Steel

Fang Zhang, Jun Peng *, Hongtao Chang and Yongbin Wang

School of Materials and Metallurgy, Inner Mongolia University of Science and Technology, Baotou 014010, China
* Correspondence: pengjun75@163.com; Tel.: +86-152-4932-0456

Abstract: In order to improve the added value of iron oxide scale and reduce the manufacturing cost of porous stainless steel, steel rolling iron oxide scale as an iron-containing raw material was used to prepare porous 316 stainless steel by high-temperature sintering under vacuum conditions, while carbon was used as a reducing agent and pore-forming agent, and the necessary metal powders were added. In our work, the specific reduction system was confirmed, including the sintering temperature, sintering time, vacuum degree and carbon amount, through thermodynamic calculation combined with experiments. Thermodynamic analysis results showed that the transformation process of the chromium element in the raw materials at 10^{-4} atm and 300~1600 °C was $FeCr_2O_4 + Cr_3O_4 \rightarrow Cr_2O_3 + Cr_3O_4 + Cr_{23}C_6 \rightarrow Cr_{23}C_6 + Cr_7C_3 + FCC \rightarrow FCC + Cr_{23}C_6 \rightarrow FCC \rightarrow FCC + BCC \rightarrow Cr(liq)$. The FCC phase with qualified carbon content could be obtained at 10^{-4} atm and 1200 °C, while 90.88 g iron oxide scale, 17.17 g carbon, 17.00 g metal chromium, 12.00 g metal nickel and 2.5 g metal molybdenum were necessary to produce 100 g porous 316 stainless steel. The porous 316 stainless steel with a carbon content of 0.025% could be obtained at 10^{-4} atm and 1200 °C for 180 min, while the chromium element underwent the transformation of metal, $Cr \rightarrow FeCr_2O_4 \rightarrow Cr_{23}C_6 \rightarrow Austenite$. The porosity of the porous 316 stainless steel was 42.07%. The maximum size of impurity particles was 5 μm when the holding time reached 180 min. Magnetic separation was an effective method to reduce impurities in the porous stainless steel.

Keywords: carbon reduction; iron oxide scale; porous 316 stainless steel; vacuum reduction sintering

Citation: Zhang, F.; Peng, J.; Chang, H.; Wang, Y. Vacuum Carbon Reducing Iron Oxide Scale to Prepare Porous 316 Stainless Steel. *Metals* 2022, 12, 2118. https://doi.org/10.3390/met12122118

Academic Editors: Denise Crocce Romano Espinosa and Petros E. Tsakiridis

Received: 25 August 2022
Accepted: 29 November 2022
Published: 9 December 2022

Publisher's Note: MDPI stays neutral with regard to jurisdictional claims in published maps and institutional affiliations.

Copyright: © 2022 by the authors. Licensee MDPI, Basel, Switzerland. This article is an open access article distributed under the terms and conditions of the Creative Commons Attribution (CC BY) license (https://creativecommons.org/licenses/by/4.0/).

1. Introduction

Iron oxide scale is a by-product of continuous casting billet or steel ingot and its rolling process, also known as iron scale, which accounts for about 1.5% of annual steel production [1]. For example, global crude steel production in 2021 was 1.95 billion tons, and the output of iron oxide scale was about 29.27 million tons, which is quite considerable [2]. Compared with other solid wastes, iron oxide scale has the advantages of high total iron content (more than 70%), low impurity content and easy purification [3]. At present, the recycling method of iron oxide scale is mainly concentrated on the production of a slagging agent, reduced iron powder and iron red pigment, and as an auxiliary iron-containing raw material for sintering, pelleting or powder metallurgy [4,5]. Thus, the above methods of iron oxide scale have a low utilization level. In order to efficiently utilize the metal components in iron oxide scale, it is necessary to carry out multi-angle research on the reduction mechanism of iron oxide scale so as to provide theoretical basis and technical support for improving the added value of its products.

Porous stainless steel has unique properties that are different from dense materials due to the presence of holes, such as its small density, large surface area, good sound absorption performance, low thermal conductivity, excellent permeability and so on [6,7]. Therefore, porous metal materials are widely used in the manufacture of filter purification materials [8], energy conversion devices [9], catalyst supports [10], sound absorbers [11] and biological transplantation materials [12–14]. Among them, porous 316 stainless steel has the advantages of high-temperature resistance, corrosion resistance and oxidation resistance, as well as good comprehensive mechanical properties, excellent biocompatibility, easy processing, et al. [15,16]. Therefore, it can be widely used as a structural material and functional material for medical-drug-carrying implant devices [13,17,18], fuel cells [19], filters [20], heat exchangers [21] and so on.

The usual preparation methods of porous stainless steel mainly include powder sintering technologies [22–24], a physical dealloying process [18], fiber felt [25,26] and so on. Most of them use stainless steel powder or stainless steel fiber as the main raw material, which is mixed with pore-forming agent, then sintered, casted or deposited in a protective gas or vacuum [27]. In addition, the stainless steel powder is produced by atomizing a molten metal with water or inert gas in centrifugal equipment or with a plasma rotary electrode [13,28]. Moreover, the production of the molten 316 stainless steel undergoes EAF (electric arc furnace)→AOD (argon oxygen decarburization furnace)→LF (ladle furnace) [29,30], which is complex, high polluting and high energy-consuming [31].

If the preparation of stainless steel powder can be combined with the molding process of the products, the process of porous stainless steel will be efficiently shortened with the raw material costs being reduced and the production efficiency being improved at the same time. In this work, steel rolling iron scale was used as an iron-containing raw material to prepare porous 316 stainless steel by high-temperature sintering under vacuum conditions, while carbon was served as a reducing agent and pore-forming agent, and other metal powders, such as alloy elements, including chromium, nickel and molybdenum. The chemical composition of 316 stainless steel is shown in Table 1. The price of iron oxide scale was 1.06~1.13 CNY per kilogram, and the average price of metal chromium powder, metal nickel powder, metal molybdenum powder was 79.2, 148.5, 275.0 CNY per kilogram, respectively [32]. The raw material cost was 39.1 CNY per kilogram according to the median component of 316 stainless steel in Table 1. The price of high-purity graphite powder was 0.6 CNY per kilogram. The commercial price of 316 stainless steel powder was 64.3~96.6 CNY per kilogram. Furthermore, the process of vacuum carbon reduction sintering provided in this paper is simpler because the stainless steel powder production is merged with the process of pore forming. Therefore, it will significantly reduce the production cost of porous 316 stainless steel. Meanwhile, the added value of the product made from iron oxide scale will been increased.

Table 1. Chemical composition of 316 stainless steel, wt %.

Component	C	Cr	Ni	Mn	Mo	S	P	Fe
Steel specification	≤0.03	16~18	10~14	≤2.0	2~3	≤0.030	≤0.035	Bal
Median	≤0.03	17	12	≤2.0	2.5	≤0.030	≤0.035	Bal

However, the metal chromium powder will be oxidized during iron oxide scale is reduced by carbon, and the reduction temperature, system pressure and carbon proportion are very crucial [33,34]. In this paper, in order to confirm the optimal preparation process of porous 316 stainless steel with iron oxide scale and metal powders, FactSage thermodynamic database combined with experimental research was implemented. In addition, the morphology and porosity of the porous 316 stainless steel were analyzed, and the occurrence state of impurities from iron oxide scale was studied to improve the purity of the product.

2. Materials and Methods

2.1. Experimental Raw Materials

The raw materials used in this paper include steel rolling iron oxide scale, metal powders and high-purity graphite powder. The chemical composition of the treated iron oxide scale is shown in Table 2. The total iron content (TFe) of the iron oxide scale was 73.17%, and the sum of oxide impurity content was 1.28%. Because the iron oxide scale comes from ordinary carbon steel, the compositions of chromium oxide, nickel oxide and molybdenum oxide were not detected.

Table 2. Chemical composition of iron oxide scale, wt %.

TFe	FeO	Fe_2O_3	SiO_2	CaO	Al_2O_3	MgO	S	P	LOI
73.17	58.63	39.38	0.30	0.36	0.23	0.39	0.02	0.01	0.68

The composition and morphology of the iron oxide scale before being treated were analyzed with a JSM-6510 scanning electron microscope equipped with an energy-dispersive X-ray spectrometer (SEM-EDS) (JEOL, Beijing, China), and the results are shown in Figure 1. In Figure 1(a), the surface of the iron oxide scale is scaly, in which the content of iron and oxygen is the highest, and that of silicon, aluminum, calcium and other impurity elements is very low, and their distributions are uneven. Meanwhile, the elements of silicon and aluminum or aluminum and calcium coexist in impurity particles on the surface of the iron oxide scale.

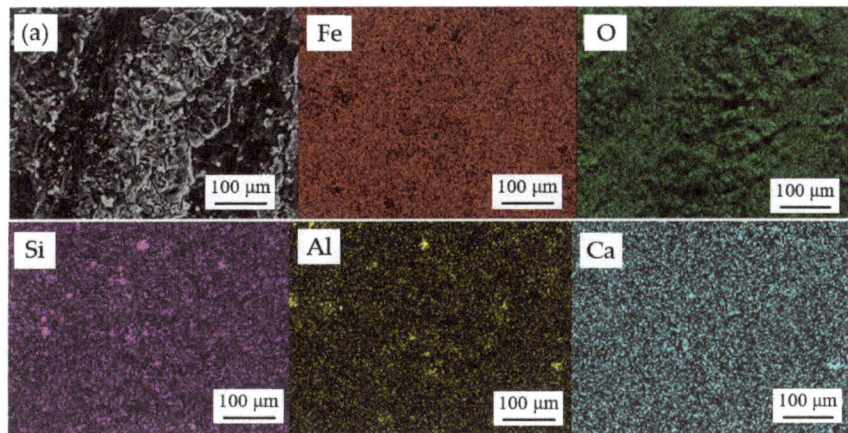

Figure 1. Microscopic morphology and elemental distribution on the surface of iron oxide scales before treatment.

The mineral composition of the iron oxide scale was analyzed by X-ray diffraction (XRD) on MiniFlex600 X-ray Diffractometer (Rigaku, Beijing, China), and the particle size of the treated iron oxide scale was tested with LS230 Laser Particle Size Analyzer (BECKMANCOULTER, Suzhou, China). The analysis results are shown in Figures 2 and 3, respectively. As can be seen from Figure 2, the main components of the iron oxide scale are FeO and Fe_3O_4, in addition to a small amount of Fe_2O_3 and Fe. The average particle size (mean) of the iron oxide scale was 9.25 μm, the surface particle size (S.D) was 6.04 μm, 10% of the particles were smaller than 0.308 μm, and 90% of the particles were smaller than 11.58 μm.

The raw materials for the preparation of the porous 316 stainless steel also included metal chromium powder, metal nickel powder, metal molybdenum powder, as well as high-purity graphite powder. Among them, the metal powders were used to adjust the alloy compositions of the porous 316 stainless steel, and the graphite powder was used as a reducing agent and a pore-forming agent. The basic information of the above raw materials is shown in Table 3.

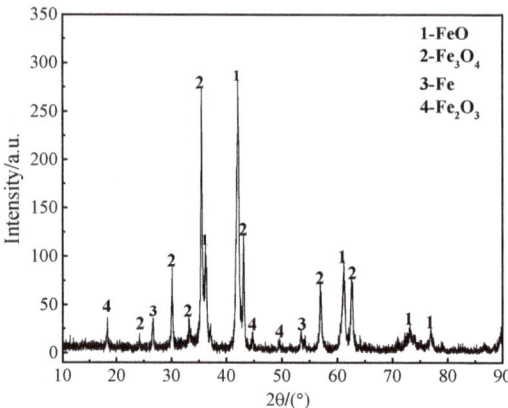

Figure 2. X-ray pattern of iron oxide scale.

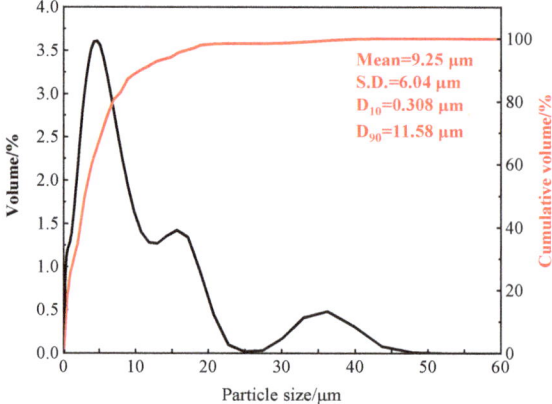

Figure 3. Particle size distribution of iron oxide scale.

Table 3. Basic information about auxiliary materials.

Auxiliary Materials	Purity	Manufacturer
Chromium metal powder	≥99.9%	Zhongmai Metal Materials Co., Ltd., Nangong, China
Nickel metal powder	≥99.9%	Zhongmai Metal Materials Co., Ltd., Nangong, China
Molybdenum metal powder	≥99.9%	Zhongmai Metal Materials Co., Ltd., Nangong, China
Graphite powder	≥99.9%	Kermel Chemical Reagent Co., Ltd., Tianjin, China

2.2. Methods

During the preparation of the porous 316 stainless steel, firstly, the iron oxide scale was crushed, sieved, cleaned for oil removal and underwent wet magnetic separation followed by being dried at 120 °C for 2 h. Then, the iron oxide scale was mixed with the appropriate amount of metal powders and high-purity graphite powder. Additionally, the mixture was press into ϕ15 × 3 mm pills. Lastly, the pill samples were sintered in a controllable atmosphere vacuum tube furnace. The preparation process of the porous 316 stainless steel is shown in Figure 4.

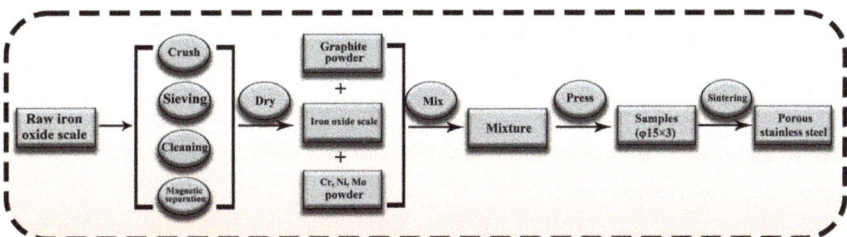

Figure 4. The preparation process of porous stainless steel.

The specification composition of the 316 stainless steel and the target composition of porous the 316 stainless steel prepared in this paper are shown in Table 1. Additionally, the median composition in Table 1 was the target component of the porous 316 stainless steel. Because metal chromium powder is bound to be oxidized by iron oxides at high temperature, it is necessary that the reduction process should be implemented under a suitable vacuum degree and sintering temperature. The specific reduction system, including the sintering temperature, vacuum degree and carbon amount, was determined through thermodynamic calculation combined with experiments. In our work, thermodynamics calculations were performed using FactSage8.1. The Equilib module was selected, and the product databases selected were FactPS, FToxid-SPINA, FToxid-CORU, FToxid-MeO, FSstel-FCC, FSstel-BCC, FSstel-M23C, FSstel-M7C3, FSStel-CEME and FSstel-Liqu. In the process of thermodynamic calculation, all of the compositions in Tables 1 and 4 were applied except sulfur and phosphorus.

Table 4. Ingredient list of raw materials for porous 316 stainless steel, g.

Iron Oxide Scale	C	Cr	Ni	Mo
90.88	16.93	17.00	12.00	2.50

In addition, in order to reveal the carbon reduction steps of the mixture of all raw materials, the major phase components of the intermediate products at different temperatures during the reduction process was analyzed by XRD with a scanning angle (2θ) from 10° to 90°, step size of 0.02° and scanning speed of 2 (°)/min. The morphology of the porous 316 stainless steel was investigated with a SEM-EDS for point and area scanning. The porosity of the porous stainless steel was measured with the water immersion method three times, and the average value of the three measurements was taken as the final result.

3. Results and Discussion
3.1. Determination of the Reduction Sintering System

The carbon addition was the most important for the preparation of quality porous 316 stainless steel, in which the carbon composition required was less than 0.03%. In addition, in order to guarantee the carbon addition was enough, it was assumed that iron oxides in the iron oxide scale were completely reduced to carbon monoxide by carbon because it was not clear that the reduction product was CO or CO_2. The dosages of iron oxide scale and alloys for the porous 316 stainless are shown in Table 4, which was confirmed with FactSage 8.1 thermodynamic database according to Tables 1 and 3. In detail, 90.88 g iron oxide scale could be reduced by 16.93 g carbon. Meanwhile, 17.00 g metal Cr, 12.00 g metal Ni and 2.50 g metal Mo also were needed for 100 g 316 stainless steel.

The function of carbon in the raw materials was to reduce the iron oxide scale, but chromium metal powder was likely oxidized by the ferric oxides from iron oxide scale before the ferric oxides were reduced by carbon. Furthermore, chromium oxides could be reduced by graphite only when the sintering temperature was raised above a certain value; that is, the intersection temperature of the oxygen potential lines of chromium oxides and of carbon oxides. However, the sintering temperature of the porous material was required to be lower than its melting point in order to effectively control its porosity. The occurrences of chromium element at 300~1600 °C and 1 atm were predicted with FactSage 8.1 thermodynamic database in order to obtain the optimal sintering conditions, and the results are shown in Figure 5. The results show that the chromium element in $FeCr_2O_4$ was the only form at 300~800 °C, and $FeCr_2O_4$ was always present in the reduction products. Then, part of $FeCr_2O_4$ converted to Cr_3O_4, as the temperature was higher than 800 °C. Most of the $FeCr_2O_4$ significantly transformed to Cr_2O_3 and a small amount of FCC as the temperature rose to 1080 °C. Then, Cr_2O_3 disappeared completely and converted to Cr_7C_3, and the liquid metal started to generate in large quantities at 1126 °C, while Cr_3O_4, Cr_7C_3 and $FeCr_2O_4$ were not reduced completely. Therefore, it could be understood that the transition sequence of the Cr-containing phase was $FeCr_2O_4 \rightarrow FeCr_2O_4 + Cr_3O_4 \rightarrow Cr_3O_4 + Cr_2O_3 \rightarrow Cr_7C_3 + Cr_3O_4 \rightarrow Cr(liq) + Cr_3O_4$. Furthermore, it was impossible to obtain the 316 stainless steel with a porous structure under 1 atm because the chromium element in Cr_3O_4, Cr_7C_3 and $FeCr_2O_4$ did not thoroughly transform into FCC before the liquid phase generated.

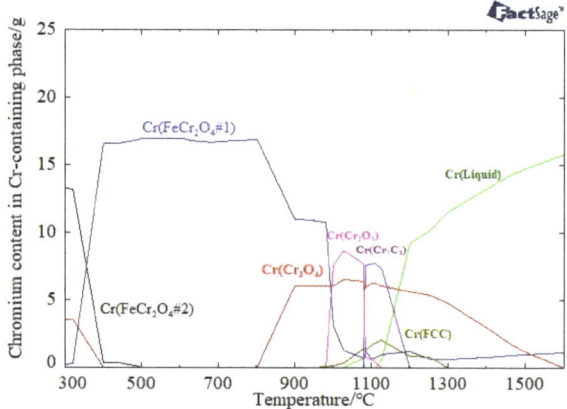

Figure 5. Occurrence of Cr element at 1 atm and different temperatures.

According to decarburization and chromium preservation theory, the chromium element in $FeCr_2O_4$, Cr_3O_4, Cr_2O_3 and Cr_7C_3 can gradually convert into FCC at lower temperatures by decreasing the system pressure, and a liquid phase does not appear at the same time [35]. Because 316 stainless steel belongs to austenitic stainless steel and the carbon specification composition in 316 stainless steel is required to be less than 0.03%; the carbon content in the FCC and the mass of FCC under 10^{-5}~1atm and 300~1600 °C were also calculated according to Tables 1 and 4, and the calculation results are shown in Figure 6. Figure 6 shows that the carbon content in FCC significantly declined at the same temperature with the decrease in the system pressures. Moreover, the equilibrium temperatures corresponding to a carbon content of 0.03% under 10^{-5}~10^{-1} atm also successively rose, and these temperatures were 793, 1080, 1226, 1385 and 1390 °C, respectively. Fortunately, the carbon content in FCC could be reduced to 0.03% when the equilibrium system was below 10^{-3} atm, which is possibly feasible for the preparation of porous 316 stainless steel.

In addition, Figure 7 shows the temperature ranges in which FCC accounted for more than 98% of the total product under 10^{-5} atm, 10^{-4} atm, 10^{-3} atm, 10^{-2} atm and 10^{-1} atm, which were 964~1100 °C, 1037~1200 °C, 1097~1228 °C, 1168~1300 °C and 1245~1300 °C, respectively. The results in Figures 6 and 7 indicate that the carbon content in FCC met the requirements of 316 stainless steel prepared at 10^{-4} atm and 1080~1200 °C. Therefore, it was also necessary to confirm the transition sequence of the Cr-containing phase under 10^{-4} atm through thermodynamic calculation.

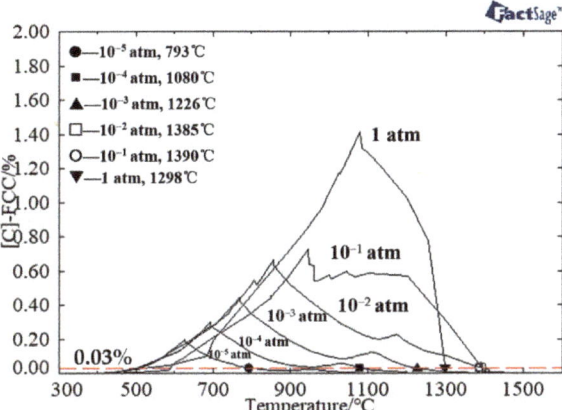

Figure 6. Carbon content in FCC under different system pressure.

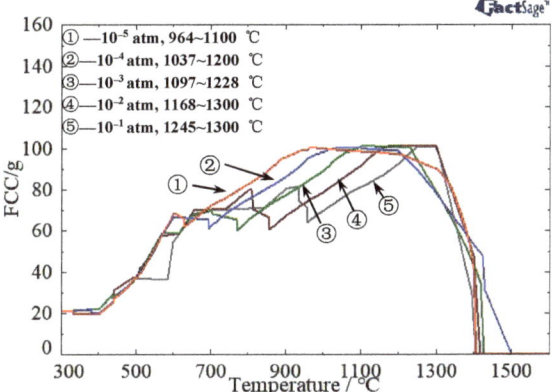

Figure 7. Temperature range of FCC existing under $10^{-5} \sim 10^{-1}$ atm.

Figure 8 shows the chromium content in the Cr-containing phase at different temperatures and under 10^{-4} atm. As can be seen from Figure 8, the transition sequence of the Cr-containing phase was $FeCr_2O_4 + Cr_3O_4 \rightarrow Cr_2O_3 + Cr_3O_4 + Cr_{23}C_6 \rightarrow Cr_{23}C_6 + Cr_7C_3 + FCC \rightarrow FCC + Cr_{23}C_6 \rightarrow FCC \rightarrow FCC + BCC \rightarrow Cr(liq)$ with the increase in the equilibrium temperature, and the liquid phase generated at 1427 °C. Meanwhile, most of the chromium element mainly existed in the form of FCC with a small amount of spinel at 1037~1200 °C. Therefore, it can be determined that the porous 316 stainless steel could be obtained under 10^{-4} atm and at 1080~1200 °C. However, $Fe_2Cr_2O_4$ always coexisted with FCC under the above conditions, which may be due to the insufficient addition of carbon. Therefore, the amount of carbon in Table 4 needs to be adjusted to ensure that $Fe_2Cr_2O_4$ is reduced completely.

In order to reduce $FeCr_2O_4$ completely, the amount of $FeCr_2O_4$ and the carbon content in the FCC at 10^{-4} atm and 1200 °C were calculated when the carbon addition was increased from 16.50 g to 17.20 g, and the results are shown in Figure 9a,b, respectively. Figure 9a shows that the amount of $FeCr_2O_4$ gradually declined with the increase in the carbon addition. It reduced to zero when the carbon addition was 17.12 g, which means that $FeCr_2O_4$ was completely reduced at this moment. In contrast, in Figure 9b, the carbon content in the FCC continuously increased with the rise of the carbon addition. It increased to 0.006% when the carbon addition was 17.12 g, which met the specified carbon content of the 316 stainless steel. Furthermore, the appropriate carbon addition should be less than 17.17 g, while the carbon content in FCC was lower than 0.03%. Therefore, a carbon addition of 17.17 g was more reasonable, while FCC was 98.95 g; the chromium content in FCC was 17.11%, and the yield of the chromium element was 99.59%.

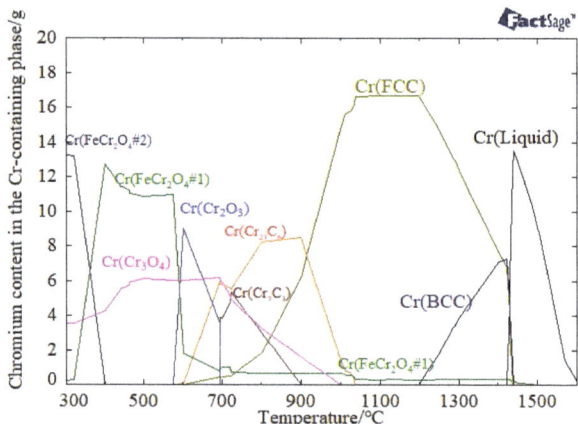

Figure 8. Content of Cr-containing phase at 10^{-4} atm and different temperature.

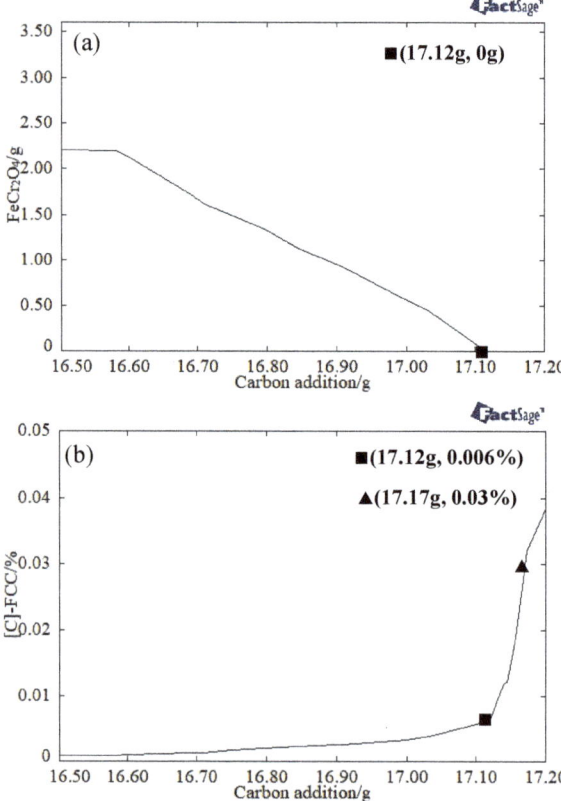

Figure 9. Effect of carbon addition at 10^{-4} atm and 1200 °C; (**a**) amount of $FeCr_2O_4$; and (**b**) carbon content in FCC.

Through the above thermodynamic analysis, it can be determined that 90.88 g iron oxide scale could be reduced to obtain the 98.95 g 316 stainless steel with 17.17 g carbon under 10^{-4} atm and 1200 °C. However, the optimal sintering time needed to be confirmed by the actual sintering experiments. A vacuum reduction sintering system is shown in Figure 10, in which the sintering samples were kept at 10^{-4} atm and 1200 °C for 120, 150, 180, 210 and 240 min, respectively.

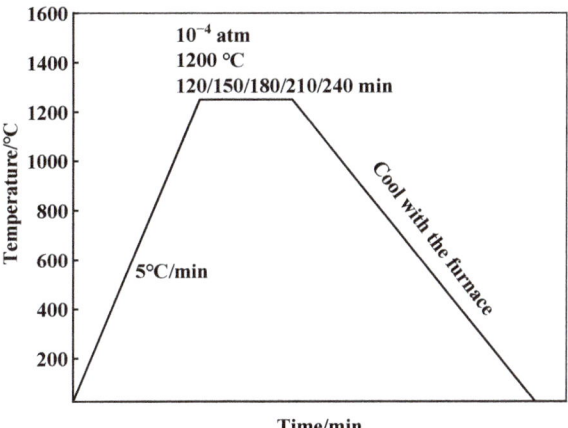

Figure 10. Vacuum reduction sintering system of porous 316 stainless steel.

By means of the vacuum reduction sintering experiments, the yield of the metal powders was confirmed, as shown in Table 5. The yield of metal chromium powder and metal molybdenum powder was 98.71% and 97.20%, respectively. The losses were caused by the evaporation of Cr_2O_3 and MoO_3 [36,37]. Every sintering sample was 5 g and held at 4 MPa for 2 min to make a sample of ϕ15 × 3 mm, and then the sintering process was carried out according to the reduction schedule in Figure 10. The weight of every sample was weighed before and after the sintering process, and the weight-loss rate and the carbon content of the sample held at 10^{-4} atm and 1200 °C is shown in Figure 11.

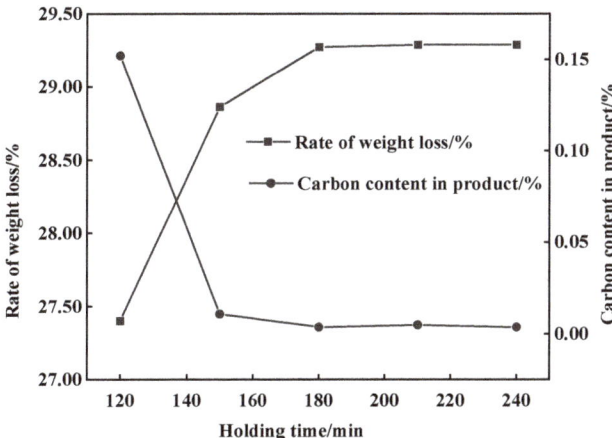

Figure 11. Rate of weight loss and carbon content in sample reduced at 10^{-4} Pa and 1200 °C.

In Figure 11, the weight loss rate and the carbon content of the sintering sample were stable after being held for 180 min under 10^{-4} atm and 1200 °C, which were 29.27% and

3.71×10^{-3}%, respectively. The carbon content met the requirement of the 316 stainless steel. Therefore, the vacuum reduction sintering system was determined to be 10^{-4} atm and 1200 °C for 180 min.

Table 5. Yield of alloy in vacuum reduction sintering process, wt %.

Raw Material	Metal Chromium	Metal Nickle	Metal Molybdenum
Yield	98.71	100.00	97.20

The chemical composition of porous 316 stainless steel prepared at 10^{-4} atm and 1200 °C for 180 min is shown in Table 6. The content of carbon, sulfur and phosphorus was 0.025%, 0.010% and 0.020%, respectively. Meanwhile the content of the alloy element was also within the specification range of the target steel.

Table 6. Chemical composition of porous 316 stainless steel, wt %.

Fe	Ni	Mn	Mo	Cr	C	O	S	P
66.89	12.00	1.98	2.49	16.60	0.025	0.20	0.010	0.020

In order to reveal the actual transformation of the chromium element, the samples being sintered at 10^{-4} atm and 700, 900, 1100 and 1200 °C for 180 min were analyzed with XRD, and the results are shown in Figure 12. The chromium element underwent the transformation of metal Cr→FeCr$_2$O$_4$→Cr$_{23}$C$_6$→Austenite at 700 °C→900 °C→1100 °C→1200 °C and 10^{-4} atm. In detail, metal chromium was oxidized to FeCr$_2$O$_4$ by the iron oxide scale at lower than 700 °C, FeCr$_2$O$_4$ changed to Cr$_{23}$C$_6$ at 1100 °C, while iron oxide scale was reduced to metal iron.

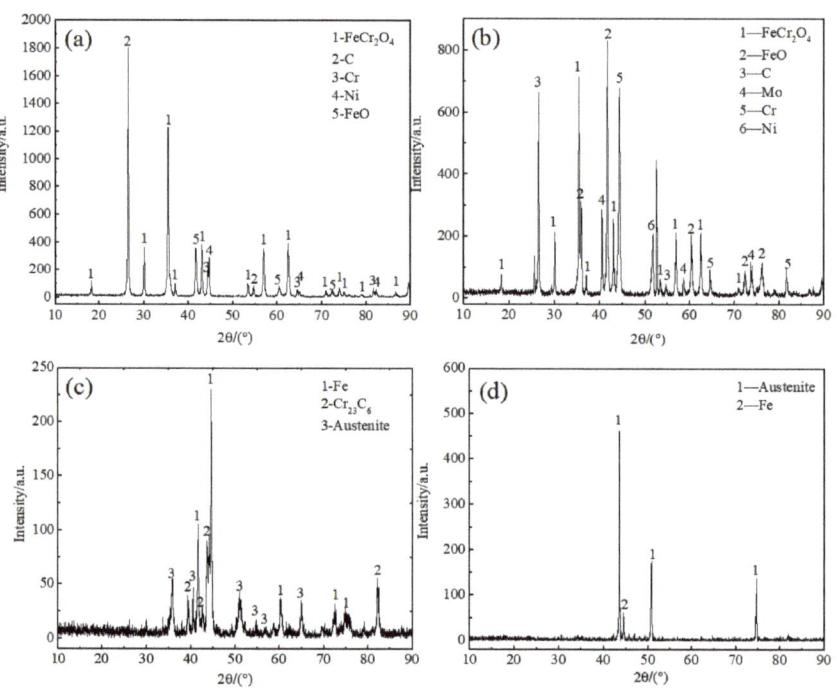

Figure 12. XRD pattern of sintering sample at 10^{-4} atm and different temperatures for 180 min; (a) 700 °C; (b) 900 °C; (c) 1100 °C; and (d) 1200 °C.

Figure 13 shows the carbon and oxygen content in the sample at 10^{-4} atm and different temperatures. As the temperature increased, the carbon and oxygen contents showed a continuous downward trend; the fast stage was at 1100~1150 °C, while the carbon in $Cr_{23}C_6$ was oxidized and removed by the oxygen in the residual ferrous oxide [38]. In the stage of 1150~1250 °C, the decline of the carbon and oxygen content in the sample was getting slower, while the carbon was dissolved into austenite, and the oxygen in impurities, such as CaO, SiO_2 and Al_2O_3, could not be removed.

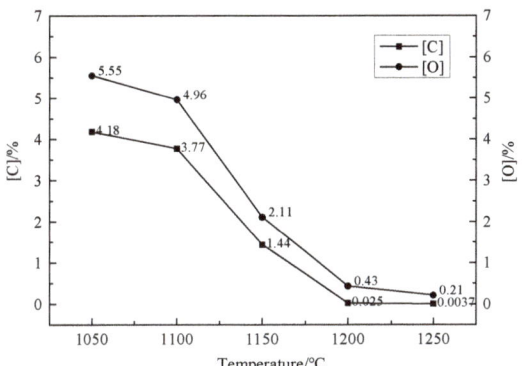

Figure 13. Carbon and oxygen content in samples at 10^{-4} atm and different temperatures.

3.2. Microstructure of Porous 316 Stainless Steel

Figure 14 shows the micromorphology of the porous 316 stainless steel prepared at 10^{-4} atm and 1200 °C for 180 min. The porous stainless steel consisted of sintering necks and pores, the size and shape of the pores were irregular, and the porosity measured by the immersion medium method was 42.07%. Zhang W.P. et al. [27] prepared porous stainless steel with a porosity of 28.21~60.16% by the vacuum melting method with 30, 40, 50 and 60 vol.% ammonium bicarbonate (NH_4HCO_3) as a pore-forming agent. In Figure 14, point ① is the matrix in which the chemical compositions included iron, chromium, nickel, molybdenum and manganese, and there was no impurity element in the matrix. At the same time, the roughly spherical particles were observed on the surface of the sintering neck, with a radius of 1~2 μm, as shown by point ②. The components of the particles mainly included Al_2O_3, SiO_2 and CaO, while the content of iron, chromium and manganese element were obviously lower than that in the matrix. It was confirmed that the particles came from the impurities in the iron oxide scale.

Figure 15 shows the sintering neck interior microstructure of the porous 316 stainless steel. As can be seen from Figure 15a, the size of the austenite grains in the sintering neck was uneven, and they were all less than 10 μm. The bright white bands or particles shown in Figure 15b were confirmed by energy spectroscopy analysis as a precipitated phase with a relatively high content of chromium and molybdenum. This precipitated phase was TCP phase, namely, σ phase, which is a hard and brittle intermetallic compound with a square lattice and is mainly composed of iron, chromium, molybdenum and other elements [39,40]. The content of molybdenum in the precipitated phase was 9.35% and much higher than that of the stainless steel at 2.49%. Meanwhile, the chromium content of the poor chromium area between the two white bands in the precipitated phase was 7.67% and far below that of the stainless steel at 16.60%.

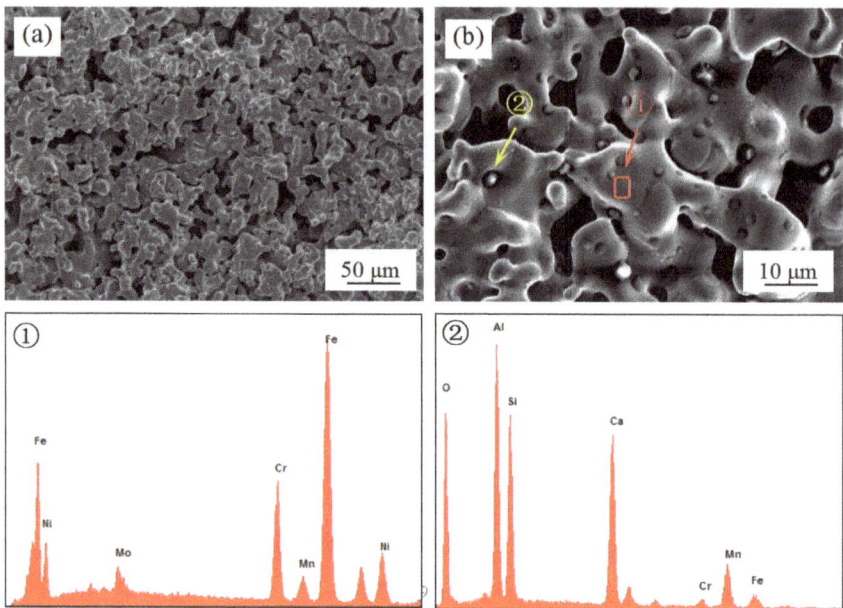

Figure 14. Micromorphology of porous 316 stainless steel at 10^{-4} atm and 1200 °C for 180 min; (**a**) macro morphology; (**b**) microscopic morphology of the matrix and impurity particles; ① matrix; and ② impurity particle.

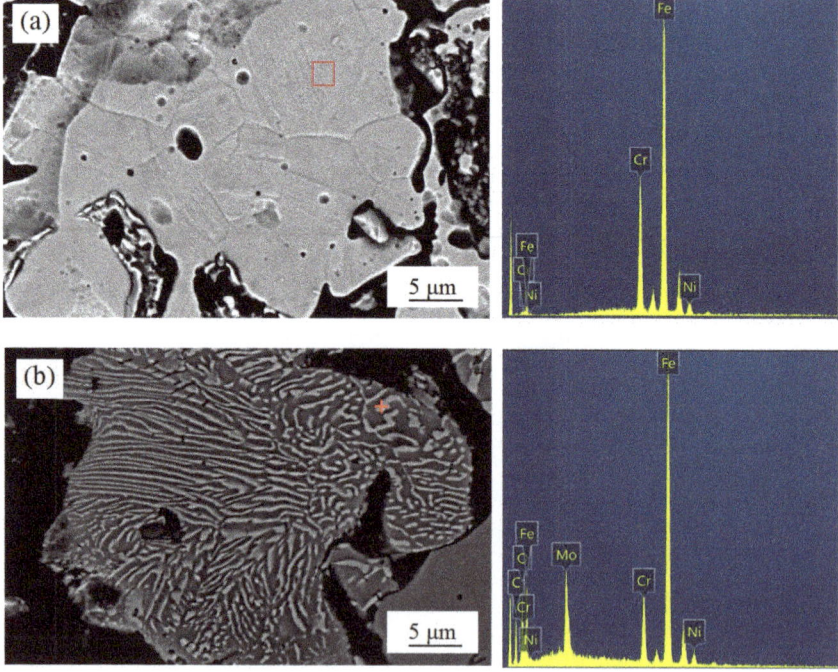

Figure 15. Microstructure of porous 316 stainless steel; (**a**) austenite grain; and (**b**) σ phase; the energy spectrum analysis is on the right side of the back-scattered electronic image.

3.3. Growing up of Impurity Particles

In order to prove the growing up of impurity particles in the iron oxide scale during reduction sintering, the morphology and composition of the particles sintered at 10^{-3} Pa and 1200 °C for 60, 120 and 180 min were analyzed with SEM-EDS. The back-scattered electronic images of the sintering neck and the particles are shown in Figure 16. The content of iron, manganese, calcium and chromium in the roughly spherical particles in Figure 16a is higher than that in Figure 16b,c. The content of iron and chromium decreased with the extension of the holding time, while the size of the impurity particle gradually increased. The maximum size of the impurity particles reached 5 µm, and their color became deepest when the holding time reached 180 min, indicating that the content of iron, manganese, chromium and other elements gradually decreased because the atomic number of these elements is greater than that of calcium, silicon and aluminum.

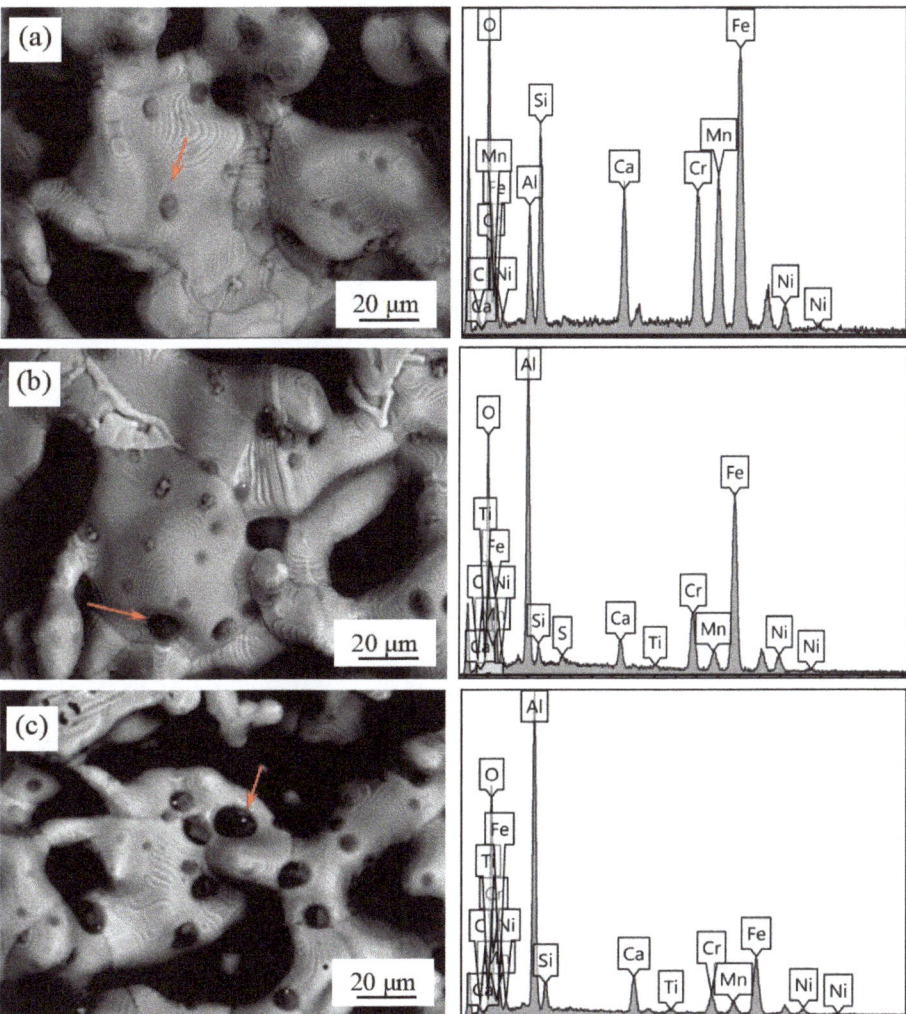

Figure 16. Back-scattered electronic image and composition analysis of particles at 10^{-4} atm and 1200 °C; (**a**) 60 min; (**b**) 120 min; and (**c**) 180 min; the energy spectrum analysis is on the right side of the back-scattered electronic image.

Because the iron oxide scale reduction process was carried out step by step, ferrous oxide was an intermediate product. In order to reveal the influence of ferrous oxide on the melting properties of impurities, the CaO-SiO$_2$-Al$_2$O$_3$ ternary phase diagram containing MgO and FeO was calculated and analyzed with FactSage 8.1 database, as shown in Figure 17. In Figure 17, there is liquid slag coexisting with several mineral phases, including Ca$_2$SiO$_4$, monoxide, Ca$_3$MgAl$_4$O$_{10}$, melilite, CaAl$_2$Si$_2$O$_8$, peridot, spinel, clinopyroxene and so on. In the calculation, the content of FeO was 20%, while that of iron oxide scale declined from 58.63% to zero. In fact, FeO has the function of reducing the melting point [41,42]. The liquid phase is a benefit to the separation of impurity oxides gradually from the iron matrix because the shrinkage of the iron phase is greater than that of the impurity particles [43]. As a result, the particles enriched on the austenitic grain boundary as roughly spherical particles.

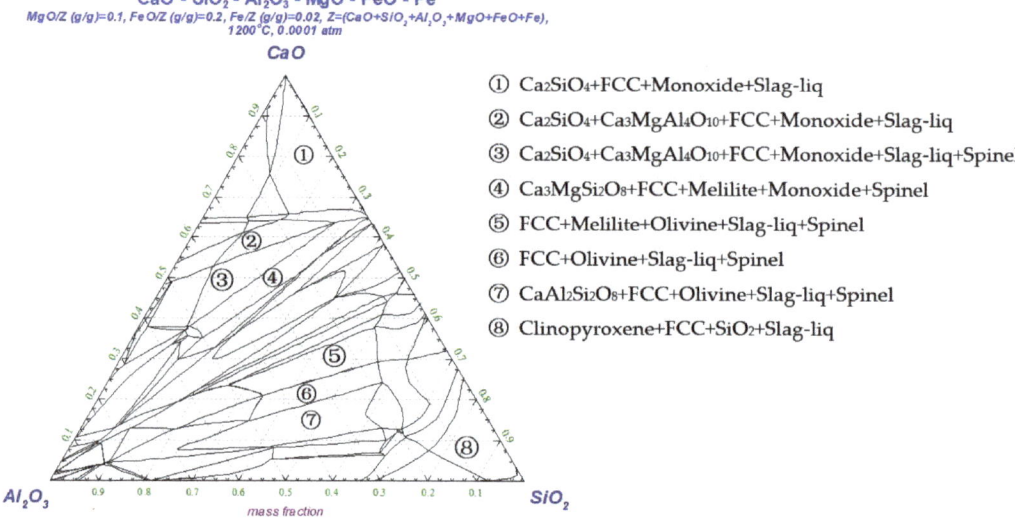

Figure 17. CaO-SiO$_2$-Al$_2$O$_3$ ternary phase diagram.

In order to reduce the impurities in the porous 316 stainless steel, the magnetic field strength was increased from 3000 Oe to 5000 Oe, and the non-magnetic substances in the iron oxide scale were further separated and removed. The SEM back scatter image of the surface and inside of the porous 316 stainless steel prepared with new raw materials and the original reduction sintering system is shown in Figure 18. Figure 18a is the surface micromorphology of the porous 316 stainless steel, and Figure 18b is that of the inside. Therefore, the impurities in the sample were significantly reduced, and magnetic separation was an effective method to reduce impurities in the porous stainless steel.

The main goal of this paper was to develop high-value-added metal materials and products with an iron oxide scale as a raw material, but the impurity in the iron oxide scale was the main factor that impacted the product quality. Therefore, it is necessary to explore effective ways to reduce the impurity content or find the appropriate application field of the above metal materials and products.

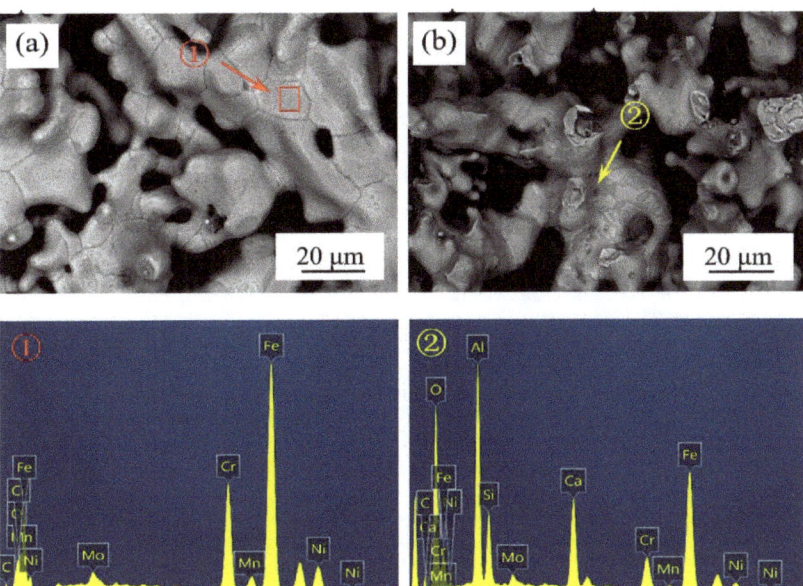

Figure 18. Back-scattered electronic image of porous 316 stainless steel with stronger magnetic separation iron oxide scale; (**a**) surface; (**b**) inside; ① the energy spectrum analysis of point ①; and ② the energy spectrum analysis of point ①.

4. Conclusions

Porous 316 stainless steel was prepared by carbon reduction under vacuum with iron oxide scale as the main raw material in this study. The specific reduction system was confirmed, including the sintering temperature, sintering time, vacuum degree and carbon amount through thermodynamic calculation combined with experiments. The characters of the intermediate products and final product were analyzed and measured. Thermodynamic analysis results showed that the conversion process of the chromium element in the raw materials was $FeCr_2O_4 \rightarrow FeCr_2O_4 + Cr_3O_4 \rightarrow Cr_3O_4 + Cr_2O_3 \rightarrow Cr_7C_3 + Cr_3O_4 \rightarrow Cr(liq) + Cr_3O_4$ at 1 atm and 300~1600 °C. The liquid phase began to generate at 1126 °C, so porous stainless steel could not be prepared at 1 atm. The transformation process of the chromium element in the raw materials at 10^{-4} atm and 300~1600 °C was $FeCr_2O_4 + Cr_3O_4 \rightarrow Cr_2O_3 + Cr_3O_4 + Cr_{23}C_6 \rightarrow Cr_{23}C_6 + Cr_7C_3 + FCC \rightarrow FCC + Cr_{23}C_6 \rightarrow FCC \rightarrow FCC + BCC \rightarrow Cr(liq)$. The FCC phase with qualified carbon content could be obtained below 10^{-4} atm and 1200 °C, while 90.88 g iron oxide scale, 17.17 g carbon, 17.00 g metal chromium, 12.00 g metal nickel and 2.50 g metal molybdenum were necessary to produce 100 g porous 316 stainless steel. The sintering experiment results showed that porous 316 stainless steel with a carbon content of 0.025% could be obtained at 10^{-4} atm and 1200 °C for 180 min, while chromium element underwent the transformation of metal, $Cr \rightarrow FeCr_2O_4 \rightarrow Cr_{23}C_6 \rightarrow Austenite$. The porosity of the porous 316 stainless steel was 42.07%. Additionally, the size of the austenite grains in the sintered neck was uneven, and they were all less than 10 μm. The σ phase appeared in the porous 316 stainless steel, in which the content of molybdenum was 9.35% and much higher than that of the stainless steel at 2.49%. Meanwhile, the chromium content of the poor chromium area was 7.67%, which was far below that of the 316 stainless steel at 16.60%. The maximum size of the impurity particles was 5 μm when the holding time reached 180 min. Magnetic separation was an effective method to reduce the impurities in the porous stainless steel.

Author Contributions: Investigation, F.Z., H.C. and Y.W.; methodology, F.Z.; data curation, F.Z., H.C. and Y.W.; writing—original draft, F.Z.; conceptualization, J.P.; supervision, J.P.; writing—review and editing, J.P. All authors have read and agreed to the published version of the manuscript.

Funding: This work was funded by the National Natural Science Foundation of China (51864041, 51874186, 51664066).

Data Availability Statement: Not applicable.

Conflicts of Interest: The authors declare no conflict of interest.

References

1. Hao, L.; Li, T.J.; Xie, Z.L.; Duan, Q.J.; Zhang, G.Y. The Oxidation Behaviors of Indefinite Chill Roll and High Speed Steel Materials. *Metals* **2020**, *10*, 1095. [CrossRef]
2. 2021 Crude Steel Production Data Analytics. China Business Intelligence Network. Available online: https://baijiahao.baidu.com/s?id=1725663902517027102&wfr=spider&for=pc (accessed on 22 October 2022).
3. Liu, J.X.; He, Y.; Xue, X.X. A new and simple route to prepare γ-Fe_2O_3 with iron oxide scale. *Mater. Lett.* **2018**, *6*, 112. [CrossRef]
4. Chen, C.; Li, Y.; Gan, W.L.; Feng, H.; He, H.Y.; Ni, H.W. Synthesis of Ni-Zn ferrite from zinc in Zn-contain dust and iron in scale. *J. Iron Steel Res.* **2021**, *33*, 911–919.
5. Yang, C. Basic characteristics and comprehensive utilization of iron-bearing solid wastes from metallurgical enterprises. *Environ. Eng.* **2012**, *30*, 287–289.
6. Zhao, C.H.; Wada, T.; Andrade, V.D.; Williams, G.J.; Gelb, J.; Li, L.; Thieme, J.; Kato, H.; Chen-Wiegart, Y.K. 3D Morphological and Chemical Evolution of Nanoporous Stainless Steel by Liquid Metal Dealloying. *ACS Appl. Mater. Interfaces* **2017**, *10*, 1021.
7. Bencina, M.; Junkar, I.; Vesel, A.; AlešIgli, M.M. Nanoporous Stainless Steel Materials for Body Implants of Synthesizing Procedure. *Nanomaterials* **2022**, *12*, 2924. [CrossRef]
8. Allioux, F.M.; Benavides, D.O.; Miren, E.; Kong, L.X.; Tanaka, D.A.P.; Dumée, L.F. Preparation of Porous Stainless Steel Hollow-Fibers through Multi-Modal Particle Size Sintering towards Pore Engineering. *Membranes* **2017**, *7*, 40. [CrossRef]
9. Mercadelli, E.; Gondolini, A.; Pinasco, P.; Sanson, A. Stainless Steel Porous Substrates Produced by Tape Casting. *Met. Mater. Int.* **2017**, *23*, 184–192. [CrossRef]
10. Abdullah, Z.; Ismail, A.; Ahmad, S. The Influence of Porosity on Corrosion Attack of Austenitic Stainless Steel. *J. Phys.* **2017**, *10*, 1088. [CrossRef]
11. Xu, X.B.; Liu, P.S.; Chen, G.F.; Li, C.P. Sound Absorption Performance of Highly Porous Stainless Steel Foam with Reticular Structure. *Met. Mater. Int.* **2021**, *27*, 3316–3324. [CrossRef]
12. Fousová, M.; Kubásek, J.; Vojtěch, D.; Fojt, J.; Čapek, J. 3D printed porous stainless steel for potential use in medicine. *Mater. Sci. Eng.* **2017**, *10*, 1088. [CrossRef]
13. Dudek, A.; Włodarczyk, R. Effect of sintering atmosphere on properties of porous stainless steel for biomedical applications. *Mater. Sci. Eng. C* **2013**, *33*, 434–439. [CrossRef] [PubMed]
14. Bender, S.; Chalivendra, V.; Rahbar, N.; Wakil, S.E. Mechanical characterization and modeling of graded porous stainless steel specimens for possible bone implant applications. *Int. J. Eng. Sci.* **2012**, *53*, 67–73. [CrossRef]
15. Kato, K.; Yamamoto, A.; Ochiai, S.; Wada, M.; Daigo, Y.; Kita, K.; Omori, K. Cytocompatibility and mechanical properties of novel porous 316 L stainless steel. *Mater. Sci. Eng. C* **2013**, *33*, 2736–2743. [CrossRef] [PubMed]
16. Groarke, R.; Danilenkoff, C.; Karam, S.; McCarthy, E.; Michel, B.; Mussatto, A.; Sloane, J.; O'Neill, A.; Raghavendra, R. 316L Stainless Steel Powders for Additive Manufacturing: Relationships of Powder Rheology, Size, Size Distribution to Part Properties. *Materials* **2020**, *13*, 5537. [CrossRef] [PubMed]
17. Bae, I.; Lim, K.S.; Park, J.K.; Song, J.H.; Oh, S.H.; Kim, J.W.; Zhang, Z.J.; Park, C.; Koh, J.T. Evaluation of cellular response and drug delivery efficacy of nanoporous stainless steel material. *Biomater. Res.* **2021**, *10*, 1186. [CrossRef] [PubMed]
18. Ren, Y.B.; Li, J.; Yang, K. Preliminary Study on Porous High-Manganese 316L Stainless Steel Through Physical Vacuum Dealloying. *Acta Metall. Sin.* **2017**, *30*, 731–734. [CrossRef]
19. Kirichenko, O.V.; Klimenko, V.N.; Shapoval, I.V.; Valeeva, I.K. Filtering properties of porous materials made of thin stainless steel fibers. *Powder Metall. Met. Cer.* **2015**, *54*, 151–155. [CrossRef]
20. Amel-Farzad, H.; Peivandi, M.T.; Yusof-Sani, S.M.R. In-body corrosion fatigue failure of a stainless steel orthopedic implant with a rare collection of different damage mechanisms. *Eng. Fail. Anal.* **2007**, *14*, 1205–1217. [CrossRef]
21. Li, W.Q.; Qu, Z.G.; Zhang, B.L.; Zhao, K.; Tao, W.Q. Thermal behavior of porous stainless-steel fiber felt saturated with phase change material. *Energy* **2013**, *5*, 846–852. [CrossRef]
22. Kazantseva, N.; Krakhmalev, P.; Åsberg, M.; Koemets, Y.; Karabanalov, M.; Davydov, D.; Ezhov, I.; Koemets, O. Micromechanisms of Deformation and Fracture in Porous L-PBF 316L Stainless Steel at Different Strain Rates. *Metals* **2021**, *11*, 1870. [CrossRef]
23. Krakhmalev, P.; Fredriksson, G.; Svensson, K.; Yadroitsev, I. Microstructure, solidification texture, and thermal stability of 316 L stainless steel manufactured by laser powder bed fusion. *Metals* **2018**, *8*, 643. [CrossRef]
24. Zhu, Y.; Lin, G.L.; Khonsari, M.M.; Zhang, J.H.; Yang, H.Y. Material characterization and lubricating behaviors of porous stainless steel fabricated by selective laser melting. *J. Mater. Process. Technol.* **2018**, *262*, 41–52. [CrossRef]

25. Feng, P.; Liu, Y.; Wang, Y.; Li, K.; Zhao, X.U.; Tang, H.P. Sintering behaviors of porous 316L stainless steel fiber felt. *J. Cent. South Univ.* **2015**, *22*, 793–799. [CrossRef]
26. Ao, Q.B.; Wang, J.Z.; Ma, J.; Ge, Y. Sound Absorption Properties of Stainless Steel Fiber Porous Materials before/after Corrosion. *Mater. Sci.* **2018**, *933*, 367–372. [CrossRef]
27. Zhang, W.P.; Li, L.J.; Gao, J.X.; Huang, J.M.; Zhang, X.K. The Effect of Porosity on Mechanical Properties of Porous FeCrN Stainless Steel. *J. Phys.* **2021**, *10*, 1088. [CrossRef]
28. Öztürk, B.; Topcu, A.; Cora, Ö.N. Influence of processing parameters on the porosity, thermal expansion, and oxidation behavior of consolidated Fe22Cr stainless steel powder. *Powder Technol.* **2021**, *382*, 199–207. [CrossRef]
29. Ma, D.; Guo, P.M.; Pang, J.M.; Zhao, P. Theoretical Analysis and Industrial Experiment of Direct Alloying of Molybdenum Oxide for 316L Steel. *Iron Steel.* **2014**, *49*, 27–30.
30. Wang, G.P.; Liu, C.J. Theoretical and experimental research of molybdenum oxide alloying for 316L stainless steel smelting process. *J. Iron Steel Rea. Int.* **2018**, *30*, 354–358.
31. Stavropoulos, P.; Panagiotopoulou, V.C.; Papacharalampopoulos, A.; Aivaliotis, P.; Georgopoulos, D.; Konstantinos, S. A Framework for CO_2 Emission Reduction in Manufacturing Industries: A Steel Industry Case. *Designs* **2022**, *6*, 22. [CrossRef]
32. Mysteel.com. Available online: https://feigang.mysteel.com/m/22/1008/10/AF407D5E375A3AA4.html (accessed on 22 October 2022).
33. Mukasheva, N.Z.; Kosdauletovb, N.Y.; Suleimen, B.T. Comparison of Iron and Chromium Reduction from Chrome Ore Concentrates by Solid Carbon and Carbon Monoxide. *Solid State Phenom.* **2020**, *299*, 1152–1157. [CrossRef]
34. Liu, Y.; Jiang, M.F.; Wang, D.Y. Material Balance Calculation of Chromium Ore Smelting Reduction and Direct Alloying Process in a Converter. *Adv. Mater. Res.* **2012**, *485*, 574–577. [CrossRef]
35. You, Y.; Wei, B.K.; You, W.; Wang, Z.L.; Jiang, B.L. Compound Control Method of Carbon Content in Argon—Oxygen Refining Ferrochromium Alloy. *Trans. Indian Inst. Met.* **2021**, *10*, 1007. [CrossRef]
36. Teng, J.W.; Gong, X.J.; Yang, B.B.; Yu, S.; Lai, R.L.; Liu, J.T.; Li, Y.P. High temperature oxidation behavior of a novel Ni-Cr-W-Al-Ti superalloy. *Corros. Sci.* **2022**, *10*, 1016. [CrossRef]
37. Wang, Y.X.; Tang, Y.J.; Wan, W.; Zhang, X. High-Temperature Oxidation Resistance of CrMoN Films. *J. Mater. Eng. Perform.* **2020**, *10*, 1007. [CrossRef]
38. Kryukova, R.E.; Goryushkina, V.F.; Bendrea, Y.V.; Bashchenkoa, L.P.; Kozyreva, N.A. Thermodynamic Aspects of Cr_2O_3 Reduction by Carbon. *Steel Trans.* **2019**, *49*, 843–847. [CrossRef]
39. Li, J.; Ren, X.; Zhang, Y.; Hou, H.; Gao, X. Effect of superplastic deformation on precipitation behavior of sigma phase in 3207 duplex stainless steel. *Mater. Int.* **2021**, *31*, 334–340. [CrossRef]
40. Liu, J.; Dang, X.T.; Peng, Y.T.; Wu, T. Microstructure and Wear Behavior of Laser Cladded CoCrNiMo$_x$ Coatings on the Low Carbon Steel. *Crystals* **2022**, *12*, 1229. [CrossRef]
41. Tu, Y.; Zhang, Y.; Su, Z.; Jiang, T. Mineralization mechanism of limonitic laterite sinter under different fuel dosage: Effect of FeO. *Powder Technol.* **2022**, *10*, 1016. [CrossRef]
42. Liao, J.; Zhao, B. Experimental Studies in Phase Equilibrium of the System "FeO-SiO_2-MgO-Al_2O_3-Cr_2O_3" at Iron Saturation. *Metal. Mater. Trans. B* **2021**, *52*, 2364–2374. [CrossRef]
43. Wang, G.; Wang, J.S.; Xue, Q.G. Kinetics of the Volume Shrinkage of a Magnetite/Carbon Composite Pellet during Solid-State Carbothermic Reduction. *Metals* **2018**, *8*, 1050. [CrossRef]

Article

Tin Removal from Tin-Bearing Iron Concentrate with a Roasting in an Atmosphere of SO₂ and CO

Lei Li [1,2,*], Zhipeng Xu [2] and Shiding Wang [1]

[1] Faculty of Metallurgical and Energy Engineering, Kunming University of Science and Technology, Kunming 650093, China
[2] College of Environmental Science and Engineering, Donghua University, Shanghai 201620, China
* Correspondence: tianxiametal1008@dhu.edu.cn; Tel.: +86-139-8761-9187; Fax: +86-21-6779-2537

Abstract: The tin could be volatilized and removed effectively from the tin-bearing iron concentrate while roasted in an atmosphere of SO_2 and CO. The reduction of SO_2 by CO occurred in preference to the SnO_2 and Fe_3O_4, and the generated S_2 could sulfurize the SnO_2 to an evaporable SnS, which resulted in the tin volatilization. However, the Fe_3O_4 could be sulfurized simultaneously, and a phase of iron sulfide was formed, retaining in the roasted iron concentrate. It decreased the quality of the iron concentrate. In addition, the formation of Sn-Fe alloy was accelerated as the roasting temperature exceeded 1100 °C, which decreased the Sn removal ratio. An appropriate SO_2 partial pressure and roasting temperature should be controlled. Under the condition of the roasting temperature of 1050 °C, SO_2 partial pressure of 0.003, CO partial pressure of 0.85, and residence time of 60 min, the tin content in the roasted iron concentrate was decreased to 0.032 wt.% and the sulfur residual content was only 0.062 wt.%, which meets the standard of iron concentrate for BF ironmaking.

Keywords: tin removal; tin-bearing iron concentrate; roasting; volatilization; SO_2 and CO

Citation: Li, L.; Xu, Z.; Wang, S. Tin Removal from Tin-Bearing Iron Concentrate with a Roasting in an Atmosphere of SO₂ and CO. *Metals* **2022**, *12*, 1974. https://doi.org/10.3390/met12111974

Academic Editors: Lijun Wang and Shiyuan Liu

Received: 12 October 2022
Accepted: 15 November 2022
Published: 18 November 2022

Publisher's Note: MDPI stays neutral with regard to jurisdictional claims in published maps and institutional affiliations.

Copyright: © 2022 by the authors. Licensee MDPI, Basel, Switzerland. This article is an open access article distributed under the terms and conditions of the Creative Commons Attribution (CC BY) license (https://creativecommons.org/licenses/by/4.0/).

1. Introduction

The Sn-bearing iron ore is a typically complex iron ore resource, with the reserve exceeding 0.5 billion tons in China [1–3]. After it is treated by traditional mineral processing technology, a Sn-bearing iron concentrate could be obtained. However, this concentrate cannot be used as a raw material for ironmaking due to the overly high tin content in it (0.3–0.8 wt %) [4]. This content exceeds the content standard of tin in the iron concentrate (<0.08 wt.%) [3,4]. To use it as a resource, much research has been carried out to remove tin from the Sn-bearing iron concentrate. Because most of the Sn phases are embedded in the iron phase at a fine-grained size, the tin could not be removed effectively through mineral processing methods [4,5]. Considering the difference between the volatility of SnO, SnS, and other phases in the Sn-bearing iron concentrate, a reduction or sulfurization roasting process has been used to remove tin from this iron concentrate [4–9]. An Fe-Sn spinel or Fe-Sn alloy was easily formed in a reduction roasting process, causing the tin removal rate to be only 80%, as reported in previous research [9,10]. By a sulfurization roasting process, the tin removal rate reached over 90% with FeS_2 [4], high-sulfur coal [6], waste tire rubber [5], or $CaSO_4$ [11] used as curing agents. The sulfurization roasting process might be suitable for treating the Sn-bearing iron concentrate.

In the phosphate rock processing, fuel and coal combustion, and non-ferrous and ferrous metals smelting [12–14], massive SO_2-containing gas was generated and would cause a serious pollution with an emission into the air. The treatment of SO_2 gas has received increasing attention worldwide. Processes, including the wet method [15,16], semi-dry method [17,18], dry method [19,20], and activated carbon adsorptive method [21,22], have been used to remove SO_2 from the flue gas. Among them, the $CaCO_3$/CaO-$CaSO_4$ wet method was mainly used in which the $CaCO_3$/CaO was firstly ground into powder, then fully mixed with water and stirred to form slurry, and at last passed into the absorption tower together with air [16]. A high desulfurization efficiency could be obtained in this wet method; however, the generated waste gypsum residue leads a secondary pollution [15,23]. A wide variety technology is required to recycle SO_2 from flue gases, especially for the gas containing low concentration SO_2.

The SO_2 could be reduced to elemental S_2 by CO, which could then sulfurize SnO_2 to SnS at a proper temperature [4,6,8]. Considering this, we proposed an innovative approach to remove Sn from the Sn-bearing iron concentrate using the SO_2 as the curing agent in a reducing atmosphere (CO) in this research. This research supplied a new thought for treating and reusing the low concentration SO_2 off-gas. Thermodynamic analysis and laboratory tests were carried out to study the feasibility of this approach. Furthermore, the reaction mechanism was elucidated through X-ray diffraction (XRD) and scanning electron microscopy coupled with energy dispersive spectrometry (SEM-EDS).

2. Materials and Methods

2.1. Materials

The tin-bearing iron concentrate used in this study was collected from an ironmaking plant in Yunnan province of China. The element analysis result shows that it contains 65.52 wt. % Fe and 0.39 wt.% Sn (Table 1). A pre-removal of Sn should be carried out before it used as an ironmaking raw material. Figure 1a shows that the main phase in this iron concentrate is Fe_3O_4, and the Sn-containing phase of cassiterite (SnO_2) is embedded in this Fe_3O_4 phase as demonstrated in the electron probe microanalysis techniques (EPMA) analysis result (Figure 1b).

Table 1. Chemical composition of the Sn-bearing iron concentrate.

Element	Fe	SiO_2	Sn	Zn	CaO	Al_2O_3	Pb	Others
Content	65.52	0.91	0.39	0.04	2.30	0.87	0.1	29.87

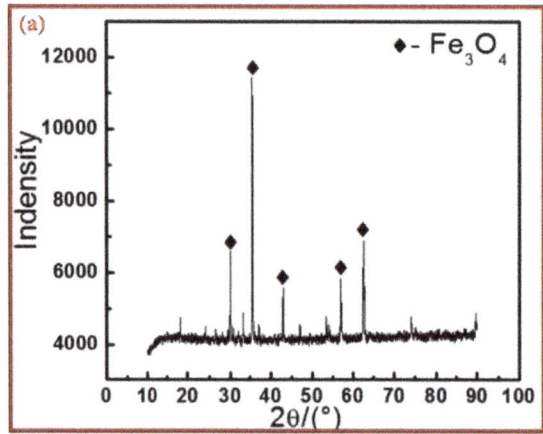
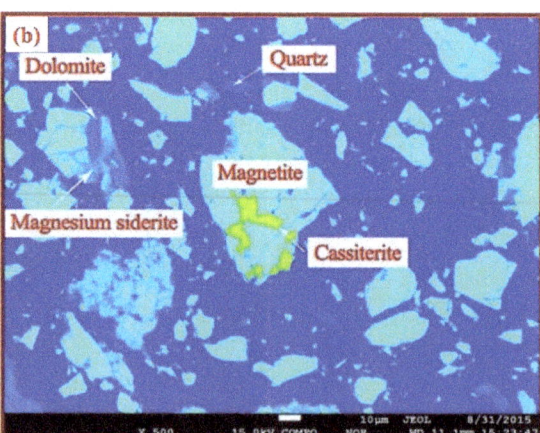

Figure 1. XRD pattern (a) and EPMA analysis (b) of the Sn-bearing iron concentrate.

The N_2 with purity of 99.99 vol.%, CO with purity of 99.99 vol.%, and mixed gas of 2 vol.% SO_2 + 98 vol.% N_2 used in this research, were supplied by the local suppliers.

2.2. Methods

The experiments were carried out in a horizontal tube furnace (GSL-1500X, Hefei Kejing Materials Technology Co. Ltd., Hefei, China), as shown in Figure 2, the temperature of which was measured by a KSY intelligent temperature controller connected to a Pt-Rh thermocouple. For the experimental procedure, the tin-bearing iron concentrate was firstly grounded to minus 0.075 mm, placed in a crucible, transferred to the horizontal tube furnace, and heated to a proper temperature under a high-purity N_2 atmosphere with a flow rate of 40 mL/min. According to previous studies, the surface area of solid particles increased as the particle size decreased, which was beneficial to improving the gas-solid reaction area. The tin-bearing iron concentrate with particle size of 0.075 μm was selected for experiment [8]. After that, the high-purity N_2 was changed into the mixed gas of (2 vol.% SO_2 + 98 vol.% N_2) and high-purity CO at a proper volume ratio with a total flow rate of 100 mL/min, and held for a certain time. After the roasting process completed and the residue cooled down to room temperature in a high-purity N_2 atmosphere at a flow rate of 40 mL/min, the roasted residue was removed and prepared for analysis.

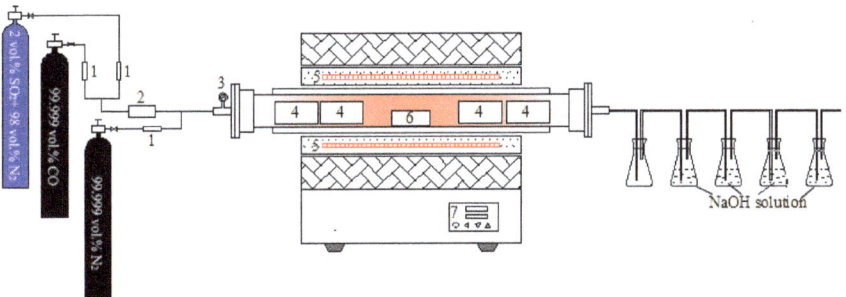

Figure 2. Schematic illustration of the experimental apparatus. (1-Mass flow meter; 2-Gas mixer; 3-Pressure gauge; 4-Filter; 5-Resistive heater; 6-Corundum reactor; 7-Temperature controller).

2.3. Characterization

Elemental composition of the samples was obtained by chemistry analytical method, all of the measurements were conducted three times and the average value was taken as the final result. The phase composition and distribution in the samples were characterized via an X-ray diffraction and EPMA techniques (JEOL, Kyoto, Japan). The XRD patterns were obtained using Cu-Kα radiation in a 2θ range of 10° to 80° with a scan step of 8°/min (Rigaku, Kyoto, Japan). In addition, FactSage 7.2 software (7.2, GTT-Technologies, Herzogenrath, Germany) was used to calculate the equilibrium phase composition during the roasting process.

Mathematical expression of the Sn volatilization ratio in this paper was defined as:

$$R = \frac{M_0 \times W_0 - M_r \times W_r}{M_0 \times W_0} \times 100\%$$

where M_0 and M_r stand for the mass of original tin-bearing iron concentrate and roasted residue, respectively, and W_0 and W_r for the Sn mass content in the original tin-bearing iron concentrate and roasted residue, respectively.

3. Thermodynamic Analysis

To investigate the effect of SO_2 (g) on the Sn volatilization rate from the tin-bearing iron concentrate under CO-SO_2 mixed atmosphere, 1 mol Fe_3O_4 and 1 mol SnO_2 were selected as the reactants to calculate the equilibrium phase composition while roasted with 2 mol CO and different amounts of SO_2 at 1100 °C using FactSage 7.2 software. Without the addition of SO_2 (g), the results in Figure 3a show that CO_2 (g), FeO, Fe_2O_3 and SnO appear, which was due to the occurrence of reactions (1)–(3). Though the decomposition of Fe_3O_4 is difficult to be carried out due to the positive value of the standard Gibbs free energy of reaction (2) at 1100 °C (Figure 3b), the occurrence of reaction (3) promoted this decomposition to happen, considering the chemical equilibrium. With the increase of the SO_2 amount from 0 to 0.4 mol, the amounts of CO (g), FeO, SnO, and Fe_2O_3 decrease, and the amounts of CO_2 (g), S_2 (g), SnS (g), SnO_2 and Fe_3O_4 increase. In Figure 3b, the Gibbs free energy for the reduction of SO_2 (g) by CO (g) (reaction (4)) at 1100 °C is minimum, causing the CO (g) amount used to reduce SnO_2, Fe_2O_3 to be decreased in the presence of SO_2 (g) and the equilibrium amount of SnO_2 and Fe_3O_4 increased. The CO (g) first reduces SO_2 (g) to form S_2 (g) by reaction (4), and then S_2 (g) reacts with SnO (s) and SnO_2 (s) through reactions (5) and (6) to form SnS (g). The SnS (g) increases in this range of SO_2 (g) amount. With the SO_2 (g) amount exceeding 0.6 mol, an excessive SO_2 (g) exists in the equilibrium composition in Figure 3a. Consequently, the reduction of SnO_2 and Fe_2O_3 was further restrained, as a result of which the formation of SnS (g) decreased and the decomposition of Fe_3O_4 occurred little.

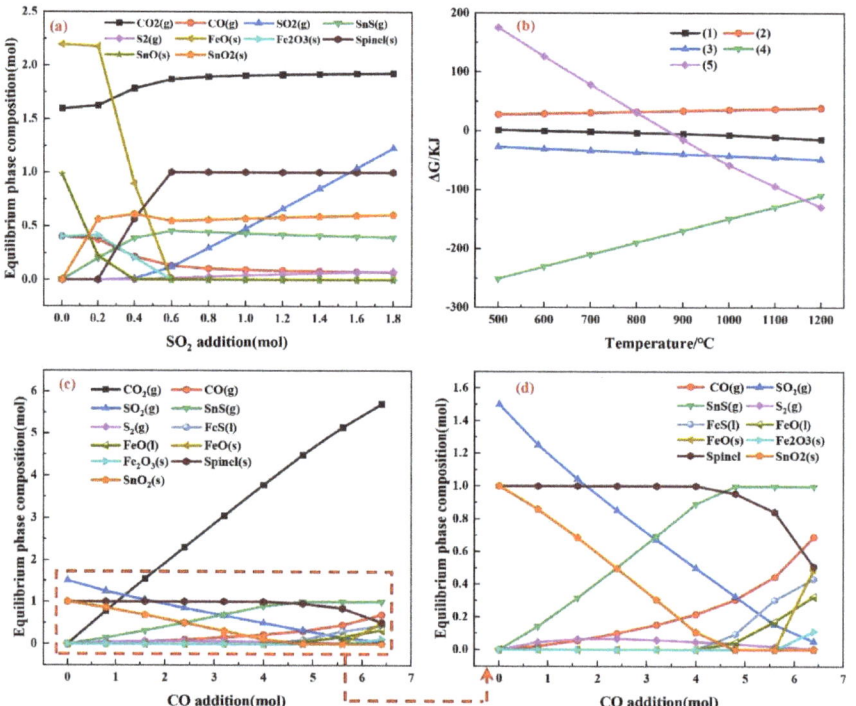

Figure 3. (**a**) Equilibrium phase composition of 2 mol CO+ 1 mol Fe_3O_4 + 1 mol SnO_2 roasted with different amounts of SO_2 at 1100 °C; (**b**) Gibbs free energy changes for reaction (1)–(5) at 500–1100 °C; (**c,d**) Equilibrium phase composition of 1.5 mol SO_2+ 1 mol Fe_3O_4 + 1 mol SnO_2 roasted with different amounts of CO at 1100 °C.

Under the condition of the Fe_3O_4 amount of 1 mol, SnO_2 amount of 1 mol, and SO_2 amount of 1.5 mol, the effect of CO (g) amount on the Sn transformation was then calculated at 1100 °C. Figure 3c,d show that with the increase of CO amount from 0 to 5 mol, the amounts of SO_2 (g) and SnO_2 decrease accompanied with the increase of SnS (g) and S_2 (g) amounts, which might be due to the occurrence of reactions (4) to (6). As the amount of CO increases over 4 mol, the amounts of Fe_3O_4 (s) and S_2 (g) decrease while the amounts of FeO (l) and FeS (l) increase, probably due to the occurrence of reactions (4) and (7).

In summary, under the CO-SO_2 mixed atmosphere, the SnO_2 can be sulfurized and volatilized in the form of SnS (g) from the tin-bearing iron concentrate. However, some sulfur might retain in the roasted iron concentrate in form of FeS at a high CO amount, which decreases the iron concentrate quality for ironmaking. A suitable SO_2 and CO partial pressures should be controlled during the roasting process.

$$CO(g) + SnO_2(s) = SnO(s) + CO_2(g) \tag{1}$$

$$Fe_3O_4(s) = Fe_2O_3(s) + FeO(s) \tag{2}$$

$$CO(g) + Fe_2O_3(s) = 2FeO(s) + CO_2(g) \tag{3}$$

$$4CO(g) + 2SO_2(g) = S_2(g) + 4CO_2(g) \tag{4}$$

$$3S_2(g) + 4SnO = 4SnS(g) + 2SO_2(g) \tag{5}$$

$$4CO(g) + S_2(g) + 2SnO_2 = 2SnS(g) + 4CO_2(g) \tag{6}$$

$$S_2(g) + Fe_3O_4 = FeS(l) + SO_2(g) + 2FeO(l) \tag{7}$$

4. Results and Discussion

4.1. Effects of the SO_2 Partial Pressure

Under the condition of the roasting temperature of 1000 °C, residence time of 40 min, tin-bearing iron concentrate particle size below 0.075 mm, and total flow rate of mixed gases of (2 vol.% SO_2 + 98 vol.% N_2) and high-purity CO of 100 mL/min, the effects of SO_2 partial pressure on the Sn volatilization ratio from the tin-bearing iron concentrate and S content in the roasted residue were focused firstly. The SO_2 partial pressure (P_{SO2}) was assumed as $P_{SO2}= V_{SO2}/(V_{SO2}+ V_{N2}+ V_{CO})$, and the CO partial pressure (P_{CO}) was assumed as $P_{CO}= V_{CO}/(V_{SO2}+ V_{N2}+ V_{CO})$. The V_{SO2}, V_{N2}, and V_{CO} corresponds to the volume fraction of SO_2, N_2, and CO in the mixed gas respectively.

Figure 4a shows the changes of CO partial pressure (P_{CO}) with the increase of SO_2 partial pressure (P_{SO2}) in this research. Based on it, the Fe_3O_4 and SnO_2 both could be sulfurized during the roasting process in this process, as presented the predominance area diagram of Fe-Sn-S-O at 1000 °C in Figure 4b. More S_2 (g) could be produced from the reduction of SO_2 (g) at a higher SO_2 partial pressure (P_{SO2}) through Equation (4), which in turn could sulfurize more cassiterite (SnO_2) to SnS (g) by Equation (6). As a result, the Sn volatilization ratio increased from 60.1% to 72.4% with the P_{SO2} from 0.001 to 0.005 as presented in Figure 4c. Furthermore, according to Figure 3c,d, the Fe_3O_4 could be sulfurized accompanied with the sulfurization of SnO_2 thermodynamically, causing the sulfur content in the roasted residue to increase with the increase of P_{SO2} as shown in Figure 4d. Figure 5a shows the phase compositions of the roasted residues under different SO_2 partial pressures. It indicated that after the roasting treatment, the Fe_3O_4 in the raw Sn-bearing iron concentrate could be reduced into FeO at the P_{SO2} of 0.003 and further reduced to metallic Fe as the P_{SO2} decreased to 0.001. The CO partial pressure increased from 0.85 to 0.95 with the decrease of P_{SO2} from 0.003 to 0.001 (Figure 4a), which promoted the further reduction of FeO to Fe. The phase of iron sulfide could not be detected in the XRD patterns of the roasted residues due to its little content. Then, an SEM-EDS analysis on the roasted residue was carried out and the result is shown in Figure 5b. In Figure 5b, the element composition of point "1" is Fe and S, which confirmed the existence of iron sulfide in the roasted residue. To increase the Sn volatilization ratio and decrease the S content in the roasted iron concentrate, the SO_2 partial pressure should be controlled at 0.003.

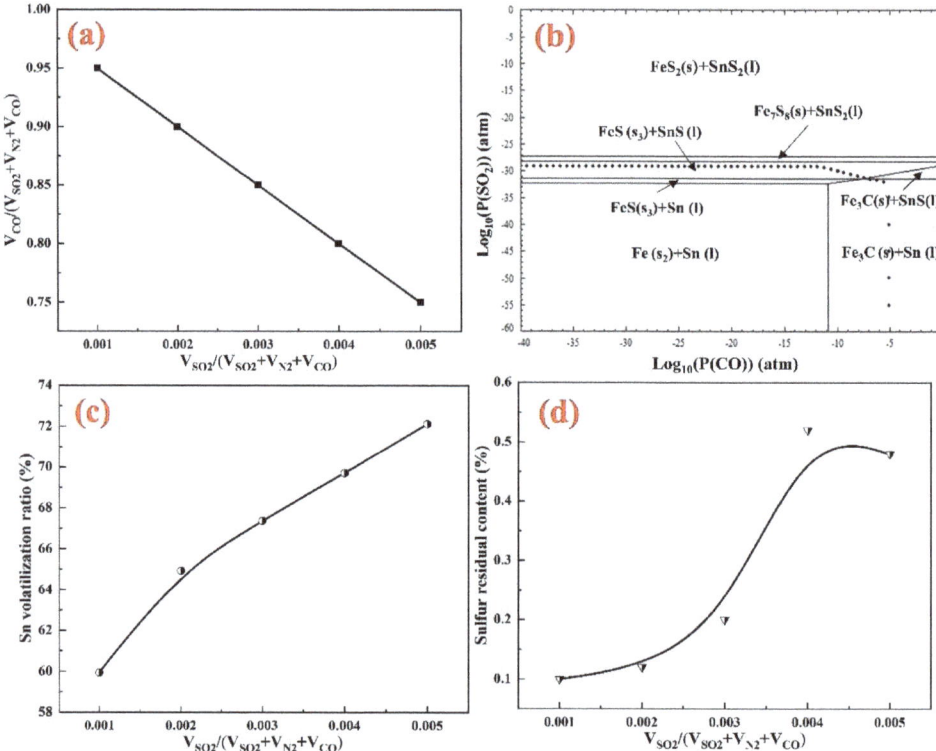

Figure 4. (a) The changes of CO partial pressure (P_{CO}) with the increase of SO_2 partial pressure (P_{SO_2}); (b) The predominance area diagram of Fe–Sn–S–O at 1000 °C; (c) Effects of SO_2 partial pressure on the Sn volatilization ratio from the tin-bearing iron concentrate; (d) The S content in the roasted iron concentrate under different SO_2 partial pressure.

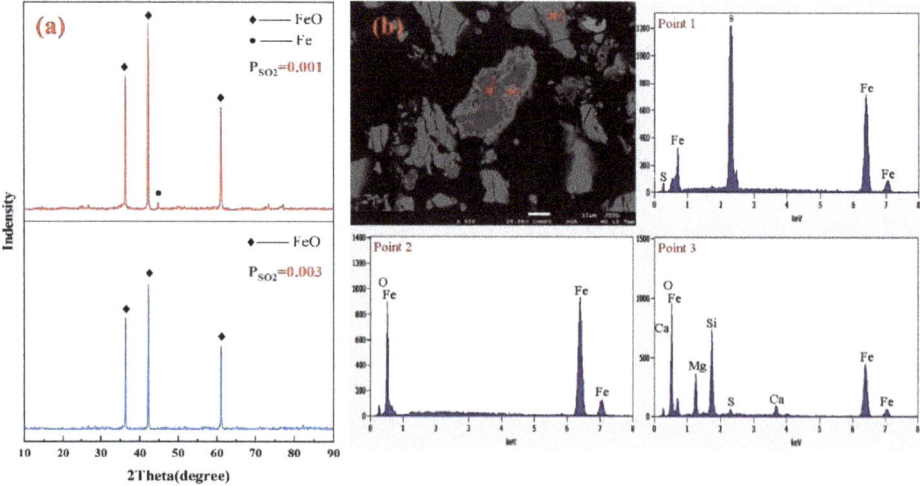

Figure 5. (a) XRD patterns of the roasted residue at the SO_2 partial pressure of 0.001 and 0.003 respectively; (b) SEM-EDS result of the roasted residue at the SO_2 partial pressure of 0.003.

4.2. Effects of Roasting Temperature

Under the condition of the SO_2 partial pressure of 0.003, residence time of 40 min, particle size of the tin-bearing iron concentrate below 0.075 mm, and total flow rate of mixed gases of (2 vol.% SO_2+ 98 vol.% N_2) and high-purity CO of 100 mL/min, the effects of roasting temperature on the Sn volatilization ratio from the tin-bearing iron concentrate and S content in the roasted residue were researched.

The sulfurization of SnO_2 through Equations (4) and (6) was accelerated at higher temperatures, and the vapor pressure of SnS (g) also increased with the increase of temperature [24]. These resulted in an increased the Sn volatilization ratio from 27.0% to 80.6% with the roasting temperature from 900 °C to 1050 °C seen in Figure 6a. However, with the roasting temperature further increased to 1100 °C, the Sn volatilization ratio decreased to 74.8%. This might be due to more generation of Fe-Sn alloy at 1100 °C, which limited the tin sulfurization and volatilization [9,10]. Comparing Figure 7a to Figure 7b, a metallic Fe phase could be detected in the roasted residue as the temperature increased from 1050 °C to 1100 °C, indicating a deeper reduction of Fe_3O_4 could be carried out and as a result more metallic Fe would be produced at a higher temperature. The generated metallic Fe might be combined with the reduced Sn to form an Fe-Sn alloy through Equation (8) [5], and more importantly the Fe content in the generated Fe-Sn alloy increased with the increase of the roasting temperature as presented in Figure 7c,d. Comparing Figure 7c to Figure 7d, the Sn content in the formed Sn-Fe alloy decreased from 97.01 wt% to 1.13 wt% as the roasting temperature increased from 900 °C to 1100 °C. The Sn activity in the Fe-Sn alloy decreased with the decrease of Sn content in it according to Raoult's law, causing the Sn sulfurization from the Fe-Sn alloy by Equation (9) to be restricted. As a result, the Sn volatilization decreased to 74.8% at 1100 °C, as shown in Figure 6a.

$$[Fe] + [Sn] = Sn\text{-}Fe\ alloy \qquad (8)$$

$$S_2 + Sn\text{-}Fe\ alloy \to SnS + Fe \qquad (9)$$

Similar to the sulfurization of SnO_2 through Equations (4) and (6), the transfer of 'S' from SO_2 to iron sulfide using Equation (7) was promoted with the increase of roasting temperature, causing the S content in the roasted residue increased with the temperature from 900 °C to 1000 °C (Figure 6b). However, as the temperature exceeded 1000 °C, the sulfur content in the roasted residue decreased. The reason might be that less S_2 (g) would be generated at higher temperatures deduced from Figure 3b, which in turn led to less 'S' fixed in the roasted residue in form of iron sulfide. In Figure 3b, the Gibbs free energy for reaction (4) increased with the temperature increase. Based on the results in Figure 6a,b, the optimum roasting temperature should be 1050 °C in order to maximize the removal of tin from the tin-bearing iron concentrate and to ensure a low sulfur content in the roasted residue.

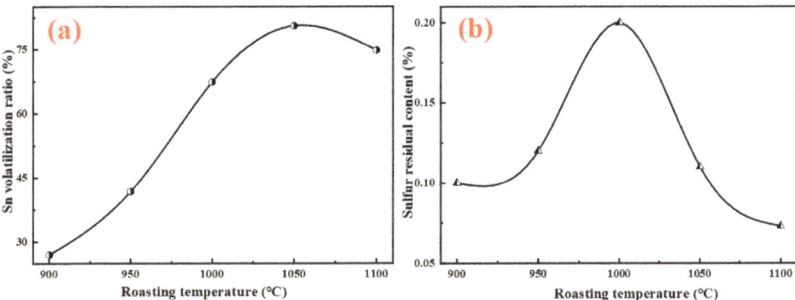

Figure 6. Effects of roasting temperature on the Sn volatilization ratio (**a**) from the tin-bearing iron concentrate and S content in the roasted iron concentrate (**b**).

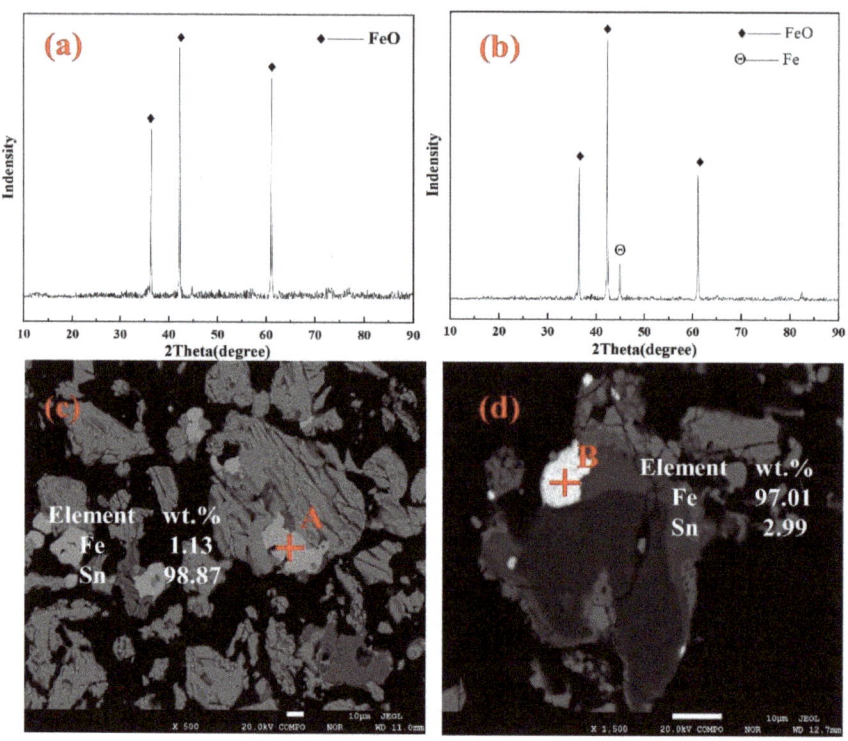

Figure 7. XRD results of the roasted residues at 1050 °C (**a**) and 1100 °C (**b**), respectively, for 40 min with the P_{SO2} of 0.003; SEM-EDS results of the roasted residues at 1050 °C (**c**) and 1100 °C (**d**), respectively, for 40 min with the P_{SO2} of 0.003.

4.3. Effects of the Residence Time

Under the condition of roasting temperature of 1050 °C, SO_2 partial pressure of 0.003, tin-bearing iron concentrate particle size below 0.075 mm and total flow rate of mixed gases of (2 vol.% SO_2 + 98 vol.% N_2) and high-purity CO of 100 mL/min, the effects of residence time on the Sn volatilization ratio from the tin-bearing iron concentrate and S content in the roasted residue are shown in Figure 8a,b respectively.

Based on Equations (4) and (6), the sulfurization reaction of Sn from the tin-bearing iron concentrate could be summarized as Equation (10) in this research. An unreacted core shrinking model was chosen to describe the sulfurization kinetics of SnO_2. The reaction process could be divided into three main steps: outer diffusion of the CO and SO_2 through the gas phase boundary layer to the reactant particle surface, internal diffusion of CO and SO_2 through gaps in the reactant particle to the gas-solid reaction interface, and interfacial chemical reaction with the SnO_2 at the reaction interface. Generally, the first step of the outer diffusion is not the rate controlling step when the gas flow exceeds 60 mL/min [25,26]. The sulfurization of SnO_2 was likely controlled by the internal diffusion, interfacial chemical reaction, or the combination of them. As reported in previous research [27–29], the kinetic equations controlled by different reaction steps could be summarized in Table 2. In Table 2, the t was the reduction time, min; the X was the Sn volatilized ratio, %; and the a, b, a_1, and b_1 were constants. With the $1-(1-X)^{1/3}$ used as Y-axis and t used as X-axis, the reaction would be controlled by the interfacial chemical reaction if there is a linear relationship between X and Y. Similarly, if there is a linear relationship between the $1-2X/3-(1-X)^{2/3}$ (Y-axis) and t (X-axis), the reaction would be controlled by the gas internal diffusion control; if there is a linear relationship between the $\left[1+(1-X)^{1/3}-2(1-X)^{2/3}\right]$ (Y-axis) and

$t/\left[1-(1-X)^{1/3}\right]$ (X-axis), the reaction would be controlled by the combination of the internal diffusion and interfacial chemical reaction.

$$4CO(g) + SO_2(g) + SnO_2 = SnS + 4CO_2(g) \qquad (10)$$

The equations listed in Table 2 were used to treat the experimental data in Figure 8a, and the results are shown in Figure 9. In comparison with Figure 9a–c, a better linear dependence between the $1 - 2X/3 - (1 - X)^{2/3}$ (Y-axis) and t (X-axis) could be seen with the residence time from 10 min to 60 min in Figure 9a. It indicated the reaction was controlled by the gas internal diffusion control. Consequently, as the residence time increased, the CO and SO_2 gas concentration at the reaction interface gradually approached the CO and SO_2 concentration in the main gas phase, causing more of the SnO_2 to be sulfurized and volatilized. With the residence time prolonged from 20 min to 60 min, the Sn volatilization rate increased from 19.9% to 92.1%. While the residence time increased further; the Sn volatilization rate increased little. The trend in sulfur content in the roasted residue is similar with that of Sn volatilization ratio as presented in Figure 8a, which raised rapidly from 20 min to 60 min and then increased slowly as the residence time continued to increase. Considering the results in Figure 8a,b, the residence time was chosen at 60 min.

Under the condition of the roasting temperature of 1050 °C, SO_2 partial pressure of 0.003, CO partial pressure of 0.85, and residence time of 60 min, the tin removal rate from the tin-bearing iron concentrate achieved 92.1% and the Sn content in the roasted iron concentrate was decreased to 0.032 wt.%. In addition, the sulfur content in the iron concentrate is only 0.062 wt.%, which meets the standard of BF ironmaking.

Table 2. Kinetic equations for different controlling steps.

Controlling Step	Kinetic Equation [26]
Interfacial chemical reaction	$t = a\,[1 - (1 - X)^{1/3}]$
Gas internal diffusion	$t = b\,[1 - 2X/3 - (1 - X)^{2/3}]$
Combination of interfacial chemical reaction and gas internal diffusion	$t = a_1\,[1 - (1 - X)^{1/3}] + b_1\,[1 - 2X/3 - (1 - X)^{2/3}]$

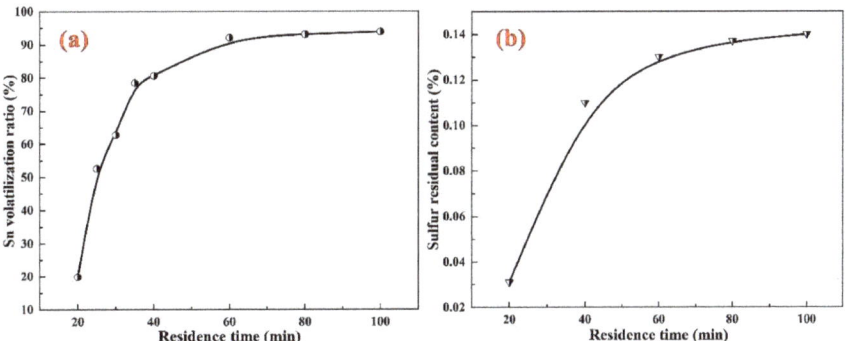

Figure 8. Effect of residence time on Sn volatilization ratio from the tin-bearing iron concentrate (a) and S content in the roasted residue (b).

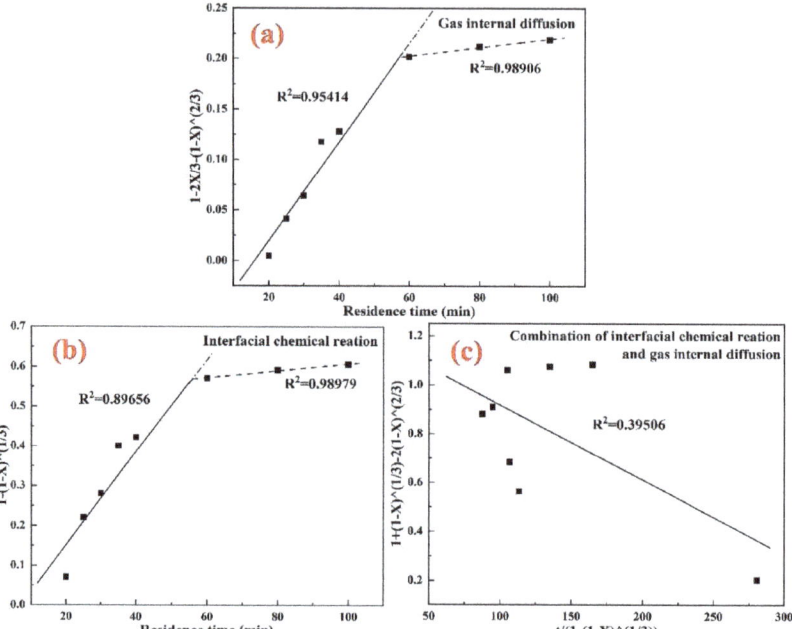

Figure 9. (a) Relationship between $[1 - 2X/3 - (1 - X)^{2/3}]$ and different residence time; (b) Relationship between $[1 - (1 - X)^{1/3}]$ and different residence time; (c) Relationship between $[1 + (1 - X)^{1/3} 2(1 - X)^{2/3}]$ and $t/[1 - (1 - X)^{1/3}]$.

5. Conclusions

The tin from the tin-bearing iron concentrate could be efficiently removed using a roasting in the mixed atmosphere of CO and SO_2 at 1050 °C. With the partial pressure of SO_2 increased from 0.001 to 0.005, more S_2 was produced from the reduction of SO_2, which in turn promoted the sulfidation and SnO_2 volatilization, resulting in the increased volatilization ratio of Sn from 60.1% to 72.4%. However, the Fe_3O_4 sulfidation also occurred simultaneously at a higher SO_2 partial pressure and an iron sulfide phase was formed, retaining in the roasted iron concentrate, due to which the residual sulfur content increased. The Sn volatilization ratio increased from 27.0% to 80.6% with the increase of roasting temperature from 900 °C to 1050 °C, but it decreased to 74.8% as the roasting temperature was further increased to 1100 °C. It was due to the more formation of a Sn-Fe alloy. The kinetics study showed that the sulfurization of SnO_2 from the concentrated tin-bearing iron was controlled by the internal gas diffusion reaction step. Under the condition of roasting temperature of 1050 °C, SO_2 partial pressure of 0.003, CO partial pressure of 0.003, and residence time of 60 min, the Sn content in the roasted iron concentration was decreased to 0.032 wt%, and the residual sulfur content was only 0.062 wt%, which meets the BF ironmaking.

Author Contributions: L.L.: Conceptualization, funding acquisition, investigation, methodology, project administration, resources, supervision, writing—original draft, writing—review and editing. Z.X.: Data curation, investigation, validation, writing—original draft. S.W.: Methodology, investigation, validation, writing—original draft. All authors have read and agreed to the published version of the manuscript.

Funding: This study is supported by the National Natural Science Foundation of China (51874153).

Data Availability Statement: All data are available in this study.

Acknowledgments: The authors wish to express their thanks to the National Natural Science Foundation of China (51874153) for their financial support of this research.

Conflicts of Interest: The authors declare no conflict of interest.

References

1. Li, C.; Sun, H.H.; Bai, J.; Li, L.T. Innovative methodology for comprehensive utilization of iron ore tailings: Part 1. The recovery of iron from iron ore tailings using magnetic separation after magnetizing roasting. *J. Hazard. Mater.* **2010**, *174*, 71–77. [CrossRef] [PubMed]
2. Wu, F.; Cao, Z.; Wang, S.; Zhong, H. Phase transformation of iron in limonite ore by microwave roasting with addition of alkali lignin and its effects on magnetic separation. *J. Alloys Compd.* **2017**, *722*, 651–661. [CrossRef]
3. Omran, M.; Fabritius, T.; Elmandy, A.M.; Abdel-Khalek, N.A.; Gornostayev, S. Improvement of phosphorus removal from iron ore using combined microwave pretreatment and ultrasonic treatment. *Sep. Purif. Technol.* **2015**, *156*, 724–737. [CrossRef]
4. Yu, Y.; Li, L.; Wang, J.Y.; Li, K.Z.; Wang, H. Phase transformation of Sn in tin-bearing iron concentrates by roasting with FeS$_2$ in CO-CO$_2$ mixed gases and its effects on Sn separation. *J. Alloys Compd.* **2018**, *750*, 8–16. [CrossRef]
5. Yu, Y.; Li, L.; Wang, J.Y.; Wang, J.C.; Li, K.Z. Sn separation from Sn-bearing iron concentrates by roasting with waste tire rubber in N$_2$ + CO + CO$_2$ mixed gases. *J. Hazard. Mater.* **2019**, *371*, 440–448. [CrossRef] [PubMed]
6. Yu, Y.; Li, L.; Sang, X.L. Removing tin from tin-bearing iron concentrates with sulfidation roasting using high sulfur coal. *ISIJ Int.* **2016**, *56*, 57–62. [CrossRef]
7. Zhang, R.J.; Li, L.; Li, K.Z.; Yu, Y.; Wang, H. Reduction and sulfurization behavior of tin phases in tin-bearing iron concentrates with sulfates in sulfur-bearing stone coal. *ISIJ Int.* **2018**, *58*, 453–459. [CrossRef]
8. Zhang, R.J.; Li, L.; Li, K.Z.; Yu, Y. Reduction roasting of tin-Bearing iron concentrates using pyrite. *ISIJ Int.* **2016**, *56*, 953–959. [CrossRef]
9. Su, Z.J.; Zhang, Y.B.; Liu, B.B.; Zhou, Y.L.; Jiang, T.; Li, G.H. Reduction behavior of SnO$_2$ in the tin-bearing iron concentrates under CO-CO$_2$ atmosphere. Part I: Effect of magnetite. *Powder Technol.* **2016**, *292*, 251–259. [CrossRef]
10. Wang, Z.W.; Wang, C.Y.; Lu, H.M. Investigation on removal of tin from Sn-Bearing iron concentrates by reduction roasting. *Mining Metall.* **2005**, *14*, 63–66. [CrossRef]
11. Yu, Y.; Li, L. Transformation behaviour of sulfur from gypsum waste (CaSO$_4$·2H$_2$O) while roasting with tin-bearing iron concentrate for tin removal and iron recovery. *ISIJ Int.* **2020**, *60*, 2291–2300. [CrossRef]
12. Pandey, R.; Biswas, R.; Chakrabarti, T.; Devotta, S. Flue gas desulfurization: Physicochemical and biotechnological approaches. *Crit. Rev. Environ. Sci. Technol.* **2005**, *35*, 571–622. [CrossRef]
13. Bates, T.; Lamb, B.; Guenther, A.; Dignon, J.; Stoiber, R. Sulfur emissions to the atmosphere from natural sources. *J. Atmos. Chem.* **1992**, *14*, 315–337. [CrossRef]
14. Ma, K.; Deng, J.; Ma, P.; Sun, C.; Zhou, Q.; Xu, J. A novel plant-internal route of recycling sulfur from the flue gas desulfurization (FGD) ash through sintering process: From lab-scale principles to industrial practices. *J. Environ. Chem. Eng.* **2021**, *10*, 106957. [CrossRef]
15. Lim, J.; Cho, H.; Kim, J. Optimization of wet flue gas desulfurization system using recycled waste oyster shell as high-grade limestone substitutes. *J. Cleaner Prod.* **2021**, *318*, 128492. [CrossRef]
16. Lim, J.; Choi, Y.; Kim, G.; Kim, J. Modeling of the wet flue gas desulfurization system to utilize low-grade limestone. *Korean J. Chem. Eng.* **2020**, *37*, 2085–2093. [CrossRef]
17. Li, H.; Zhang, H.; Li, L.; Ren, Q.; Yang, X.; Jiang, Z.; Zhang, Z. Utilization of low-quality desulfurized ash from semi-dry flue gas desulfurization by mixing with hemihydrate gypsum. *Fuel* **2019**, *255*, 115783. [CrossRef]
18. Fang, D.; Liao, X.; Zhang, X.; Teng, A.; Xue, X. A novel resource utilization of the calcium-based semi-dry flue gas desulfurization ash: As a reductant to remove chromium and vanadium from vanadium industrial wastewater. *J. Hazard. Mater.* **2018**, *342*, 436–445. [CrossRef]
19. Matsushima, N.; Li, Y.; Nishioka, M.; Sadakata, M.; Qi, H.; Xu, X. Novel dry-desulfurization process using Ca(OH)$_2$/fly ash sorbent in a circulating fluidized bed. *Environ. Sci. Technol.* **2004**, *38*, 6867–6874. [CrossRef]
20. Li, Y.; You, C.; Song, C. Adhesive carrier particles for rapidly hydrated sorbent for moderate-temperature dry flue gas desulfurization. *Environ. Sci. Technol.* **2010**, *44*, 4692–4696. [CrossRef]
21. Zhang, X.; Li, Y.; Zhang, Z.; Nie, M.; Wang, L.; Zhang, H. Adsorption of condensable particulate matter from coal-fired flue gas by activated carbon. *Sci. Total Environ.* **2021**, *778*, 146245. [CrossRef] [PubMed]
22. Zhao, Y.; Dou, J.; Duan, X.; Chai, H.; Oliveira, J.; Yu, J. Adverse effects of inherent CaO in coconut shell-derived activated carbon on its performance during flue gas desulfurization. *Environ. Sci. Technol.* **2009**, *54*, 1973–1981. [CrossRef] [PubMed]
23. Yao, S.; Cheng, S.; Li, J.; Zhang, H.; Jia, J.; Sun, X. Effect of wet flue gas desulfurization (WFGD) on fine particle (PM2.5) emission from coal-fired boilers. *J. Environ. Sci.* **2019**, *77*, 32–42. [CrossRef]
24. Zhang, Y.Y. Behaviour Pattern of Stannous Sulfide under Vacuum and High Temperature and Experimental Study of Vacuum Distillation of Sulfur Slag. Master's. Thesis, Kunming University of Science and Technology, Kunming, China, 2014.
25. Guo, Y.F. Study on Strengthening of Solid-State Reduction and Comprehensive Utilization of Vanadiferous Titanomagnetite. Ph.D. Thesis, Central South University of Technology, Changsha, China, 2007.

26. Zhang, G.H.; Chou, K.C.; Zhao, H.L. Reduction kinetics of FeTiO$_3$ powder by hydrogen. *ISIJ Int.* **2012**, *52*, 1986–1989. [CrossRef]
27. Mo, D.C. *Kinetics for Metallurgy*, 1st ed.; Central South University Press: Changsha, China, 1987; pp. 154–196.
28. Padilla, R.; Sohn, H.Y. The reduction of stannic oxide with carbon. *Metall. Mater. Trans. B* **1979**, *10*, 109–115. [CrossRef]
29. Zhang, C.F.; Peng, B.; Peng, J. Electric arc furnace dust non-isothermal reduction kinetics. *Trans. Nonferrous Met. Soc. China* **2000**, *10*, 524–530.

Article

Synthesis of Ferroalloys via Mill Scale-Dross-Graphite Interaction: Implication for Industrial Wastes Upcycling

Praphaphan Wongsawan, Weerayut Srichaisiriwech and Somyote Kongkarat *

Faculty of Sciences and Technology, Thammasat University, Pathum Thani 12120, Thailand
* Correspondence: ksomyote@tu.ac.th; Tel.: +66-2-564-4444; Fax: +66-2-564-4484

Abstract: Mill scale and aluminum dross are the industrial wastes from steel and aluminum industries, which have high concentrations of Fe_2O_3 and Al_2O_3, respectively. This paper reports the conversion of reducible metal oxides in scale and dross into an alloy via carbothermic reduction at 1550 °C. Scale and dross were mixed with graphite into three different C/O molar ratios of 1, 1.5, and 2 to produce a pellet. The pellets were heated at 1550 °C for up to 6 h under an argon atmosphere. By this method, carbothermic reductions were found to proceed and formed Fe–Si–Al–C alloy that consists of Fe_3Al and Fe_3Si phases. The presence of Si in the alloy came from the reduction of SiO_2 in aluminum dross. Levels of Al and Si in the alloy increase with increasing C/O molar ratios. However, the Si level in the alloy was found to stabilize since 3 h, while the Al level increases with increasing time up to 6 h. Unreacted oxides in the wastes had an insignificant effect on the ferroalloy formation. These results provide evidence for carbothermic reduction of the Fe_2O_3-Al_2O_3-SiO_2 system at 1550 °C and show the novel method to upcycling aluminum dross and mill scale toward a circular economy.

Keywords: industrial wastes; mill scale; aluminum dross; ferroalloys

1. Introduction

The industrial sector had been developed and expanded rapidly in recent decades, leading to the generation of waste. Steel and aluminum making are heavy industries that produce tremendous amounts of waste and by-products, such as slag, dust, mill scale, and dross. Mill scale is categorized as a by-product, generated from the hot rolling of semi-finished steel products, such as slab, bloom, and billet. Mill scale contains >70 wt% of metallic iron or >90 wt% of iron oxides, and thus it had been widely interested in many applications [1–10]. It was found to be recycled in the smelting process by mixing with coal or coke and used as a charge material to replace iron ore or scrap iron [1,2]. The direct reduction of mill scale had also been investigated using several reducing agents, such as biomass [3,4] and reducing gas [5,6]. The obtained products from the direct reduction of mill scale were iron powder and iron-bearing compound [7–10]. Aluminum dross is a by-product of the aluminum industry. It is generated during the aluminum melting process. The dross is mostly re-melted to extract the metallic aluminum using a rotary kiln. Aluminum dross generated after the re-melted process is secondary dross, which mainly composes Al_2O_3 and salts [11]. It was mostly buried or landfilled causing soil and groundwater pollution. Aluminum dross is reactive with water which can generate various gaseous species, such as NH_3, CH_4, and H_2S [12]. Thus, it is considered hazardous industrial waste, and the movement or transfer of the waste needs to comply with the Basel Convention [13]. The deposal of dross is a high cost for aluminum melting industries. Therefore, a good management method is needed for the tremendous increase in the volume of industrial waste. Aluminum dross had been utilized via several processes and can be classified into wet, dry, and without process [14–24]. The wet process was carried out using acid and base leaching for Al recovery [15], absorbent/catalyst [16,17], and alumina [18,19] productions. The dry process was conducted by heating at a temperature above 1000 °C for the production of composite [20], refractory [21], and hercynite [22]. Aluminum dross can also utilize directly without processing for the production of geopolymers [23] and filler [24].

Synthesis of various metals using wastes and non-waste materials had been reported previously [25–32]. Jamieson et al. [25] separated red sand into a high iron oxides fraction, silica, and a mixture of iron and silica by using low and high magnetic separators. Zhu et al. [26] reported the extracting of various metals from red mud using acid leaching, solvent extraction, polymerization process, alkali leaching with pressure, and the aging process. Jadhav and Hocheng [27] have reported on extracting metals from several types of wastes using sulfuric/citrus/oxalic acids and microbiological leachants. Wei et al. [28] utilized silicon cutting waste (SCW) from the diamond wire sawing process by mixing it with aluminum powder and heated at temperatures of 1000–1500 °C. The production of Al–Si alloys can be synthesized by a one-step smelting process [28]. Khanna et al. [29] have reported on recovering multiple metals from various industrial wastes including fly ashes, red mud, mill scales, water treatment residues, and biomass. Carbothermic reduction of Al_2O_3-Fe_2O_3-C and Al_2O_3-Fe_2O_3-SiO_2-C systems at the temperatures of 1450–1700 °C for up to 2 h had been widely studied. The reduction of Fe_2O_3 and SiO_2 was observed at 1450 °C, leading to the formation of Fe–Si–C alloys, while the reduction of Al_2O_3 was seen to start at 1550 °C and complete at 1600–1700 °C [29]. Synthesis of ferrosilicon alloys from rice husk and rubber tree bark had also been reported [30,31]. Boonyaratchinda et al. [31] employed rice husk as a silica source and rubber tree bark as a carbon source for the recovery of ferroalloy. By this, the Fe–Si alloy was observed from a carbothermic reduction in the SiO_2-Fe_2O_3-C system at 1550 °C after 30 min of interaction with Si content in the ferrosilicon alloy of 45.32 wt% [31].

Khanna et al. [32] investigated interaction of Al_2O_3-C-Fe system at 1550 °C under inert argon atmosphere using pure alumina, graphite and a piece of steel. Carbothermic reduction reaction of Al_2O_3 by carbon in the presence of liquid steel was observed. The formation of Fe–Al–C ferroalloys was reported for the system of Al_2O_3-22.82 %C substrate interacted with a piece of Fe (0.6 %C). In this system, alumina fibrous was observed after 2–3 h of reaction and then the Fe–Al–C ferroalloy was formed after 5.5 h and completely transformed after 6 h. The synthesized Fe–Al–C ferroalloy composes of a metallic Fe_3Al phase along with a carbon phase in the form of graphite flake. The presence of liquid steel in the system acts as a metallic solvent to trap the reduced Al in the system [32]. These studies on extracting or recovering metal from waste, non-waste, and industrial byproducts are still to a moderate extent [25–32]. Some of these used several steps and/or high temperatures to extract the metals, and thus have a small or moderate economic worthiness.

Aluminum dross and mill scale contain high amounts of Al_2O_3 and Fe_2O_3, respectively. The utilization of aluminum dross as a source of alumina for extracting Al had not been reported previously. Therefore, it is quite new to extract Al and the other metal from aluminum dross, which could be one of the novel management methods rather than buried or landfilled. The present study aims to evaluate the possibility of utilizing aluminum dross and mill scale as a source of Al_2O_3 and Fe_2O_3 to synthesize ferroalloys at 1550 °C. The carbothermic reduction reaction of the Fe_2O_3-Al_2O_3-C system will be investigated. The influence of carbon content in the system and interaction times will be reported and investigated.

2. Materials and Methods
2.1. Sample Preparation
2.1.1. Aluminum Dross

Aluminum dross used in the present study was supported by TOP Five manufacturing, the aluminum melting industries located in the eastern province of Thailand. The dross was sieved into a powder of <180 µm. XRF analysis of aluminum dross is presented in Table 1, showing 69.94 wt% of Al_2O_3 and 5.01 wt% of SiO_2 as the major components.

Table 1. Composition of aluminum dross.

Oxides (wt%)													
Al_2O_3	SiO_2	Fe_2O_3	CaO	K_2O	MgO	MnO	Na_2O	SO_3	CuO	TiO_2	ZnO	Others
69.94	5.01	0.54	1.0	0.76	4.91	0.15	10.65	2.46	0.37	0.17	0.25	3.79

2.1.2. Mill Scale

Mill scale was collected from UMC Metal Co., Ltd., Chonburi, Thailand. The scale was ground in a ring mill and sieved into a powder of <180 μm. Table 2 shows XRF analysis of the mill scale used in this study, containing 93.66 wt% of Fe_2O_3 as the main components.

Table 2. Composition of the mill scale.

Oxides (wt%)								
Fe_2O_3	SiO_2	Al_2O_3	CaO	SO_3	TiO_2	K_2O	P_2O_5
93.66	1.42	0.82	0.17	0.08	0.04	0.02	0.04

2.1.3. Pellet Preparation

Aluminum dross, mill scale, and graphite were blended homogeneously into three different C/O molar ratios using rolling mill, as shown in Table 3. C is the total moles of carbon from graphite in the blends, while O is total moles of oxygen from Al_2O_3 and Fe_2O_3 in the dross and mill scale, respectively. Graphite used in the experiments contains 98.5 wt% of carbon (Cat. No.17046-02) obtained from Kanto Chemical Co., Inc., Tokyo, Japan. Some water was added to the blends to make a spherical composite pellet with the weight of approximately 5 g. The pellets were dried in the oven at 90 °C for 48 h and used for high temperature experiments.

Table 3. Composition in the blend samples.

Blend	Dross (wt%)	Scale (wt%)	Graphite (wt%)	C/O Ratios
A	41.34	48.44	10.21	1
B	39.37	46.09	14.57	1.5
C	37.52	43.95	18.53	2

2.2. High-Temperature Interactions

The overview of experimental procedures is given in Figure 1. The high-temperature interactions were investigated in a tube furnace. The pellet was put in a crucible and then inserted into the cold zone of the furnace for 5 min to prevent thermal shock. The crucible was then inserted into the hot zone where the temperature was 1550 °C for 1, 2, and 3 h. High-purity argon (99.99%) was purged into the tube furnace at the rate of 1 L/min at all times to prevent oxidation of the pellet. The quenched pellets were collected for further analysis.

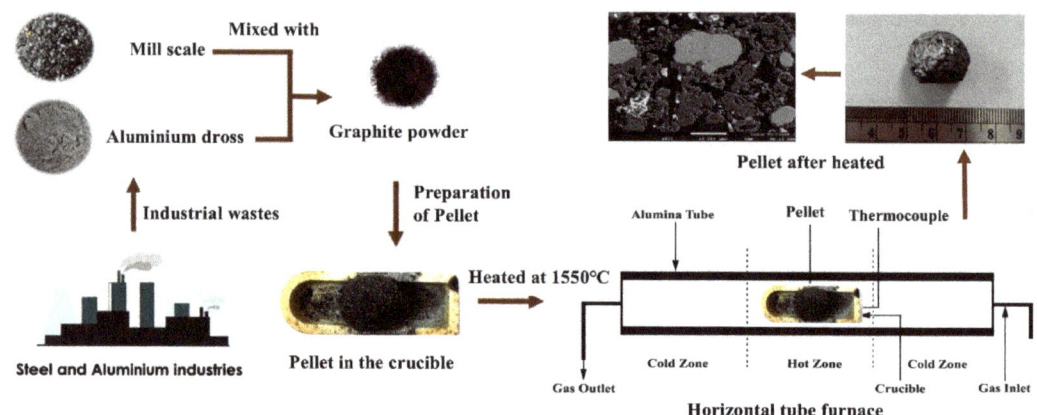

Figure 1. Overview of samples preparation and experimental procedure.

2.3. Analysis

The pellets were mounted in the resin and cross-sectioned for SEM analysis. SEM and EDS analysis were used to observe the formation of ferroalloy and its compositions. Some of the quenched pellets were ground into powder for XRD analysis. XRD was employed to investigate the phases that occurred in the synthesized ferroalloys.

3. Results and Discussion

3.1. Carbothermic Reduction Reactions

Figure 2 shows the pellets after they were heated at 1550 °C as a function of times. The outer surface of the pellets was seen to crack and some of them were covered by white materials, which is fibrous alumina [32]. The presence of fibrous alumina on the pellets indicates the occurrence of a carbothermic reduction reaction of alumina [32]. The SEM micrograph of the cross-sectioned pellets is shown in Figure 3. The formation of metal droplets was observed as a white–grey phase and clearly separated from the unreacted-oxides phase. These indicate the occurrence of carbothermic reduction reactions of Fe_2O_3 in mill scale, Al_2O_3, and SiO_2 in aluminum dross at 1550 °C.

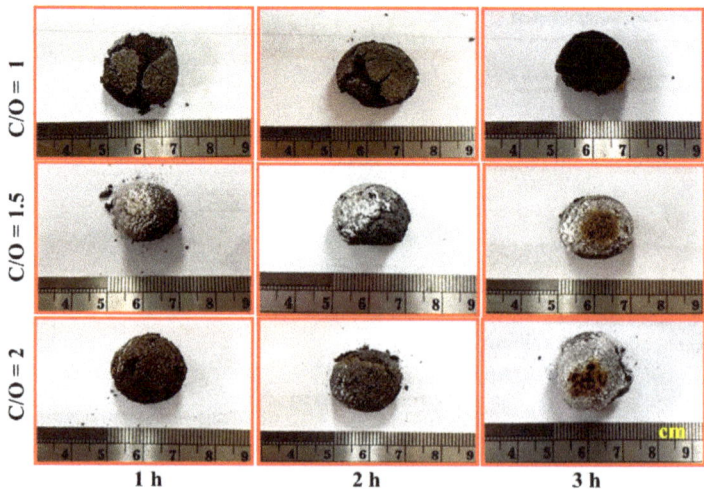

Figure 2. Pellet samples after heating at 1550 °C for up to 3 h.

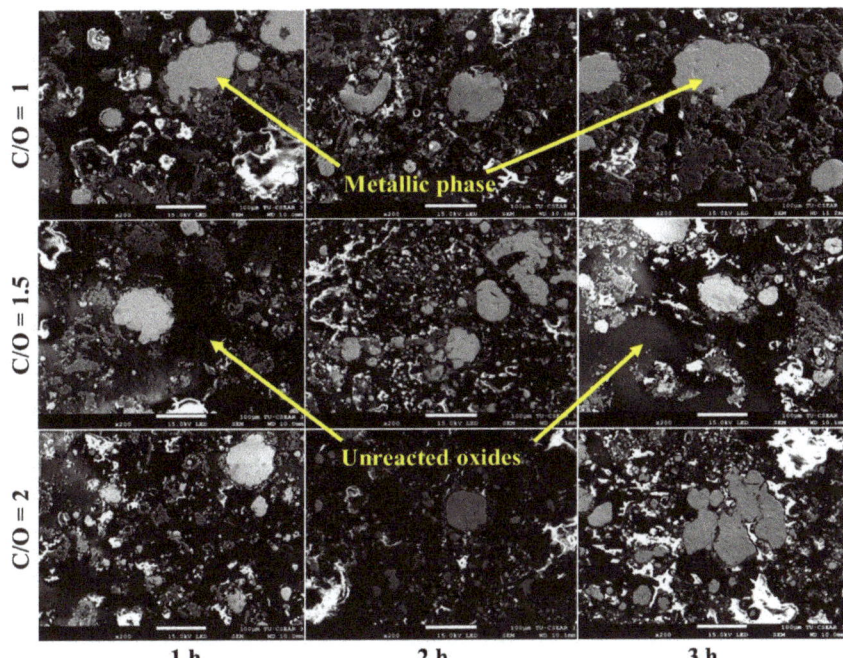

Figure 3. SEM micrograph of cross-sectioned pellets after heating at 1550 °C for up to 3 h.

Figures 4–6 show XRD spectra of the pellets with C/O ratios of 1, 1.5, and 2 after heating at 1550 °C, respectively. XRD spectra confirmed that the synthesized alloys produced in the systems were iron aluminide (Fe_3Al) and iron silicide (Fe_3Si). The other phases were carbon and unreacted Al_2O_3 and SiO_2 remain in the system. Fe_3Al peak occurs at 2theta of approximately 43.4°, while the major peak of Fe_3Si occurs at 2theta of 44.8°. For the pellets with C/O = 1, intensity of Fe_3Al peak was higher than that of Fe_3Si peak, and this trend happened for 3 h of interaction. With increasing C/O ratios, the intensity of Fe_3Si peak was observed to increase and get stronger than that of Fe_3Al peak in the case of the pellet with C/O = 2.

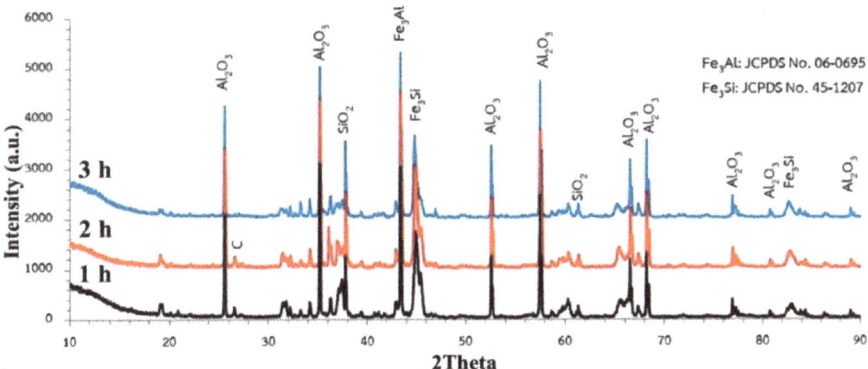

Figure 4. XRD patterns of the pellet (C/O = 1) after heating at 1550 °C for up to 3 h.

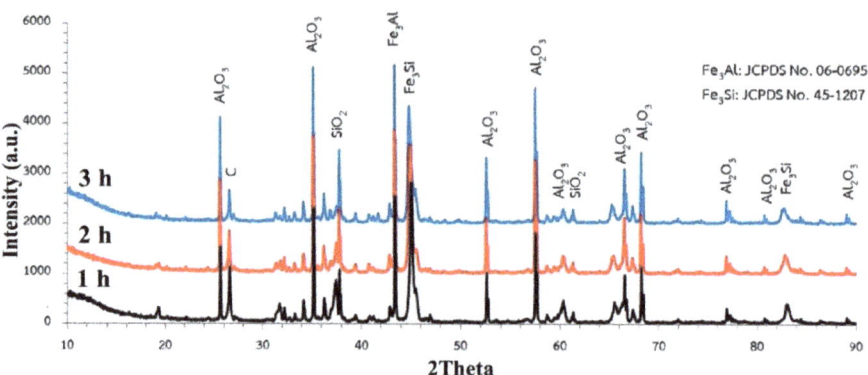

Figure 5. XRD patterns of the pellet (C/O = 1.5) after heating at 1550 °C for up to 3 h.

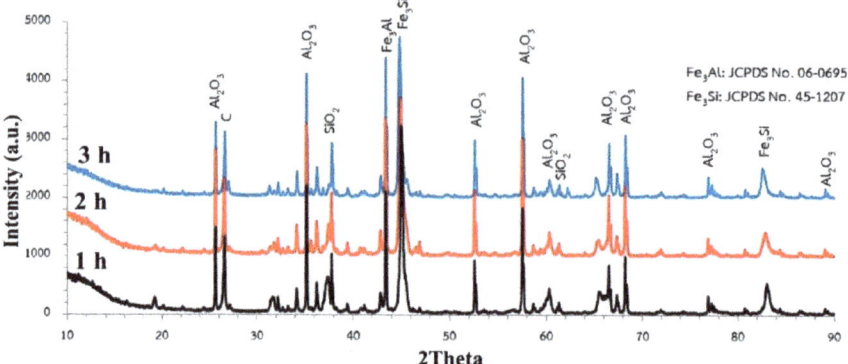

Figure 6. XRD patterns of the pellet (C/O = 2) after heating at 1550 °C for up to 3 h.

The formation of Fe$_3$Al and Fe$_3$Si can be explained via the carbothermic reduction reaction of Fe$_2$O$_3$, Al$_2$O$_3$, and SiO$_2$ at 1550 °C under an argon atmosphere. The reduction of Fe$_2$O$_3$ by solid carbon atom can occurred via Equations (1)–(3), and produce liquid Fe and CO into the system [32,33]. The produced CO could possibly react with Fe$_2$O$_3$ as a reducing agent and produce liquid Fe and CO$_2$ into the system, as shown in Equations (4)–(6) [32,33]. At 1550 °C, standard Gibbs free energy (ΔG°) for Equations (3) and (6) is −124.46 kJ and −272.71 kJ, respectively.

$$3Fe_2O_3\,(l) \;+\; C\,(s) \;=\; 2Fe_3O_4\,(l) \;+\; CO\,(g) \tag{1}$$

$$Fe_3O_4\,(l) \;+\; C\,(s) \;=\; 3FeO\,(l) \;+\; CO\,(g) \tag{2}$$

$$FeO\,(l) \;+\; C\,(s) \;=\; Fe\,(l) \;+\; CO\,(g) \tag{3}$$

$$3Fe_2O_3\,(l) \;+\; CO\,(g) \;=\; 2Fe_3O_4\,(l) \;+\; CO_2\,(g) \tag{4}$$

$$Fe_3O_4\,(l) \;+\; CO\,(g) \;=\; 3FeO\,(l) \;+\; CO_2\,(g) \tag{5}$$

$$FeO\,(l) \;+\; CO\,(g) \;=\; Fe\,(l) \;+\; CO_2\,(g) \tag{6}$$

The overall carbothermic reduction of Al$_2$O$_3$ as shown in Equation (7), is known to occur at the temperature of over 2200 °C at 1 atm [34]. This reaction could produce Al$_2$O gas, Al vapor, and CO gas via the reaction pathway, Equations (8) and (9). However, the reduction reaction of Al$_2$O$_3$ by solid carbon had also been reported to start at a temperature of 1450 °C with a small amount of Al vapor and gaseous species of AlO and Al$_2$O [35]. Equations (7)–(9) had also been reported to proceed at 1550 °C in the presence of metallic solvent [29,32,34]. Thus, the carbothermic reduction of Al$_2$O$_3$ in the present study could occur in the presence of liquid Fe as the metallic solvent.

$$Al_2O_3(s) + 3C(s) = 2Al(l) + 3CO(g) \quad (7)$$

$$Al_2O_3(s) + 2C(s) = Al2O(g) + 2CO(g) \quad (8)$$

$$Al_2O(g) + C(s) = 2Al(g) + CO(g) \quad (9)$$

$$SiO_2(s) + C(s) = SiO(g) + CO(g) \quad (10)$$

$$SiO(g) + 2C(s) = SiC(s) + CO(g) \quad (11)$$

$$SiO(g) + SiC(s) = 2Si + CO(g) \quad (12)$$

$$SiO_2(g) + 2SiC(s) = 3Si + 2CO(g) \quad (13)$$

For SiO$_2$, the possible reaction could proceed through the reduction of SiO$_2$ into Si or SiC, as shown in Equations (10)–(13). The carbothermic reaction of SiO$_2$ has reaction rates at temperature above 1400 °C, and thus could occur at the experimental temperature of 1550 °C [29,30]. The carbothermic reduction reactions of Fe$_2$O$_3$, Al$_2$O$_3$, and SiO$_2$ at 1550 °C produce Fe, Al and Si into the system. The produced Al could have high affinity for liquid iron and thus transfer into liquid iron phase and form iron aluminide (Fe$_3$Al) [36], while Si could be removed from the reaction zone through dissolution into liquid iron phase and form iron silicide (Fe$_3$Si) [30]. During the presence of liquid iron, carbon dissolution into liquid steel could proceed due to the contact between liquid iron and solid carbon from graphite. Therefore, the final product was the Fe–Al–Si–C alloy.

3.2. Effect of Carbon

SEM and EDS analysis were employed for the metallic phase that obtained from the pellets of C/O ratios of 1, 1.5, and 2 after heating for 3 h, as shown in Figures 7–9. The composition of the ferroalloys was roughly determined using EDS analysis and given in Table 4. The metallic phase was found to be Fe–Al–Si–C alloys for all cases. For C/O = 1, the alloy composes of 0.69 at% of Al and 4.15 at% of Si with the others being iron and carbon. With increasing carbon to C/O = 1.5, Al and Si levels have increased to 7.30 and 9.94 at%, respectively. However, there was no further increase in Si level with increasing carbon to C/O = 2, while the Al level kept rising from 7.3 at% to 8.59 at%.

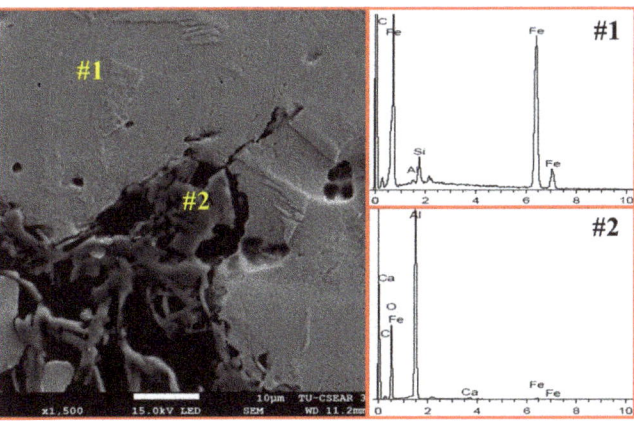

Figure 7. SEM micrograph and EDS spectra of the cross-sectioned pellet (C/O = 1) after heating at 1550 °C for 3 h.

Figure 8. SEM micrograph and EDS spectra of the cross-sectioned pellet (C/O = 1.5) after heating at 1550 °C for 3 h.

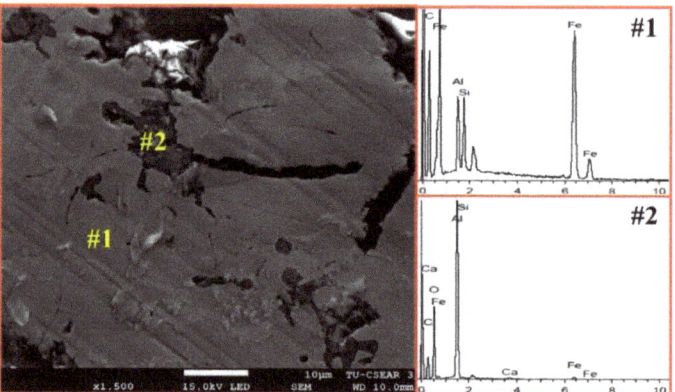

Figure 9. SEM micrograph and EDS spectra of the cross-sectioned pellet (C/O = 2) after heating at 1550 °C for 3 h.

Table 4. Composition of the synthesis metals after heating for 3 h.

Blend	C/O Ratios	Atomic (at%)			
		Fe	Al	Si	C
A	1	69.32	0.69	4.15	25.84
B	1.5	56.08	7.30	9.94	26.68
C	2	48.19	8.59	8.29	34.93

Figure 10 shows the variation of Al and Si concentration in the ferroalloy as a function of carbon content in the system (C/O ratios). It is clearly seen that the carbothermic reduction of SiO_2 precedes the reduction of Al_2O_3 when carbon content in the system has reached C/O = 1.5, and vice versa when increasing carbon to C/O = 2. These indicate that the reduction of SiO_2 can reach completion at 1550 °C, and carbon content in the system of C/O > 1.5 is excessive. On the other hand, the reduction of Al_2O_3 still carries on and cannot achieve complete reduction at this state. The lower carbon in the system of C/O = 1 was likely to have an inadequate carbon atom for Al_2O_3 reduction. This is because carbon will be consumed for Fe_2O_3 reduction first and then SiO_2 reduction. The remaining carbon will be for Al_2O_3 reduction due to the lowest driving force of the reaction at 1550 °C.

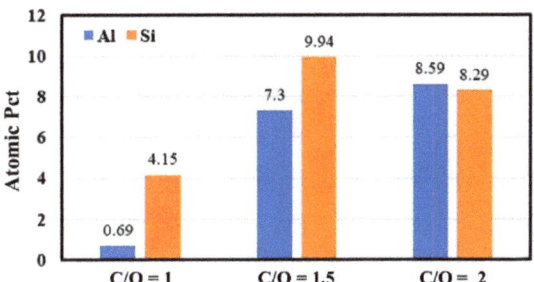

Figure 10. Comparison of elemental atomic percent in the synthesized metal after heating at 1550 °C for 3 h.

3.3. Effect of Time

SEM and EDS techniques were used to analyze the metallic phase that obtained from the pellets of C/O = 2 after heating at 1550 °C for 1, 2, 3 and 6 h, as shown in Figure 11. Composition of the alloy was EDS analyzed and given in Table 5. With high carbon concentration in the system, the carbothermic reduction of Al_2O_3 precedes the reduction of SiO_2 since 1 h of interaction. The level of Al in the alloy was 6.53 at% at 1 h and keep pace to reach 14.62 at% after 6 h, while Si level was 2.17 at% at 1 h and found to stabilize between 6–8 at% since 2 h until 6 h. These indicate the faster complete reduction of SiO_2 than that of Al_2O_3.

Table 5. Composition of the synthesis metals for the sample (C/O = 2).

Heating Time (h)	Atomic (at%)			
	Fe	Al	Si	C
1	55.49	6.53	2.17	35.81
2	42.11	8.7	6.06	43.13
3	48.19	8.59	8.29	34.93
6	48.02	14.62	7.99	29.37

Figure 12 shows the variation of Al and Si concentration in the ferroalloy as a function of time for the pellet with C/O = 2. It looked like the carbothermic reduction of Fe_2O_3, Al_2O_3, and SiO_2 in this system had proceeded almost concurrently, but with different kinetic rates depending on the driving force for each reaction. This is because the excess carbon of C/O = 2 in the pellet, provide adequate carbon atom for the carbothermic reduction of each reducible oxide. Even though SiO_2 is known to proceed at a faster rate than Al_2O_3, Al concentration in the alloy was higher than Si because the higher amount of Al_2O_3 (69.94 wt%) in the dross compared to SiO_2 content (5.01 wt%).

For the synthesis of Fe–Al–Si–C alloys, carbon need to be provided adequately for each reducible oxide, such as C/O = 2. An interaction time of 3 h was more suitable than 6 h due to an economic point of view. Figure 13 shows the SEM micrograph and EDS elemental distribution in the bulk metal obtained from the pellets of C/O = 2 after heating at 1550 °C for 3 h. It can be observed that Al, Si, and C distribute over the entire Fe matrix. The present studies have shown that industrial wastes such as mill scale and aluminum dross can be successfully utilized or valorized as a source of Fe_2O_3, Al_2O_3, and SiO_2 for the synthesis of Fe–Al–Si–C alloys.

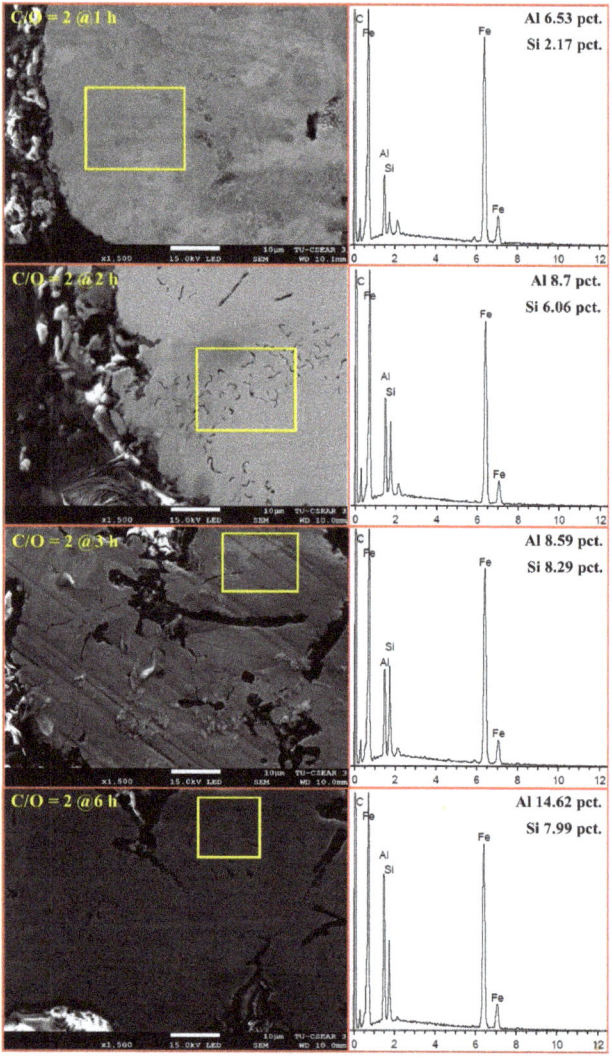

Figure 11. Comparison of SEM micrograph and EDS spectra of the cross-sectioned pellet (C/O = 2) after heating at 1550 °C for 1, 2, 3, and 6 h.

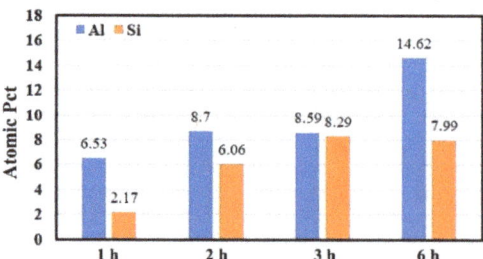

Figure 12. Comparison of elemental atomic percent in the synthesized metal from the pellet (C/O = 2) after heating at 1550 °C for 1, 2, 3, and 6 h.

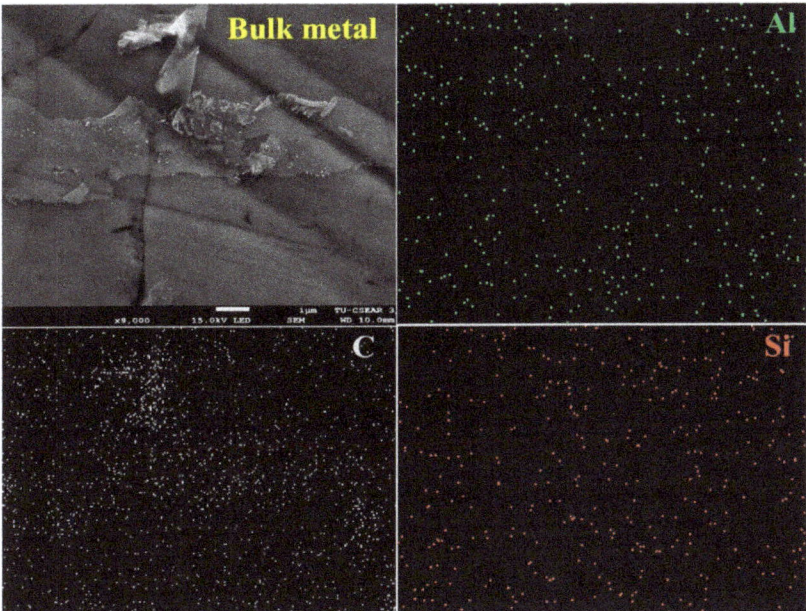

Figure 13. SEM micrograph and EDS elemental contour of the cross-sectioned metal droplet obtained from pellet (C/O = 2) after heating at 1550 °C for 3 h.

4. Conclusions

Synthesis of ferroalloys at 1550 °C using a mill scale and aluminum dross as a source of metal oxides was successfully investigated. The experimental results can be concluded as below.

1. Carbothermic reduction reactions of Fe_2O_3 in mill scale, Al_2O_3, and SiO_2 in aluminum dross can proceed at 1550 °C. The formation of metal droplets was observed and clearly separated from the unreacted-oxides phase. The synthesized alloys produced in the systems were Fe–Al–Si–C ferroalloys consisting of iron aluminide (Fe_3Al) and iron silicide (Fe_3Si) phases. Carbothermic reduction of Al_2O_3 in the present study can occur in the presence of liquid Fe as the metallic solvent.
2. For low carbon content in the system (C/O = 1), carbon was inadequate for Al_2O_3 reduction because it will be consumed by Fe_2O_3 reduction first and then SiO_2 reduction. The remaining carbon will be for Al_2O_3 reduction due to its lowest driving force of the reaction at 1550 °C.
3. For high carbon content in the system (C/O = 2), it looked like the carbothermic reduction of Fe_2O_3, Al_2O_3, and SiO_2 in this system had occurred almost concurrently, but different kinetic rates depend on the driving force for each reaction. The excess carbon in the pellet will provide adequate carbon atoms for the carbothermic reduction of each reducible oxide.
4. Carbothermic reduction of SiO_2 1550 °C can complete within 2–3 h, while a longer time is needed for the carbothermic reduction of Al_2O_3 to reach completion. However, the conditions of C/O = 2 and interaction time of 3 h were suitable for the synthesis of Fe-Al-Si-C due to the economic point of view. Further investigation is essential for the mechanical properties of the synthesized ferroalloys.
5. The innovation of this study was to extract Fe, Al, and Si from metal oxides bearing industrial wastes at temperatures as low as 1550 °C in one step process. The final product was in the form of Fe–Al–Si–C alloys. This research increases the possible methods

for industrial waste management, decreases the negative effect on the environment, and enhances sustainability for materials processing toward a circular economy.

Author Contributions: Conceptualization, S.K. and W.S.; methodology, P.W., W.S. and S.K.; validation, S.K.; formal analysis, P.W. and S.K.; investigation, P.W., W.S. and S.K.; resources, W.S. and S.K.; data curation, P.W.; writing—original draft preparation, P.W., W.S. and S.K.; writing—review and editing, S.K.; visualization, P.W. and S.K.; supervision, S.K.; project administration, S.K.; funding acquisition, W.S. All authors have read and agreed to the published version of the manuscript.

Funding: This research was funded by Faculty of Sciences and Technology, Contract No. SciGR14/2564 and Thammasat University Research Fund, Contract No. TUFT 045/2564.

Institutional Review Board Statement: Not applicable.

Informed Consent Statement: Not applicable.

Data Availability Statement: Not applicable.

Conflicts of Interest: The authors declare no conflict of interest.

References

1. Martin, M.I.; Lopez, F.A.; Torralba, J.M. Production of sponge iron powder by reduction of rolling mill scale. *Ironmak. Steelmak.* **2012**, *39*, 155–162. [CrossRef]
2. Sen, R.; Dehiya, S.; Pandel, U.; Banerjee, M.K. Utilization of low grade coal for direct reduction of mill scale to obtain sponge iron: Effect of reduction time and particle size. *Procedia Earth Planet. Sci.* **2015**, *11*, 8–14. [CrossRef]
3. Ye, Q.; Zhu, H.; Peng, J.; Khannan, C.S.; Chen, J.; Dai, L.; Liu, P. Preparation of Reduce Iron Powders from Mill Scale with Microwave Heating: Optimization Using Response Surface Methodology. *Metall. Mater. Trans. B* **2013**, *44B*, 1478–1485. [CrossRef]
4. Khaerudini, D.S.; Chanif, I.; Insiyanda, D.R.; Destyorini, F.; Alva, S.; Premono, A. Preparation and characterization of mill scale industrial waste reduced by biomass-based carbon. *J. Sustain. Metall.* **2019**, *5*, 510–518. [CrossRef]
5. Shi, J.; Wang, D.R.; He, Y.D.; Qi, H.B.; Wei, G. Reduction of oxide scale on hot-rolled Strip steels by carbon monoxide. *Mater. Lett.* **2008**, *62*, 3500–3502.
6. Guan, C.; Li, J.; Tan, N.; He, Y.; Zhang, S. Reduction of oxide scale on hot-rolled steel by hydrogen at low temperature. *Int. J. Hydrog. Energy* **2014**, *39*, 15116–15124. [CrossRef]
7. Benchiheub, O.; Mechachti, S.; Serrai, S.; Khalifa, M.G. Elaboration of iron powder from mill scale. *J. Mater. Environ. Sci.* **2010**, *1*, 267–276.
8. Mechachti, S.; Benchiheub, O.; Serrai, S.; Shalabi, M.E.H. Preparation of iron Powders by Reduction of Rolling Mill Scale. *Int. J. Sci. Eng. Res.* **2013**, *4*, 1467–1472.
9. Joshi, C.; Dhokey, N.B. Study of Kinetics of Mill Scale Reduction: For PM Applications. *Trans. Indian Inst. Met.* **2015**, *68*, 31–35. [CrossRef]
10. Sista, K.S.; Dwarapudi, S.; Nerune, V.P. Direct reduction recycling of mill scale through iron powder synthesis. *ISIJ Int.* **2019**, *59*, 787–794. [CrossRef]
11. Satish Reddy, M.; Neeraja, D. Aluminium residue waste for possible utilization as a material: A review. *Indian Acad. Sci.* **2015**, *43*, 124.
12. Das, B.R.; Dash, B.; Tripathy, B.C.; Bhattacharya, I.N.; Das, S.C. Production of η-alumina from waste aluminium dross. *Miner. Eng.* **2006**, *20*, 252–258. [CrossRef]
13. Available online: http://www.basel.int/ (accessed on 11 August 2022).
14. Lukita, M.; Abidin, Z.; Riani, E.; Ismail, A. Utilization of hazardous waste of black dross aluminum: Processing and application-a review. *J. Degrad. Min. Lands Manag.* **2022**, *9*, 3265–3271. [CrossRef]
15. Yoldi, M.; Fuentes-Ordoñez, E.G.; Korili, S.A.; Gil, A. Efficient recovery of aluminum from saline slag wastes. *Miner. Eng.* **2019**, *140*, 1–8. [CrossRef]
16. Gil, A.; Arrieta, E.; Vicente, M.A.; Korili, S.A. Synthesis and CO_2 adsorption properties of Hydrotalcite like compounds prepared from aluminum saline slag wastes. *Chem. Eng. J.* **2018**, *334*, 1341–1350. [CrossRef]
17. Santamaría, L.; López-Aizpún, M.; García-Padial, M.; Vicente, M.A.; Korili, S.A.; Gil, A. Zn-Ti-Al layered double hydroxides synthesized from aluminum saline slag wastes as efficient drug adsorbents. *Appl. Clay Sci.* **2020**, *187*, 1–14. [CrossRef]
18. Nguyen, T.H.; Nguyen, T.T.N.; Lee, M.S. Hydrochloric acid leaching behavior of mechanically activated black dross. *J. Korean Inst. Resour. Recycl.* **2018**, *27*, 78–85.
19. Nguyen, T.T.N.; Lee, M.S. Synthesis of magnesium aluminate spinel powder from the purified sodium hydroxide leaching solution of black dross. *Processes* **2019**, *7*, 612. [CrossRef]
20. Zawrah, M.F.; Taha, M.A.; Abo Mostafa, H. In situ formation of Al_2O_3/Al core-shell from waste material: Production of porous composite improved by graphene. *Ceram. Int.* **2018**, *44*, 10693–10699. [CrossRef]

21. Ramaswamy, P.; Ranjit, S.; Bhattacharjee, S.; Gomes, S.A. Synthesis of high temperature (1150 °C) resistant materials after extraction of oxides of Al and Mg from aluminum dross. *Mater. Today Proc.* **2019**, *19*, 670–675. [CrossRef]
22. Chobtham, C.; Kongkarat, S. Synthesis of hercynite from aluminum dross at 1550 °C: Implication for industrial waste recycling. *Mat. Sci. Forum* **2020**, *977*, 223–228. [CrossRef]
23. Font, A.; Soriano, L.; Monzó, J.; Moraes, J.C.B.; Borrachero, M.V.; Payá, J. Salt slag recycled by-products in high insulation alternative environmentally friendly cellular concrete manufacturing. *Constr. Build. Mater.* **2020**, *231*, 117114. [CrossRef]
24. Udvardi, B.; Géber, R.; Kocserha, I. Investigation of aluminum dross as a potential asphalt filler. *Int. J. Eng. Manag. Sci.* **2019**, *4*, 445–451. [CrossRef]
25. Jamieson, E.; Jones, A.; Cooling, D.; Stockton, N. Magnetic separation of Red Sand to produce value. *Miner. Eng.* **2006**, *19*, 1603–1605. [CrossRef]
26. Zhu, X.; Niu, Z.; Li, W.; Zhao, H.; Tang, Q. A novel process for recovery of aluminum, iron, vanadium, scandium, titanium and silicon from red mud. *J. Environ. Chem. Eng.* **2020**, *8*, 103528. [CrossRef]
27. Jadhav, U.U.; Hocheng, H. A review of recovery of metals from industrial waste. *J. Achiev. Mater. Manuf. Eng.* **2012**, *54*, 156–167.
28. Wei, D.; Kong, J.; Gao, S.; Zhou, S.; Zhuang, Y.; Xing, P. Preparation of Al–Si alloys with silicon cutting waste from diamond wire sawing process. *J. Environ. Manag.* **2021**, *290*, 112548. [CrossRef]
29. Khanna, R.; Konyukhov, Y.V.; Ikram-ul-hag, M.; Burmistrov, I.; Cayumil, R.; Belov, V.A.; Rogachev, O.; Leybo, D.V.; Mukherjee, P.S. An innovative route for vararising iron and aluminium oxide rich industrial wastes: Recovery of multiple metals. *J. Environ. Manag.* **2021**, *295*, 113035. [CrossRef]
30. Kongkarat, S.; Boonyaratchinda, M.; Chobtham, C. Formation of ferrosilicon alloy at 1550 °C via carbothermic reduction of SiO_2 by coal and graphite: Implication for rice husk ash utilization. *Solid State Phenom.* **2021**, *315*, 16–24. [CrossRef]
31. Boonyaratchinda, M.; Kongkarat, S. Fundamental investigation of ferrosilicon production using rice husk and rubber tree bark at 1550 °C: Implication for utilization of agricultural waste in steelmaking industry. *Mat. Sci. Forum* **2020**, *977*, 171–177. [CrossRef]
32. Khanna, R.; Kongkarat, S.; Seetharaman, S.; Sahajwalla, V. Carbothermic Reduction of Alumina at 1823K in the Presence of Molten Steel: A Sessile Drop Investigation. *ISIJ Int.* **2012**, *52*, 992–999. [CrossRef]
33. Dankwah, J.R.; Koshy, P.; Saha-Chaudhury, N.; O'Kane, P.; Skidmore, C.; Knights, D.; Sahajwalla, V. Reduction of FeO in EAF steelmaking slag by metallurgical coke and waste plastics blends. *ISIJ Int.* **2011**, *51*, 498–507. [CrossRef]
34. Frank, R.A.; Finn, C.W.; Elliott, J.F. Physical chemistry of the carbothermic reduction of alumina in the presence of a metallic solvent: Part II. Measurements of kinetics of reaction. *Metall. Mater. Trans. B* **1989**, *20*, 161–173. [CrossRef]
35. Cox, J.H.; Pidgeon, L.M. An investigation of the Aluminium-Oxygen-Carbon system. *Can. J. Chem.* **1963**, *27*, 671. [CrossRef]
36. Akinlade, O.; Singh, R.N.; Sommer, F. Thermodynamics of liquid Al–Fe alloys. *J. Alloys Comp.* **2000**, *299*, 163–168. [CrossRef]

Article

Selenium and Tellurium Separation: Copper Cementation Evaluation Using Response Surface Methodology

Seyedreza Hosseinipour, Eskandar Keshavarz Alamdari * and Nima Sadeghi

Department of Materials and Metallurgical Engineering, Amirkabir University of Technology (Tehran Polytechnique), Tehran 15875-4413, Iran
* Correspondence: alamdari@aut.ac.ir; Tel.: +98-2164542971

Abstract: In recent years, high demands for Se and Te in the solar panels and semiconductors industry have encouraged its extraction from primary and secondary sources. However, the two elements' similar chemical and physical properties make pure element production, Se or Te, arduous. This work is aimed to investigate the significant factors of Se and/or Te recovery in the copper cementation process using the response surface methodology. The test was carried out in two series, for Te and Se, so that H_2SO_4, $CuSO_4$, Te(or Se) concentration, and temperature are the factors of experimentation. According to response surface methodology (RSM) results for both test series (i. e. Se and Te), 50 g/L H_2SO_4, 15 g/L Cu, and 35 °C, 3000 mg/L Se (or 750 mg/L Te) was specified for higher Se recovery (97%), and the lowest Te extraction (2%) as an optimum condition, so that could make a suitable separation process. Hence, the cementation test was conducted in the simultaneous presence of Se and Te, so the separation index became 5291. Moreover, the cementation test was carried out in the pregnant leach solution of copper anode slime, and the separation factor was measured to be 606. On the other hand, the thermodynamic evaluation and XRD patterns of the process's sediments confirm that Se is precipitated as Cu_2Se and $Cu_{1.8}Se$, whereas no Te components are detected in the sediments.

Keywords: separation; cementation; selenium; tellurium; response surface methodology (RSM)

1. Introduction

Selenium is a metalloid element found in the sulfide minerals and copper anode slime alongside precious metals, tellurium, copper, silver, and nickel. Selenium as a metalloid has broad applications in solar cell fabrication [1], semiconductor manufacturing [2], pharmaceuticals and biomedical uses [3], pigments for ceramics, glasses, and plastics [4,5], metallurgical applications [4], and agriculture uses [6]. Se is usually observed as a red-colored powder in amorphous form and metallic gray in crystalline form, with intermediate properties between tellurium and sulfur [7].

On the other hand, *tellurium* is another semi-metallic element that has specific characteristics that make it helpful for energy conversion [7,8], chemical reaction catalysis [9], alloying, and semiconductors [8]. The electrolytic copper refinery slimes contain gold and precious metals alongside selenium and tellurium, periodically gathered for valuable metals recovery [10,11]. The main purpose of copper anode slime treatment is the extraction of precious metals and gold. However, Se and Te recovery are of secondary importance, so various methods are raised for metals recovery [4]. The chemical and physical specifications of selenium are akin to tellurium, which is an arduous purification process [7,10–12].

Conventionally, selenium fumes were recovered from the exhausted gas of roasting furnaces. However, a portion of selenium and tellurium remain in the residue, sent to an acidic or basic leaching process [13]. Moreover, selenium gas may not be entirely gathered in the filters and causes enormous ecological problems, such as air pollution by heavy metals, so it must be diminished in the coming years [4]. Additionally, there is a commercial process based on roasting copper slimes with soda ash to convert both selenium and tellurium

compounds to a +6 oxidation state [14]. This way, a part of selenium is recovered by the natural leaching process (pH: 7), and tellurium has to be retrieved in chloric or sulfuric acid solutions. The pregnant tellurium solution also contains a significant quantity of selenium that should be separated [15–17].

Acid roasting technology is another practical method to recover Se from copper anode slime based on selective volatilization of selenium compound from slimes [18]. Although selenate, Se VI, compounds are recovered in the roasting process, tellurite and some selenite compounds remain in the sulfated slimes that should be separated [4]. Moreover, selenium and tellurium could also be recovered in oxidative sulfuric acid leaching [19–21]. Under the optimal condition, the concentration of Se and Te is 3.7 and 1.1 g/L, whereas the concentration of copper is approximately 15 g/L [19]. Thus, Se and Te were dissolved in leaching media simultaneously, so the two elements' separation can become a vital process to produce pure metals.

Solvent extraction is one of the popular separation processes. Accordingly, ketone, phosphate, and ether extractants [22,23] were used for selenium extraction; whereas, phosphate [24] and phosphine oxides [25] were proposed for tellurium extraction from sulfuric acid solutions [26].

On the other hand, selenium and tellurium can be precipitated by chemical agents, such as cuprous ions [27], chromous ions [28], hydrazine hydrate [29], sulfur dioxide [30], sodium metabisulphite [31], and sodium sulfite [32] from sulfuric acid media. Copper is another cost-efficient reducer, reported for tellurium [33,34] and tellurium/selenium [35] cementation from the aqueous solution. Although some work has been conducted on Te and Se cementation, no solid report was found for Se/Te separation from copper anode slime liquor. Thus, an accurate study that investigates Se cementation should be done. The tests should be discretely carried out for Se and Te to eliminate interaction between Se and Te cementation, and the results will be compared with dual cementation of Se and Te.

The Cu cementation process is carried out based on the electrochemical reaction that different species of tellurium or selenium can precipitate through a redox reaction. Different parameters, such as pH, electrochemical potential, and the presence of other ionic species in solution, can thermodynamically affect Se/Te cementation. Thus, a thermodynamic evaluation should have been conducted for Se and Te cementation. Moreover, Se or Te cementation was discretely studied in a synthetic solution, similar to Se or Te concentration in pregnant leaching solution, to figure out each factor's effect on metal cementation exclusively. After finding each factor's influence on the discrete Se and Te cementation model, the test should be conducted to find the optimum position of Se/Te separation in synthetic and pregnant leach solutions. Finally, a cost-effective method will be designed to generate a high-purity product in this approach.

2. Materials and Methods
2.1. Precipitation Procedure

Initially, a pregnant leach solution was provided by leaching in sulfuric acid and oxygen peroxide solution. The elemental analysis of the solution is brought in Table 1. According to this, synthetic solutions were provided at different levels of elements and H_2SO_4 concentration. After an optimal condition of cementation was achieved, Se, Te, or H_2SO_4 concentration could be made up by changing the S/L ratio or H_2SO_4 concentration in the copper anode slime leaching. Then, the cementation test was carried out in PLS to calculate Se/Te separation index.

Table 1. Analysis of primitive pregnant solution from copper anode slime leaching.

Elements	Cu (g/L)	Se (mg/L)	Te (mg/L)	As (mg/L)	Pb (mg/L)	Ag (mg/L)	Pd (mg/L)
concentration	13.85	2910	723	185	3.2	132	<1

The experimental procedure of cementation was divided into two branches, illustrated in Figure 1. In the first stage, the effect of parameters was investigated on the Te cementation, and the optimum condition was specified. In the next section, Se cementation, in which Se concentration is four times Te concentration, was investigated to find influential factors in the process. These two sections are entitled discrete cementation for Te and Se cementation. Afterward, the dual selenium and tellurium cementation was conducted based on the optimum condition for Se and Te separation in the synthetic solution. This test was replicated in the solution of copper anode slime leach too. Finally, the obtained solution from Se cementation in pregnant leaching solution has been sent to Te precipitation process.

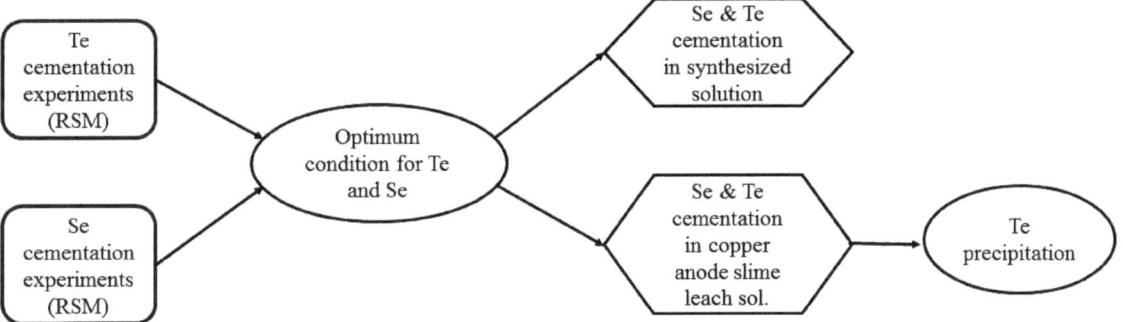

Figure 1. Flowchart for Se and Te cementation process by solid copper.

The batch extraction experiments were carried out in the Erlenmeyer flask to recover Se or Te from synthetic solutions. Initially, a specific volume of Te and/or Se was poured into the volumetric flask (100 mL, class A) from stock solutions, and then copper sulfate and sulfuric acid solutions were added according to the experiment design. Finally, the volume of the solution was brought to the required volume by distilled water. After solution preparation, the samples were transferred into the Erlenmeyer flask and placed in a bain-marie bath to reach the target temperature. Two grams of copper chops were cast to the 100 mL solution, 20 g/L copper chops density, in the flask at the desired temperature, and the mixture was agitated by a mechanical shaker at 500 rpm for 2 h. After filtration, samples were taken for analysis. The remaining metallic ion concentration in the solution was determined by AAS (AA240, Varian, Palo Alto, CA, USA) analytical instrument. The extraction efficiency was expressed as extraction percentage (%E) as defined in Equation (1).

$$\%E = \frac{[C_{i,Me} - C_{f,Me}]}{C_i} \times 100 \tag{1}$$

Moreover, the distribution coefficient was considered, as a criterion, to assess the process as follows:

$$D_{Me} = \frac{C_{i,Me} - C_{f,Me}}{C_f} \tag{2}$$

$C_{i,Me}$, $C_{f,Me}$, and D_{Me} are the initial, final elements concentration and distribution coefficient, respectively. Regarding this approach, another scale can be applied, which can be a helpful tool to survey the separation capability of the proposed process. This criterion is called separation index and is defined as Equation (3):

$$\beta = \frac{D_{Se}}{D_{Te}} \tag{3}$$

2.2. Materials and Apparatus

All chemicals were of analytical reagent grade, and all solutions were prepared with deionized water. Stock solutions for Te (10 g/L) and Cu (70 g/L) were separately prepared by dissolving a certain amount of K_2TeO_3 (Sigma-Aldrich, A.R., St. Louis, Mo, USA), and $CuSO_4 \cdot 5H_2O$ (Neutron, Tehran, Iran) in 0.1 M H_2SO_4 (Ghatranshimi, Tehran Iran) solutions, respectively. Moreover, 35 g/L Se (IV) solution was prepared in 0.25 M HNO_3+0.1M H_2SO_4 solution by adding pure Se (Umicore, Brussels, Belgium, technical grade). Then, the samples were made by adding a specific volume of the stock solution and sulfuric acid solution to the volumetric flask. Afterward, the obtained solutions were allowed to stand for more than 24 h at ambient temperature. Pure copper chop (99.99, National Iranian Copper Industries Co. (NICICO, Tehran, Iran), with a size < 200 µm, was used as a reducing agent that was directly added to the sulfate solution. Sodium hydroxide (Merck KGaA, Darmstadt, Germany) solution was used for acid analysis and pH adjustment of the sulfate solutions. In order to detect selenium and tellurium concentration, atomic absorption spectroscopy (AAS and AA240, Varian, Palo Alto, CA, USA) was used, and pH and ORP (oxidation-reduction potential) of solutions were measured by a pH meter (InoLab 7110, WTW, Weilheim, Germany).

2.3. Optimization Procedure

The parametric approximation models are widely exploited through the design of experiments to figure out optimum conditions on the pilot and industrial scale [9]. In this way, the independent parameters' influence on the experiments' outcomes as dependent variables can be achieved using the least number of tests. According to computer technology progress, even complicated problems can be solved with a minimum cost and time through optimization methods [36].

Response surface methodology (RSM) is a well-arranged technique to conduct systematic investigations of complicated systems via statistical and mathematical techniques such as central composite design (CCD). The main purpose of this procedure is to discover more effective factors and the exact optimum condition with a reasonable number of runs by extension of an empirical correlation between the controlled variables (X) and response (Y) [37]. Thus, the experimental design was carried out by Design Expert software (Version 12) developed by Stat-Ease company (Minneapolis, Min, USA). The CCD model presents the second-order polynomial equation in Equation (4). This relation can be exploited to recognize curvature in a response function.

$$Y = \beta_0 + \sum_{i=1}^{k} \beta_i X_i + \sum_{i=1}^{k} \beta_{ii} X_i^2 + \sum\sum_{i<j} \beta_{ij} X_i X_j + \varepsilon \quad (4)$$

where X_i and X_j are the independent factors, β_0 and β_i are constant value and linear coefficient, β_{ii} and β_{ij} are squared, and interaction coefficients, respectively, and ε is the random experimental error [38]. The second-order response equation discovers the effect of one factor with their quadratic and interactions over the responses.

Some rough tests were carried out to figure out effective parameters. Temperature, pH, Te and/or Se concentration, and copper sulfate concentration were selected as more effective parameters. RSM is comprehensive and can specify the order of factors on the response(s) and calculate interactions between factors. This method can establish the relation between response(s), independent variables, and the probable interactions between variables can be established. In a continuous operation, the numeric factors can be put on any desired amount, presented at five levels in Table 2.

As an appropriate model for industrial functions, the quadratic polynomial model can precisely estimate the interconnection between the independent variables and the response [37]. After attaining the quadratic polynomial model based on five studied levels, analysis of variance (ANOVA) was applied to validate the provided model.

Table 2. The main factors and the corresponding levels.

Parameters	Unit	Factor Code	Level of Factors				
			−2	−1	0	1	2
Temperature	°C	X1	15	35	55	75	95
H_2SO_4 concentration	g/L	X2	25	50	75	100	125
Te concentration	mg/L	X3	500	750	1000	1250	1500
Se concentration			2000	3000	4000	5000	6000
$CuSO_4$ concentration	g/L	X4	5	15	25	35	45

3. Results and Discussion

3.1. Thermodynamic Evaluation for Se and Te Cementation

Thermodynamic simulation can always provide reliable insight into experimental design and implementation. Thus, some thermodynamic analyses were carried out for the precipitation process via FactSage^{TM} thermochemical software (version 6.0, Aachen, Germany) [39] and other thermodynamic databases, so the evaluation results are calculated for both Te and Se in 1 L of solution. As can be seen in Figure 2a, the solid phases of Se are presented, Cu_2Se is stable in a lower concentration of H_2SO_4, but the amount of Cu_2Se falls higher than 0.51 mol (50 g). In contrast, CuSe and $CuSe_2$ species rise in the range. Even though pure Se can become stable in the higher 0.7 mol (68.6 g) range, the previous work [40] reports $Cu_{2-x}Se$ as a middle phase that can form in the deposits.

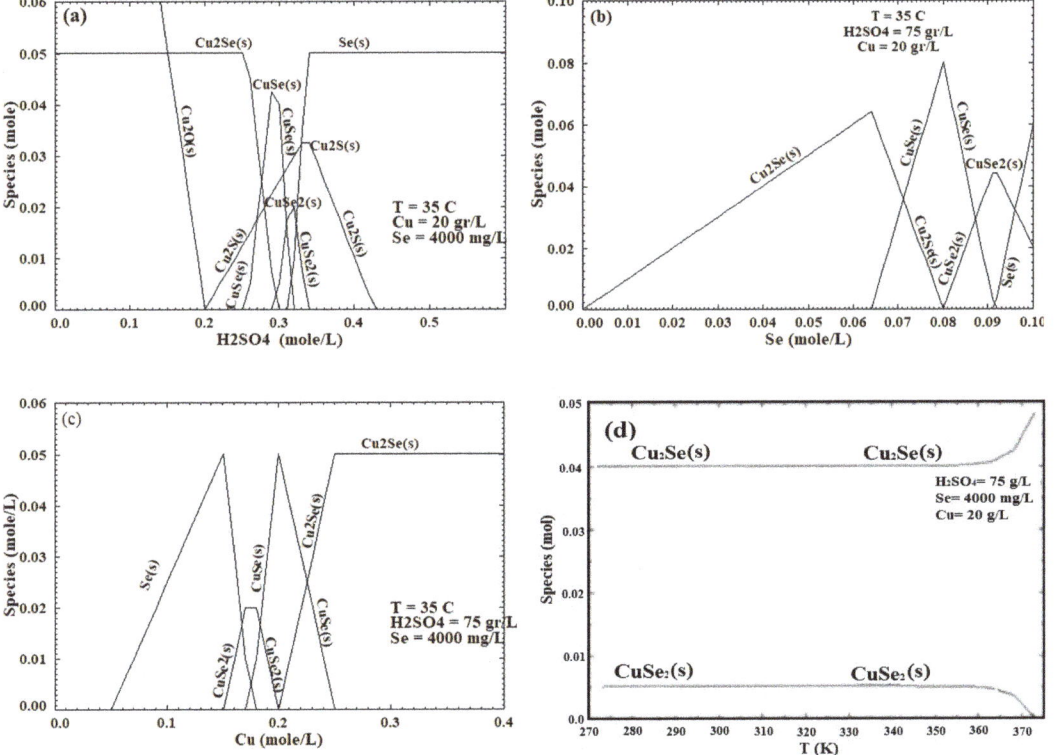

Figure 2. Se species in different (a) H_2SO_4; (b) Se concentration; (c) solid copper values; (d) temperature at mean quantity of other factors.

Furthermore, Figure 2b depicts Se ions values on various Se ionic species. Cu_2Se is persistent in lower 0.065 mol (5.13 g), whereas CuSe and $CuSe_2$ can stabilize in a higher concentration range. Figure 2c exposes the solid copper amount at the mean number of other parameters. As can be detected, the copper contributes to the precipitation reaction at higher than 0.15 mol (9.53 g) and adding more copper to the system escalates the copper contribution in the deposited phase. In conclusion, Se concentration and Cu chops can escalate the Se cementation process; in contradiction, H_2SO_4 declines CuSe cementation efficiency. Finally, Figure 2d exhibits that rising temperature does not change Cu_2Se and $CuSe_2$ species until 363 K (90 °C). However, the $CuSe_2$ phase has diminished higher than 90 °C, while Cu_2Se species extends in the system.

On the other hand, Te solid phases have been illustrated in Figure 3. As observed, Figure 3a presents that temperature could not influence the Cu_2Te values, whereas Figure 3b indicates that the initial amount of Te increases Te sediments in the system. Although Cu_2Te sediments accumulated at 0–0.065 mol (8.29 g) Te, TeO_2 has formed more than 0.065 mol Te, and an amount of Cu_2Te is decayed in the system. Moreover, Figure 3c exposes that solid copper value does not affect Cu_2Te formation, even in a system with no copper metal additive. Moreover, thermodynamic results show that despite H_2SO_4 variation in the system, Cu_2Te is the dominant species in 0–0.8 mol H_2SO_4.

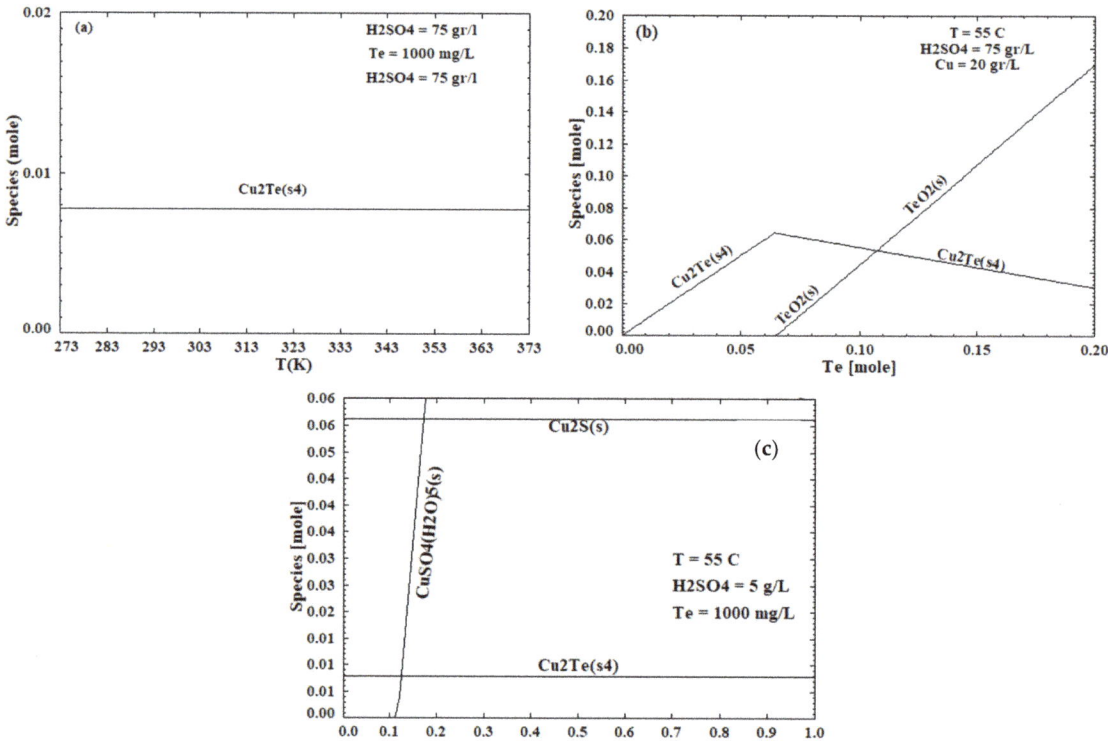

Figure 3. Te species in different amount of (**a**) H2SO4; (**b**) tellurium; (**c**) copper sulfate at mean quantity of other factors.

3.2. Optimization through CCD model

As mentioned in Table 2, a five-level design for four different variables was provided through the central composite design (CCD) illustrated in Table 3. These tests were separately conducted for Se and Te, but Table 3 is presented for both elements to summarize the contents. Accordingly, the experiments were randomly conducted to diminish the influence of uncontrolled variables [41]. Different conditions for each experiment set (e.g., Se or Te) cementation were provided with sixteen cube points and eight axial points with six center points in one cube. The quadratic polynomial model was utilized based on the responses in Table 3, through which the regression coefficients were achieved.

Table 3. Design matrix for CCD experiments and responses.

Run	X_1 T °C	X_2 C_{H2SO4} g/L	X_3 C_{Te} mg/L	C_{Se}	X_4 C_{CuSO4} g/L	R_1 Te Recovery %	Se Recovery
1	55	75	1000	4000	25	4.61	98.203
2	35	50	750	3000	35	11.99	98.00
3	35	50	1250	5000	35	3.98	99.91
4	55	75	500	2000	25	6.78	79.36
5	55	75	1000	4000	45	1.07	98.46
6	75	50	750	3000	15	1.51	90.00
7	15	75	1000	4000	25	8.96	88.48
8	75	100	750	3000	35	1.84	91.89
9	35	100	750	3000	35	4.00	86.96
10	55	75	1000	4000	25	7.33	93.93
11	55	25	1000	4000	25	4.72	98.55
12	95	75	1000	4000	25	7.00	96.31
13	35	100	1250	5000	35	3.97	89.93
14	35	100	750	3000	15	0.67	85.00
15	55	75	1000	4000	25	7.10	94.14
16	75	100	1250	5000	35	3.61	99.22
17	55	75	1000	4000	25	6.01	99.24
18	55	125	1000	4000	25	3.46	91.11
19	55	75	1000	4000	25	7.88	98.71
20	75	50	750	3000	35	4.83	93.00
21	75	100	750	3000	15	7.54	89.97
22	55	75	1500	6000	25	10.61	90.00
23	35	50	1250	5000	15	7.28	98.98
24	35	50	750	3000	15	6.98	98.52
25	75	50	1250	5000	35	2.01	99.89
26	35	100	1250	5000	15	8.80	87.47
27	55	75	1000	4000	5	5.33	94.22
28	75	100	1250	5000	15	15.91	91.57
29	75	50	1250	5000	15	3.25	99.02
30	55	75	1000	4000	25	7.33	98.07

Based on the responses in Table 3, a statistical model via the CCD model was achieved for both selenium and tellurium extraction. The coefficient of determination (R^2), adjusted R-square (adj. R^2), and the analysis of variance (ANOVA) tests were used to estimate the goodness-of-fit of the suggested model. As observed in Tables 4 and 5, the determination coefficient for Se(IV) cementation is 0.910, and the determination coefficient for Te(IV) cementation is 0.917, demonstrating the appropriate efficiency of the suggested models. Moreover, the predicted R^2 for Te and Se is 0.7034 and 0.6418, respectively, and the differences between adjusted and predicted R^2 are less than 0.2 for Te and Se. In general, the higher level of F-Values in the model increases the unity of the model, and a proposed model becomes significant because of the higher F-values. In contradiction, a considerable amount of *p*-values, or lack of fit, can make a model insignificant [33]. As can be observed, the *p*-values of models are negligible, whereas the criteria are insignificant for both Se and Te. Moreover, the lower pure error can make a convenient model to fit experimental outcomes. Therefore, the proposed Se and Te extraction model can be accepted as a feasible and practical tool.

Table 4. Analysis of variance (ANOVA) and coefficient of determination for the suggested quadratic polynomial model for Se.

Source	Sum of Squares	Degree of Freedom	Mean Square	F-Value	p-Value	Description
Model	706.34	13	54.33	11.11	<0.0001	Significant
Residual	78.25	16	4.89			
Lack of Fit	52.41	11	4.76	1.9219	0.5795	not significant
Pure Error	25.84	5	5.17			
R^2						0.910
Adjusted R^2						0.8192
Predicted R^2						0.6418
A-Temperature	25.73		54.33	11.11	<0.0001	
B-Sulfuric Acid	206.75		25.73	5.26	0.0357	
C-Se Concentration	121.70		206.75	42.27	<0.0001	
D-Cu Concentration	30.30		121.70	24.88	0.0001	
AB	85.03		85.03	17.39	0.0007	
AC	18.28		18.28	3.74	0.0711	
AD	4.55		4.55	0.9298	0.0034	
BC	1.10		8.10	3.2258	0.0064	
BD	5.68		5.68	12.16	0.0029	
CD	2.13		2.13	0.4360	0.5185	
A^2	11.03		11.03	2.26	0.1637	
B^2	0.1485		0.1485	0.0285	0.8682	
C^2	187.46		189.46	38.74	<0.0001	
D^2	2.76		2.76	0.5633	0.4638	

Table 5. Analysis of variance (ANOVA) and coefficient of determination for the suggested quadratic polynomial model for Te.

Source	Sum of Squares	Degree of Freedom	Mean Square	F-Value	p-Value	Description
Residual	29.56	18	29.36	17.88	<0.0001	Significant
Lack of Fit	25.19	13	1.64			
Pure Error	4.37	5	1.94	2.22	0.1945	not significant
R^2			0.8734			
Adjusted R^2						0.917
Predicted R^2						0.8649
A-Temperature	8.05		8.05	4.90	0.0400	
B-Sulfuric Acid	0.8791		0.8791	0.5354	0.04738	
C-Te concentration	9.00		9.00	5.48	0.0309	
D-Cu cobcentration	30.00		30.00	18.27	0.0005	
AB	62.81		62.81	38.25	<0.0001	
AD	19.58		19.58	11.92	0.0028	
BC	51.48		51.48	31.35	<0.0001	
BD	29.98		29.98	18.26	0.0005	
CD	52.93		52.93	32.23	<0.0001	
B^2	25.54		25.54	15.56	0.0010	
D^2	39.00		39.00	23.75	0.0001	

Regarding inputs and the models provided via response surface methodology and CCD, a semi-empirical relation for Se extraction containing interactions between the existing parameters, is defined as Equation (5):

$$\%E_{Se} = 85.26090 - 0.4018101(T) - 0.38856(C_{H2SO4}) + 0.019998(C_{Se}) \\ + 0.51612(C_{Cu}) - 0.005159(T)(C_{H2SO4}) - 0.002671(T)(C_{Cu}) \\ + 0.000011(C_{Se})(C_{H2SO4}) - 0.00238(C_{Cu})(C_{H2SO4}) \\ + 0.000037(C_{Se})(C_{Cu}) - 0.00157(T)^2 - 2.6 \times 10^{-6}(C_{Se})^2 \\ + 0.003137\,(C_{Cu})^2 \quad (5)$$

The semi-empirical equation for Te precipitation percentage is found as Equation (6):

$$\begin{aligned}\%E_{Te} = &-5.019280 - 0.187851(T) - 0.132960(C_{H2SO4}) + 0.000887(C_{Te}) \\ &+ 1.96010(C_{Cu}) - 0.003962(T)(C_{H2SO4}) - 0.005531(T)(C_{Cu}) \\ &+ 0.000287(C_{Te})(C_{H2SO4}) - 0.005475(C_{Cu})(C_{H2SO4}) \\ &+ 0.000728(C_{Te})(C_{Cu}) - 0.001516(C_{H2SO4})^2 \\ &- 0.011171(C_{Cu})^2\end{aligned} \quad (6)$$

Positive terms express a synergistic effect, whereas negative terms designate antagonism. Moreover, the interactions between some factors were not significant, so these interactions were eliminated in the suggested models.

Figure 4a,b illustrate the validity of the suggested models for precipitation percentages of Se and Te alongside the experimental extraction percentages presented in Table 3. As can be seen, the validity of the predicted models for both elements versus the actual outputs is acceptable. Moreover, the determination coefficients for Se and Te cementation are 0.912 and 0.92, respectively, which confirm the significant efficiency of the achieved models. These models (Equations (5) and (6)) can be used for the prediction of Se and Te precipitation in sulfuric acid media. Although the models may not always present the accurate rate of the process, but these models can undoubtedly be a helpful index to estimate Se or Te concentration in the Cu cementation process.

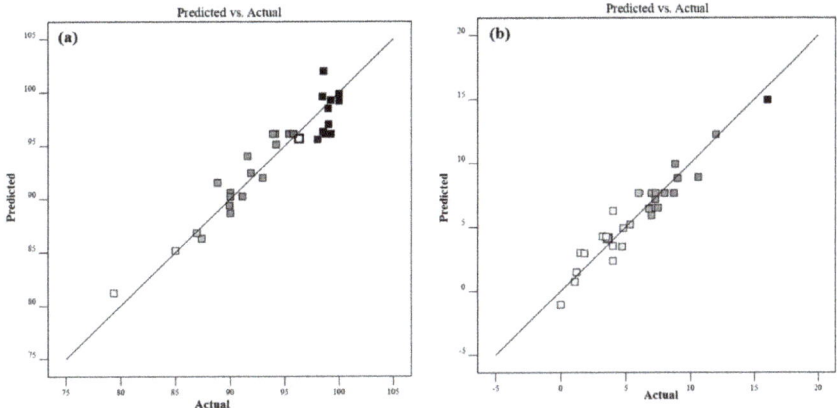

Figure 4. Predicted extracted percent versus actual extraction percent for (**a**) Se and (**b**) Te.

3.3. Three-Dimensional (3D) Response Surface Graphs

Three-dimensional surface plots of the parameters influencing the cementation of Se (IV) are illustrated in Figure 5. The three-dimensional surface graph in Figure 5a demonstrates the Se recovery as a function of temperature and initial H_2SO_4 concentration, which are both practical parameters in the separation process at a constant Se concentration (4000 mg/L) and initial Cu concentration of 25 g/L. As observed, the minor level of temperature and H_2SO_4 concentration led to the highest Se recovery value (99.4%). Nevertheless, rising temperature slightly reduces Se recovery at a minimum concentration of H_2SO_4, whereas the temperature escalates the criterion at a higher level of H_2SO_4 and reduces the negative influence of H_2SO_4. Additionally, the detrimental effect of H_2SO_4 at 75 °C is more conspicuous than at lower temperatures. According to thermodynamic analysis, Figure 5, despite the temperature effect not changing the Se cementation, H_2SO_4 reduces Se solid phase stability.

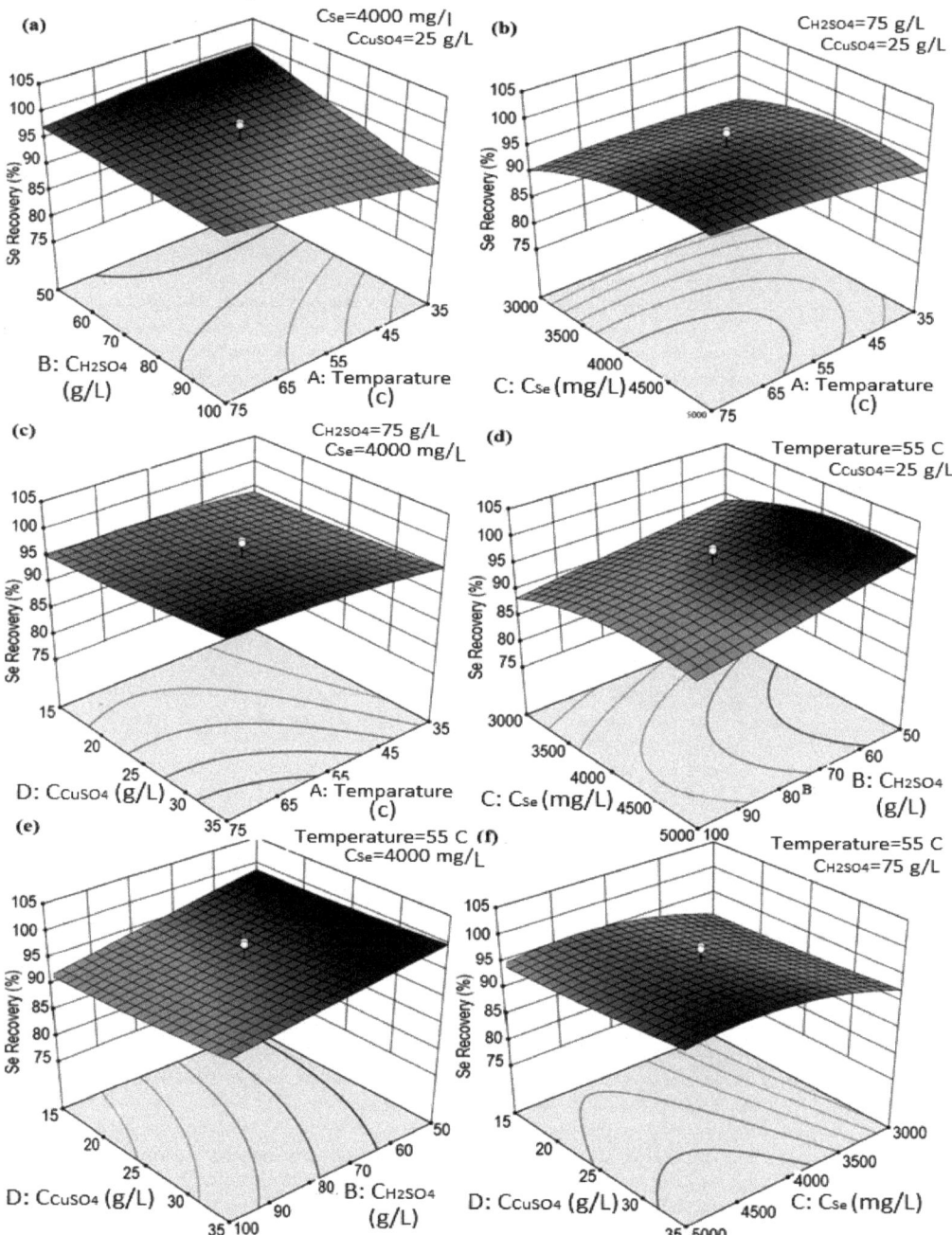

Figure 5. Response surface graphs for interactions of parameters of Se (IV) cementation by solid copper. (**a**) Effect of H_2SO_4 concentration and temperature; (**b**) Effect of Temperature and Se concentration; (**c**) Effect of Cu concentration and temperature; (**d**) Effect of of H_2SO_4 concentration and Se concentration; (**e**) Effect of H_2SO_4 concentration and Cu concentration; (**f**) Effect of Se concentration and Cu concentration.

Figure 5b illustrates the simultaneous effect of temperature and Se concentration on Se recovery at a constant level of Cu, 25 g/L, and H_2SO_4, 75 g/L. As the thermodynamic evaluation also confirms in Figure 5b, Se recovery has enhanced with Se concentration at a constant temperature. In this way, the selenious acid (H_2SeO_3) can be reduced based on an electrochemical reaction expressed in Equation (7) [40]:

$$H_2SeO_3 \text{ (aq)} + 2Cu^{2+} 4H^+ 8e^- = Cu_2Se(s) + 3H_2O \qquad (7)$$

According to the above reaction, Mokemeli [40] expressed that Stewart et al. [42] suggested a desirable effect of initial Se concentration for Se cementation by copper. Moreover, this behavior was confirmed in other studies [40], as the Se concentration order in the kinetic equation was specified between 1 and 1.8. In addition, Figure 5c shows the effects of temperature and Cu concentration on Se recovery. Temperature is a reluctant parameter in a lower Cu concentration, as shown in the thermodynamic survey, but enhancing Cu concentration promotes Cu_2Se phase formation.

Figure 5d exhibits the Se recovery as the function of sulfuric acid concentration and Se concentration at a certain temperature, 55 °C, and Cu concentration, 25 g/L. Based on Figure 5a,b, the higher concentration of H_2SO_4 diminishes Cu_2Se, while the higher Se concentration extends Se cementation. H_2SO_4 increases the copper dissolution affinity, decreasing Cu_2Se stability and recovery. Nevertheless, Se concentration can escalate the Se cementation reaction, Equation (8), and extend the Cu_2Se precipitation rate. Figure 5e shows the interaction between H_2SO_4 and Cu concentration that confirms a futile effect of Cu concentration on Se recovery because of Cu_2Se amount promotion, while H_2SO_4 diminishes Se recovery.

Moreover, Figure 5f illustrates the effect of Se and Cu concentration on Se recovery. As can be seen, the concentration of Se definitely increases the Se recovery percentage. In contrast, Cu concentration has a limited effect on the Se cementation efficiency.

The prime purpose of the work is to determine the functional condition of the cementation process. As observed, adjusting the different parameters could lead to the desired Se cementation by copper. Nevertheless, as mentioned in previous reports [12,34], tellurium is able to be precipitated by the copper cementation method. However, our results exhibit that tellurium slightly precipitates in this temperature range, and the extraction percentage is low. The main reason for restricted tellurium cementation is the instability of Cu^+ at a temperature lower than 75 °C [27], illustrated in the thermodynamic analysis. Nevertheless, Figure 5 was brought to explore the interaction between some parameters, e.g., initial Cu concentration, H_2SO_4 concentration, and temperature, on the Te precipitation process.

As seen in Figure 6a, both temperature and sulfuric acid concentration slightly increase Te precipitation, indicating the synergism effect of both factors. Regarding thermodynamic analysis, rising temperature is favorable for the endothermic cementation reaction leading to higher recovery. Moreover, Jennings et al. [12] expressed that the tellurium precipitation reaction happens at least 75 °C, and the process rate at a lower temperature is too slow. In another way, H_2SO_4 promotes Te cementation reaction to form Cu_2Te as follows:

$$H_2TeO_3 + 3Cu + H_2SO_4 = CuSO_4 + Cu_2Te + 3H_2O \qquad (8)$$

As can be observed, sulfuric acid leads to higher Se recovery, as Cooper [43] reported that at least 50 g/L of sulfuric acid is needed to accelerate the precipitation of tellurium. Figure 6b presents the effect of Cu concentration and temperature on Te extraction percent at 1000 mg/L Te and 75 g/L H_2SO_4. These data indicate that the influence of temperature on Te recovery is not desirable, whereas raising the level of Cu concentration hurts Te recovery. In the Te cementation process, the cupric species can reach an equilibrium between cuprous and Cu_2Te, which is expressed as follows [40]:

$$Cu_2Te + Cu^{2+} = 2Cu^+ + CuTe \qquad (9)$$

However, CuTe is an unstable component that can be disassociated from Cu_2Te to Te and Cu, decreasing Te extraction efficiency. However, by adding more Te to the solution, the detrimental effect of cupric ions is declined in the model, which may confirm the occurrence of the Cu_2Te dissociation. It should be noted that the interaction of other parameters is quite limited, so they were not discussed in the section to summarize the content.

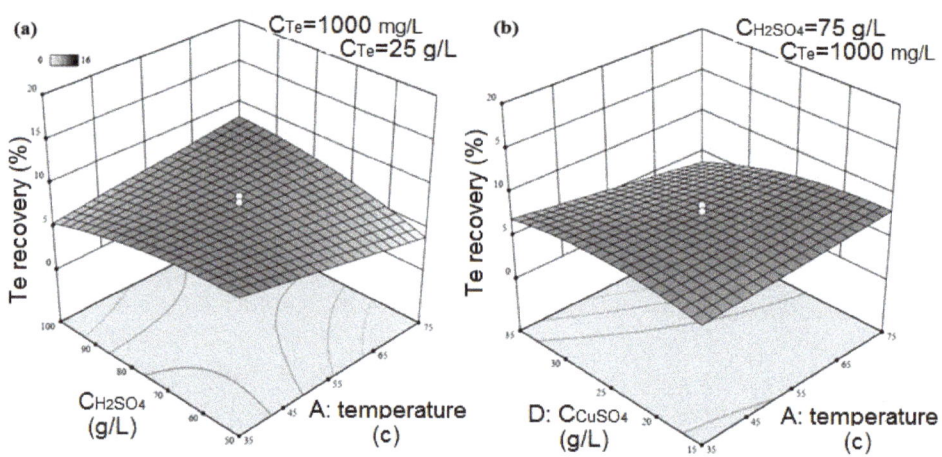

Figure 6. Response surface graphs for interactions of parameters of Te (IV) cementation by solid copper. (a) Effect of H_2SO_4 concentration and temperature; (b) Effect of Temperature and Cu concentration.

3.4. Validation of the Proposed Models

Additional experiments should confirm the Se (Equation (5)) and Te (Equation (6)) precipitation equation. Hence, three points, which have more Se extraction and less Te extraction, were chosen to validate the cementation equations in Table 6. The first row corresponds to the discrete precipitation of Se and Te, and the third row belongs to the dual extraction of both elements.

Table 6. Experiments for models' validations of Se and Te extraction.

T °C	H_2SO_4 g/L	Se mg/L	Te mg/L	Cu g/L	Predicted %E				Expe. %E		D		β
					Se	Std D.	Te	Std D.	Se	Te	Se	Te	
35	50	3000	-	15	97.88	2.21	-	-	97.21	-	34.8	-	1659
35	50	-	750	15	-	-	2.33	1.29	-	2.12	-	0.02	
35	50	3000	750	15	-	-	-	-	98.46	1.37	63.5	0.01	5291

As shown in Table 6, the extraction efficiency difference between the predicted model and experiment results is less than the standard deviations validating the achieved models in this work. Thus, the predicted value is plausible with the experimental outputs, which have less than 2% standard error. Moreover, the results of the first test, e.g., 35 °C, 50 g/L H_2SO_4, and 15 g/L Cu have values of separation factor greater than the two other ones, being more desirable for the separations process.

On the other hand, the selenium and tellurium extraction were carried out in co-presence, and the result was brought at the third line of the test. The separation indexes in the dual cementation process have been better improved than the discrete process. The thermodynamic evaluation [44] demonstrates that tellurium can reduce selenite anions according to Equation (10), leading to a more Se and Te separation index which is more desirable in separation processes.

$$SeO_3^{2-}{}_{aq} + Te_s = TeO_3^{2-}{}_{aq} + Se_S \quad \Delta G_{298}^0 = -66.86 \tag{10}$$

3.5. Se and Te Separation in Copper Anode Slime Leaching Solution

Liquor, obtained from copper anode slime, contains different impurities, Fe, Pd, Ag, As, Sb, and Pb, disturbing the separation process. It should be mentioned that the synthetic solution was prepared based on industrial conditions, and the optimum level of factors in Section 2 can be exploited for Se cementation in the copper anode slime liquor. Hence, the precipitation process for selenium or tellurium is carried out in the liquor, which has a chemical composition presented in Table 7.

Table 7. Analysis of pregnant solution according to optimal condition of Se/Te separation.

Elements	Cu (g/L)	Se (mg/L)	Te (mg/L)	As (mg/L)	Pb (mg/L)	Ag (mg/L)	Pd (mg/L)
concentration	15.05	2980	783	300	3.2	163	<1

The results are in Table 8 after 0.5, 1, 2, and 4 h, and the separation indexes are reported. The extraction efficiency at 30 min, 1, 2 and 4 h is 34.78, 76.304%, 95.480% and 97.304%, respectively. The co-extraction of impurities, such as As, may slightly diminish the selenium cementation by copper metal [45]. The outputs indicate that although the extraction percentage and separation index is diminished in copper anode slime liquor compared to the synthetic solution, the proposed process can still be efficient for Se and Te separation in industrial operations. Moreover, extending the process time can slightly enhance Se extraction at four hours, but the co-extraction of tellurium restricts the separation index.

Table 8. Selenium and tellurium cementation by 20 g/L Cu chop.

Time (h)	Se		Te		As		$\beta_{Se,Te}$
	%E	D	%E	D	%E	D	
0.5	34.782	0.5346	1.212	0.0125	1.88	0.019	42.768
1	76.304	3.2218	2.224	0.0258	2.8	0.028	124.876
2	95.480	21.12	3.366	0.0348	3.25	0.034	606.896
4	97.304	25.479	11.422	0.1290	5.1	0.054	197.511

Moreover, if results obtained from copper anode slime leaching are compared with results from the statistical model, e.g., Equations (5) and (6), we will conclude that the presented models can predict the range of Se or Te recovery percent in the cementation process. Thus, these models can be useful for a practical process design on a pilot or industrial scale.

3.6. Characterization of the Process Sediments

X-ray diffraction (XRD) is one of the technics that could provide useful data about the sediments of the process. The XRD pattern for sediments was obtained according to optimum conditions, expressed in Table 6, which are presented in Figure 7. As observed, $Cu_{1.8}Se$ and Cu_2Se are the dominant phases in the sediment that approves Se has been cemented in the system, whereas tellurium phases are not detectible in the condition. Additionally, thermodynamic assessments, Figure 2, present CuSe and $CuSe_2$ as the equilibrium phases in the Se-Cu-H2O system, and the blend of these phases are represented as $Cu_{1.8}Se$.

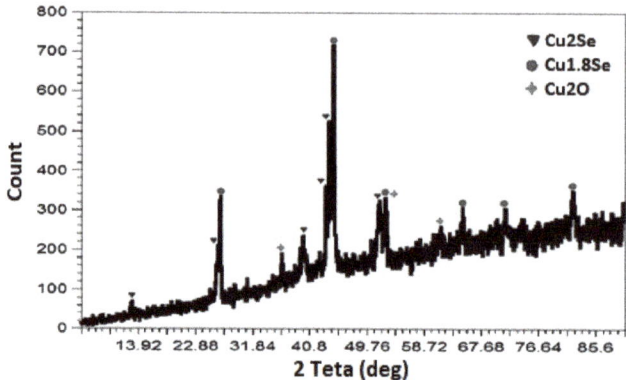

Figure 7. XRD pattern of sediments for dual cementation of Se and Te at 4000 mg/L Se, 1000 g/L Te, 75 g/L H_2SO_4 and 15 g/L $CuSO_4$ and 35 °C.

Furthermore, the XRD patterns for the discrete experiment of Se and Te cementation were brought in Figure 8a,b, respectively. Although selenium phases are $Cu_{1.8}Se$ and Cu_2Se, as recognized in Figure 8a, tellurium is not found in the XRD histogram. In addition, pure copper and Cu_2O are recognized as prime components in the Te cementation sediments. As shown in Table 6, Te recovery percent is less than 2.5%, which is too weak alongside staple Cu phase peaks.

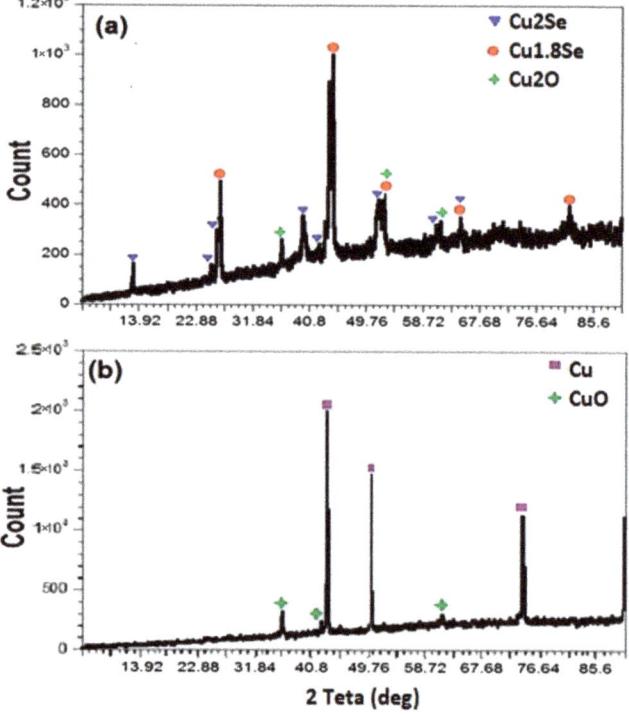

Figure 8. XRD pattern for discrete cementation of (**a**) Se at 4000 mg/L Se, 75 g/L H_2SO_4 and 15 g/L $CuSO_4$ and 35 °C and (**b**) Te at 1000 g/L Te, 75 g/L H_2SO_4 and 15 g/L Cu SO_4 and 35 °C.

4. Conclusions

The cementation of Se and Te by copper metal was surveyed using response surface methodology (RSM) as a tool for experiment design and thermodynamic analysis. The results presented that copper sulfate concentration and temperature diminished Se extraction percent in the 5–45 g/L Cu and 15–95 °C range, while temperature and sulfuric acid can slightly increase Te extraction efficiency. The optimum condition is 35 °C, 50 g/L H_2SO_4, 3000 mg/L Se, 750 mg/L Te, and 15 g/L $CuSO_4$ in which the separation index (β) is 5291 in synthetic solution and 606 in liquor of copper anode slime leaching. Although there is a significant difference between separation index (β) in synthetic and pregnant solutions, the presented models can specify the Se or Te recovery range in the sulfuric media. Moreover, the separation indexes demonstrate that the proposed method can efficiently separate these elements, e.g., Se and Te. Moreover, the XRD patterns approve copper selenide formation in the sediments. In contrast, a negligible amount of Te is extracted in the sulfate solution. Finally, a practical process from copper anode slime has been proposed via the copper cementation process.

Author Contributions: S.H., investigation, methodology, chemical, formal analysis and data curation, funding acquisition, writing the original draft; E.K.A., supervision, conceptualization, methodology, data curation, review and editing; N.S., supervision, conceptualization, methodology, data curation, review and editing. All authors have read and agreed to the published version of the manuscript.

Funding: This research received no external funding.

Data Availability Statement: Restrictions apply to the availability of these data. Data were obtained from Amirkabir University of technology and are available from Eskandar Keshavarz Alamdari with the permission of Amirkabir University of Technology.

Acknowledgments: Administrative and technical support from Rafsanjan non-ferrous metals recycling company is gratefully acknowledged.

Conflicts of Interest: The authors declare no conflict of interest.

References

1. Panahi-Kalamuei, M.; Salavati-Niasari, M.; Hosseinpour-Mashkani, S.M. Facile microwave synthesis, characterization, and solar cell application of selenium nanoparticles. *J. Alloys Compd.* **2014**, *617*, 627–632. [CrossRef]
2. Champness, C.H.; Chan, A. Relation between barrier height and work function in contacts to selenium. *J. Appl. Phys.* **1985**, *57*, 4823–4825. [CrossRef]
3. Khurana, A.; Tekula, S.; Saifi, M.A.; Venkatesh, P.; Godugu, C. Therapeutic applications of selenium nanoparticles. *Biomed. Pharmacother.* **2019**, *111*, 802–812. [CrossRef] [PubMed]
4. Hoffmann, J.E.; King, M.G. Selenium and selenium compounds. In *Kirk-Othmer Encyclopedia of Chemical Technology*; J. Wiley: Hoboken, NJ, USA, 2000; pp. 1–36. [CrossRef]
5. Naumov, A.V. Selenium and tellurium: State of the markets, the crisis, and its consequences. *Metallurgist* **2010**, *54*, 197. [CrossRef]
6. Adnan, M. Application of Selenium a Useful Way to Mitigate Drought Stress: A Review. *Open Access J. Biog. Sci. Res.* **2020**, *3*, 39. [CrossRef]
7. Knockaert, G. Tellurium and Tellurium Compounds. In *Ullmann's Encyclopedia of Industrial Chemistry*; J. Wiley: Hoboken, NJ, USA, 2000. [CrossRef]
8. Capper, P.; Garland, J.; Kasap, S.; Willoughby, A. *Mercury Cadmium Telluride: Growth, Properties and Applications*; J. Wiley: Hoboken, NJ, USA, 2011.
9. Surai, P.F.; Taylor-Pickard, J.A. *Current Advances in Selenium Research and Applications*; Wageningen Academic Publishers: Wageningen, The Netherlands, 2008; Volume 1.
10. Lu, D.-K.; Chang, Y.-F.; Yang, H.-Y.; Xie, F. Sequential removal of selenium and tellurium from copper anode slime with high nickel content. *Trans. Nonferrous Met. Soc. China* **2015**, *25*, 1307–1314. [CrossRef]
11. Nassar, N.T.; Graedel, T.E.; Harper, E.M. By-product metals are technologically essential but have problematic supply. *Sci. Adv.* **2015**, *1*, e1400180. [CrossRef]
12. Jennings, P.H.; Themelis, N.J.; Stratigakos, E.S. A continuous-flow reactor for the precipitation of tellurium. *Can. Metall. Q.* **1969**, *8*, 281–286. [CrossRef]
13. Li, Z.; Deng, J.; Liu, D.; Jiang, W.; Zha, G.; Huang, D.; Deng, P.; Li, B. Waste-free separation and recovery of copper telluride slag by directional sulfidation-vacuum distillation. *J. Clean. Prod.* **2022**, *335*, 130356. [CrossRef]
14. Hoffmann, J.E. Recovering selenium and tellurium from copper refinery slimes. *Jom* **1989**, *41*, 33–38. [CrossRef]

15. Sadeghi, N.; Alamdari, E.K. Selective extraction of gold (III) from hydrochloric acid-chlorine gas leach solutions of copper anode slime by tri-butyl phosphate (TBP). *Trans. Nonferrous Met. Soc. China* **2016**, *26*, 3258–3265. [CrossRef]
16. Xu, Z.; Guo, X.; Li, D.; Tian, Q. Leaching kinetics of tellurium-bearing materials in alkaline sulfide solutions. *Miner. Process. Extr. Metall. Rev.* **2020**, *41*, 1–10. [CrossRef]
17. Saeedi, M.; Sadeghi, N.; Alamdari, E.K. Modeling of Au Chlorination Leaching Kinetics from Copper Anode Slime. *Min. Metall. Explor.* **2021**, *38*, 2559–2568. [CrossRef]
18. Fan, J.; Wang, G.; Li, Q.; Yang, H.; Xu, S.; Zhang, J.; Chen, J.; Wang, R. Extraction of tellurium and high purity bismuth from processing residue of zinc anode slime by sulfation roasting-leaching-electrodeposition process. *Hydrometallurgy* **2020**, *194*, 105348. [CrossRef]
19. Rao, S.; Liu, Y.; Wang, D.; Cao, H.; Zhu, W.; Yang, R.; Duan, L.; Liu, Z. Pressure leaching of selenium and tellurium from scrap copper anode slimes in sulfuric acid-oxygen media. *J. Clean. Prod.* **2021**, *278*, 123989. [CrossRef]
20. Shi, G.; Liao, Y.; Su, B.; Zhang, Y.; Wang, W.; Xi, J. Kinetics of copper extraction from copper smelting slag by pressure oxidative leaching with sulfuric acid. *Sep. Purif. Technol.* **2020**, *241*, 116699. [CrossRef]
21. Kurniawan, K.; Lee, J.-C.; Kim, J.; Kim, R.; Kim, S. Leaching Kinetics of Selenium, Tellurium and Silver from Copper Anode Slime by Sulfuric Acid Leaching in the Presence of Manganese (IV) Oxide and Graphite. *Mater. Proc.* **2021**, *3*, 16.
22. Matsuo, N.; Oshima, T.; Ohe, K.; Otsuki, N. Extraction behavior of arsenic, selenium, and antimony using cyclopentyl methyl ether from acidic chloride media. *Solvent Extr. Res. Dev. Jpn.* **2019**, *26*, 81–89. [CrossRef]
23. Sattari, A.; Kavousi, M.; Alamdari, E.K. Solvent Extraction of Selenium in Hydrochloric Acid Media by Using Triisobutyl Phosphate and Triisobutyl Phosphate/Dodecanol Mixture. *Trans. Indian Inst. Met.* **2017**, *70*, 1103–1109. [CrossRef]
24. Chowdhury, M.R.; Sanyal, S.K. Separation by solvent extraction of tellurium (IV) and selenium (IV) with tri-n butyl phosphate: Some mechanistic aspects. *Hydrometallurgy* **1993**, *32*, 189–200. [CrossRef]
25. Mhaske, A.A.; Dhadke, P.M. Separation of Te (IV) and Se (IV) by extraction with Cyanex 925. *Sep. Sci. Technol.* **2003**, *38*, 3575–3589. [CrossRef]
26. Havezov, I.; Jordanov, N. Separation of tellurium (IV) by solvent extraction methods. *Talanta* **1974**, *21*, 1013–1024. [CrossRef]
27. Mokmeli, M.; Dreisinger, D.; Wassink, B. Modeling of selenium and tellurium removal from copper electrowinning solution. *Hydrometallurgy* **2015**, *153*, 12–20. [CrossRef]
28. Mohammadi, M. Selenium Removal from Waste Waters by Chemical Reduction with Chromous Ions. Ph.D. Thesis, University of British Columbia, Vancouver, BC, Canada, 2019.
29. Zhang, F.-Y.; Zheng, Y.-J.; Peng, G.-M. Selection of reductants for extracting selenium and tellurium from degoldized solution of copper anode slimes. *Trans. Nonferrous Met. Soc. China* **2017**, *27*, 917–924. [CrossRef]
30. Bello, Y.O. Tellurium and Selenium Precipitation from Copper Sulphate Solutions. Master's Thesis, Stellenbosch University, Stellenbosch, South Africa, 2014.
31. Anderson, M.D.; Thomas, T.R. *Separation of Tellurium and Iodine from Other Fission Products: Application to Loft Samples*; Atomic Energy Div., Phillips Petroleum Co.: Idaho Falls, ID, USA, 1965.
32. Guo, X.; Xu, Z.; Li, D.; Tian, Q.; Xu, R.; Zhang, Z. Recovery of tellurium from high tellurium-bearing materials by alkaline sulfide leaching followed by sodium sulfite precipitation. *Hydrometallurgy* **2017**, *171*, 355–361. [CrossRef]
33. Hashemi, M.; Mousavi, S.M.; Razavi, S.H.; Shojaosadati, S.A. Comparison of submerged and solid state fermentation systems effects on the catalytic activity of Bacillus sp. KR-8104 α-amylase at different pH and temperatures. *Ind. Crops Prod.* **2013**, *43*, 661–667. [CrossRef]
34. Shibasaki, T.; Abe, K.; Takeuchi, H. Recovery of tellurium from decopperizing leach solution of copper refinery slimes by a fixed bed reactor. *Hydrometallurgy* **1992**, *29*, 399–412. [CrossRef]
35. Wang, S.; Wesstrom, B.; Fernandez, J. A novel process for recovery of Te and Se from copper slimes autoclave leach solution. *J. Miner. Mater. Charact. Eng.* **2003**, *2*, 53–64. [CrossRef]
36. Fjodorova, N.; Novič, M. Searching for optimal setting conditions in technological processes using parametric estimation models and neural network mapping approach: A tutorial. *Anal. Chim. Acta* **2015**, *891*, 90–100. [CrossRef]
37. Farzam, S.; Feyzi, F. Response surface methodology applied to extraction optimization of gold (III) by combination of imidazolium-based ionic liquid and 1-octanol from hydrochloric acid. *Sep. Sci. Technol.* **2020**, *55*, 1133–1145. [CrossRef]
38. Khuri, A.I.; Mukhopadhyay, S. Response surface methodology. *WIREs Comput. Stat.* **2010**, *2*, 128–149. [CrossRef]
39. GTT-Technologies; Bale, C.; Chartrand, P.; Harvey, J.P.; Pelton, A.; Decterov, S.; Robelin, C.; Gheribi, A.; Jin, L.; Bélisle, E.; et al. *FactSage*; CRTC & GTT: Montreal, QC, Canada, 2007.
40. Mokmeli, M. Kinetics of Selenium and Tellurium Removal with Cuprous Ion from Copper Sulfate-Sulfuric Acid Solution. Ph.D. Thesis, University of British Columbia, Vancouver, BC, Canada, 2014.
41. Sereshti, H.; Khojeh, V.; Samadi, S. Optimization of dispersive liquid-liquid microextraction coupled with inductively coupled plasma-optical emission spectrometry with the aid of experimental design for simultaneous determination of heavy metals in natural waters. *Talanta* **2011**, *83*, 885–890. [CrossRef] [PubMed]
42. Stewart, D.A.; Tyroler, P.; Stupavsky, S. The Removal of selenium and tellurium from copper electrolyte at INCO's copper refinery electrowinning department. In Proceedings of the 15th Annual Hydrometallurgical Meeting, Vancouver, BC, Canada, 11–14 August 1985; pp. 18–22.
43. Cooper, W.C. *Tellurium*; Van Nostrand Reinhold Company: New York, NY, USA, 1971.

44. McPhail, D.C. Thermodynamic properties of aqueous tellurium species between 25 and 350. *Geochim. Cosmochim. Acta* **1995**, *59*, 851–866. [CrossRef]
45. Wu, L.-K.; Xia, J.; Zhang, Y.-F.; Li, Y.-Y.; Cao, H.-Z.; Zheng, G.-Q. Effective cementation and removal of arsenic with copper powder in a hydrochloric acid system. *RSC Adv.* **2016**, *6*, 70832–70841. [CrossRef]

Article

Selective Disintegration–Milling to Obtain Metal-Rich Particle Fractions from E-Waste

Ervins Blumbergs [1,2,3], Vera Serga [4], Andrei Shishkin [3,5], Dmitri Goljandin [6], Andrej Shishko [3], Vjaceslavs Zemcenkovs [3,5], Karlis Markus [3,7], Janis Baronins [3,8,*] and Vladimir Pankratov [9]

1. Institute of Physics, University of Latvia, 32 Miera Street, LV-2169 Salaspils, Latvia
2. Faculty of Civil Engineering, Riga Technical University, 21/1 Azenes Street, LV-1048 Riga, Latvia
3. ZTF Aerkom SIA, 32 Miera Street, LV-2169 Salaspils, Latvia
4. Institute of Materials and Surface Engineering, Faculty of Materials Science and Applied Chemistry, Riga Technical University, P. Valdena Street 3/7, LV-1048 Riga, Latvia
5. Rudolfs Cimdins Riga Biomaterials Innovations and Development Centre of RTU, Institute of General Chemical Engineering, Faculty of Materials Science and Applied Chemistry, Riga Technical University, 3 Pulka Street, LV-1007 Riga, Latvia
6. Department of Mechanical and Industrial Engineering, Tallinn University of Technology, Ehitajate Tee 5, 19086 Tallinn, Estonia
7. Life Sciences and Technologies Department, Latvia of University, Svetes Street, LV-3001 Jelgava, Latvia
8. Latvian Maritime Academy, Flotes Street 12 k-1, LV-1016 Riga, Latvia
9. Institute of Solid State Physics, University of Latvia, 8 Kengaraga Street, LV-1063 Riga, Latvia
* Correspondence: janis.baronins@gmail.com

Abstract: Various metals and semiconductors containing printed circuit boards (PCBs) are abundant in any electronic device equipped with controlling and computing features. These devices inevitably constitute e-waste after the end of service life. The typical construction of PCBs includes mechanically and chemically resistive materials, which significantly reduce the reaction rate or even avoid accessing chemical reagents (dissolvents) to target metals. Additionally, the presence of relatively reactive polymers and compounds from PCBs requires high energy consumption and reactive supply due to the formation of undesirable and sometimes environmentally hazardous reaction products. Preliminarily milling PCBs into powder is a promising method for increasing the reaction rate and avoiding liquid and gaseous emissions. Unfortunately, current state-of-the-art milling methods also lead to the presence of significantly more reactive polymers still adhered to milled target metal particles. This paper aims to find a novel and double-step disintegration–milling approach that can provide the formation of metal-rich particle size fractions. The morphology, particle fraction sizes, bulk density, and metal content in produced particles were measured and compared. Research results show the highest bulk density (up to 6.8 g·cm^{-3}) and total metal content (up to 95.2 wt.%) in finest sieved fractions after the one-step milling of PCBs. Therefore, about half of the tested metallic element concentrations are higher in the one-step milled specimen and with lower adhered plastics concentrations than in double-step milled samples.

Keywords: disintegration; e-waste; e-waste mechanical pretreatment; e-waste milling; precious metals; printed circuit boards

Citation: Blumbergs, E.; Serga, V.; Shishkin, A.; Goljandin, D.; Shishko, A.; Zemcenkovs, V.; Markus, K.; Baronins, J.; Pankratov, V. Selective Disintegration–Milling to Obtain Metal-Rich Particle Fractions from E-Waste. *Metals* **2022**, *12*, 1468. https://doi.org/10.3390/met12091468

Academic Editors: Lijun Wang and Shiyuan Liu

Received: 21 July 2022
Accepted: 26 August 2022
Published: 1 September 2022

Publisher's Note: MDPI stays neutral with regard to jurisdictional claims in published maps and institutional affiliations.

Copyright: © 2022 by the authors. Licensee MDPI, Basel, Switzerland. This article is an open access article distributed under the terms and conditions of the Creative Commons Attribution (CC BY) license (https:// creativecommons.org/licenses/by/ 4.0/).

1. Introduction

Recyclability and reusability of the materials are highly relevant to modern trends and manufacturing technologies [1]. All industries must reduce any waste significantly by implementing a computer-controlled manufacturing approach. These technologies radically minimize waste by reusing powders and filaments [2]. These benefits decrease manufacturing costs from micron-sized equipment manufacturing up to large-volume industries such as mining [3], shipbuilding [4], and civil engineering [5,6]. However, manufacturers typically produce electronic products and components from ecologically unfriendly materials [7,8]. Therefore, researchers meet the high demand for a novel technique to ensure rapid and cost-effective recycling of electronic waste (e-waste).

About 53.6 Mt of e-waste was generated in 2019, as reported by the Global E-waste Monitor 2020. From these, waste printed circuit boards (PCBs) represent the most economically attractive portion and account for about 3% of the total e-waste [9]. Therefore, PCBs recycling is a business opportunity with a high potential to obtain revenue from growth in the extraction and reuse of precious and base metals, such as gold, silver, and copper [10]. In addition, PCBs contain one of the highest concentrations of rare and precious metals (RPM). Therefore, such waste has the potential to become a sustainable source of RPM for the manufacturing of future generation electronic equipment [11].

The traditional treatment of PCBs includes cutting processes with the help of cutting/shredder mills or a combination of low-intensity impacts, shear, and abrasion with hammermills. However, both methods have significant drawbacks [12,13]. The composition of PCB is complex, wear-resistant, and creates highly abrasive particles during cutting processes. Cutting blades passes through all layers of the PCB composites, which consist not only of epoxy resins and fiberglass but also of robust and malleable metals and alloys, as well as ceramics. The presence of such components leads to high wear of the cutting edges [14]. Worn equipment reduces the efficiency of separating PCB components and requires costly maintenance. Reducing the size of the pieces does not change the structure of the composite, which remains predominantly solid/unbroken [15] with a relatively small area of uncovered precious metals.

As opposed to traditional methods, the high-intensity impact generates high stresses in the structure of PCBs. In addition, these impacts destroy bonding between adhered layers such as resin and fiberglass (less mechanically resistant materials). Additionally, the rapid release and uncovering of the metallic fraction (MF) and non-metallic fraction (NMF) phases [16] occurs. Therefore, the extractor can achieve more efficient further separation by releasing high-quality metal concentrate or exposing a large surface area for the increased possibility/acceleration of chemical reactions [17]. Thus, impact selectivity can achieve a high fragmentation level and becomes the main factor for the mechanical enrichment of target metals. Furthermore, such selective disassembly is less energy consuming and saves the remaining components of the PCB composite from excessive grinding and conversion into technological emissions [18].

PCBs are complex composite materials that consist directly of a multilayer PCB plate, solder, and PCB components [19]. The PCB's plate generally consists of three layers that are heat laminated together into a single layer. Typically, these are silkscreen, solder mask, copper, and substrate [20]. PCB components are a general term for various components, such as capacitors, resistors, transistors, and other electronic devices. These components include connectors, contacts, fasteners, and many other components attached and connected to a PCB [21].

The typical substrate of the PCB is made of fiberglass and is also known as FR4 (letters FR mean "fire-retardant"). FR-4 glass epoxy is a popular and versatile high-pressure thermoset plastic laminate with an excellent strength-to-weight ratio. This substrate layer provides a solid base for PCBs. However, the thickness may vary depending on the expected PCB's application and service conditions. The standard thickness of four-layer boards is about 1.6 mm [22].

The second layer of PCB is copper, typically laminated onto the substrate by supplying heat to the adhesive. The copper layer is relatively thin to ensure high electrical conductivity with the lowest possible heat generation. Several boards contain the sandwich of two copper layers on opposite sides of the substrate. Manufacturers usually produce cheaper electronic devices of single copper layered PCBs. The standard level of copper thickness on plane layers is about 35 μm [19].

The PCB's solder mask provides the visually observable green color. However, sometimes solder masks are designed to give the appearance in other colors, such as brown, red, or blue. The solder mask is also known as a liquid photo imageable solder mask [23]. The solder mask's purpose is to prevent molten leakage [19]. The metal that facilitates the transfer of current between the board and any attached components is solder, which also serves a dual purpose due to its adhesive properties [24].

The main task for the mechanical processing of PCBs is the separation of PCBs into constituent elements by following or immediate isolation from each other. Therefore, destroying the mechanical bonds between these elements is essential to increase the separation process efficiency [25]. This separation is required to create concentrates of metals and non-metals to facilitate the access of reagents to the exposed surfaces. The most significant and protruding parts (metal and ceramic components) are primarily exposed to abrasion and shear. This effect results in the waste of additional energy and material from the wear of grinding media. This factor accompanies the unnecessary grinding of metal and ceramic components and materials' ineffective mixing/heating [26]. One of the most widely used e-waste recycling methods is pyrolysis. However, high brominated antipyrine concentration causing a release of toxic gases is the method's main disadvantage [27]. Currently, manufacturers add the compound PCB to reduce the flammability of computer components in case of fire [28].

PCBs in common could contain up to 30 wt.% polymers, 30 wt.% ceramics, and 40 wt.% metals [29]. Detailed composition by fractions was investigated by Roberto et al. and is represented in Table 1 [30].

Table 1. Typical PCB composition attributed to the general groups of materials, data from [30].

Metals (~40 wt.% in Total)	wt.% of Metals	Ceramics (~30 wt.% in Total)	wt.% of Ceramics	Plastics (~30 wt.% in Total)	wt.% of Plastics
Cu	6–27	SiO_2	15–30	Polyethylene	10–16
Fe	1.2–8	Al_2O_3	6–9.4	Polypropylene	4.8
Al	2–7.2	Alkali-earth oxides	6	Polystyrene	4.8
Sn	1–5.6	Titanates-micas	3	Epoxy resin	4.8
Pb	1–4.2			Polyvinyl chloride	2.4
Ni	0.3–5.4			Polytetrafluoroethylene	2.4
Zn	0.2–2.2			Nylon	0.9
Sb	0.1–0.4				
Au (ppm)	250–2050				
Ag (ppm)	110–4500				
Pd (ppm)	50–4000				
Pt (ppm)	5–30				
Co (ppm)	1–4000				

However, certain PCB elements contain several precious metals at much higher concentrations. For example, the content of Au, Ag, and Pd in the contact group, connection slots, interfaces, and the board surface range from 180 to 3695 mg·kg^{-1}, from 809 to 12,321 mg·kg^{-1}, and from 96 to 118 mg·kg^{-1}, respectively [31]. As shown above, PCBs contain a significant concentration of valuable and expensive metals. However, along with them, a typical PCB contains up to 70 wt.% non-metallic components made of ceramics, plastics, and fiberglass. These components are part of the textolite. Therefore, developing an efficient method for the preliminary separation of these components is necessary. In addition, material recyclers are interested in preventing the formation of large amounts of liquid or gaseous phases during waste pyrolysis.

Industry experts frequently research and implement new methods for more efficient PCB pretreatment for valuable metals extraction. Reviewed literature shows a general division of PCB pretreatment by mechanical and solvent-based methods [27,32–34].

Y. Zhou and K. Qiu, in their publication, have reported a new process of "centrifugal separation +vacuum pyrolysis" to recover solder and organic materials from wasted PCBs [34]. This approach has exhibited the relatively complete separation of solder from PCBs with the help of centrifugal equipment, heated at 240 °C, and the rotating drum set at 1400 rpm for 6 min intermittently. The results of vacuum pyrolysis showed that the PCB without solder pyrolyzed to form an average of 69.5 wt.% solid residue, 27.8 wt.%

oil, and 2.7 wt% gaseous phase [34]. This method effectively separates microchips and other functional elements from PCB by removing the solder. However, researchers have admitted that multilayer PCBs avoid complete Cu extraction by further processing steps. Additionally, the pyrolysis process generates significant volumes of toxic organic gases.

M. Tatariants et al. have described a ball milling process to produce a powder exhibiting high fineness from the crushed PCBs [32]. The authors have set the ball mill's frequency at 20 Hz for 60 min. Such an approach resulted in micro-scaled PCBs powder. Obtained powder consisted of three phases: metal particles with adhered fragments of epoxy resin, fiberglass particles partially covered with epoxy resin, and fiberglass–metal–epoxy composite agglomerates.

Researchers have reported the use of organic solvents, such as dimethylformamide [32], dimethyl sulfoxide (DMSO), and N, N-dimethyl pyrrolidone [33] as efficient tools for brominated epoxy resins removal from PCB's structure. However, these methods also require other subsequential treatment methods, such as milling, air-flow separation, etc.

Size reduction by disintegration–milling is one of the state-of-the-art PCBs mechanical pretreatment approaches. Various solutions can be mentioned as effective analogues, which still demand improvements or alternative solutions to achieve reliable pretreatment result from e-wastes with variable compositions.

A 3.25 mm fraction particle production from PCBs with the help of a rotary cutting shredder and a subsequent three-stage grinding process in the ceramic ball mill allows for manufacturing particles with sizes down to 125 μm [35]. However, the two-stage PCBs crushing into a rotary cutting shredder down to 3.35 mm, then size reduction down to about 1 mm in a four-bladed rotary cutting shredder, and final grounding with the help of an ultra-centrifugal mill (Retsch ZM 200, Retsch GmbH, Haan, Germany,) allowed for the production of particles with sizes up to 250 μm prior to use for leaching tests [36]. At the same time, the PCB hammer-crushing and grounding with the help of an ultra-centrifugal mill (Retsch ZM 200) provides the production of particles with fraction sizes of 4 mm–212 μm) [37]. Finally, the grounding of PCBs, preliminarily cut into 2 mm pieces, with the help of an LM1-M ring mill (LabTechnics Australia, Victoria, Australia), allowed for the production of particles sieved into <365, 365–500, and 500–750 μm with the help of the Retsch AS200 control sieve shaker (Retsch GmbH, Haan, Germany) [38]. All these methods characterize an inefficient multi-step milling approach with high-energy and human workload consumption.

The present article aims to reveal the milling rate's effect on the properties of mechanically disintegrated PCBs. The idea is to apply mechanical disintegration as the pre-treatment method to increase the content of recovered valuable metals from specific fractions of disintegrated PCBs. The studied milling process provides the formation of a wide range of particle sizes. The work demonstrates the dependence of common and valuable metal contents on the size of obtained disintegration–milled particles. Therefore, entrepreneurs can use this approach in various manufacturing technologies for further processing the extracted valuable metals. Furthermore, recycling companies can scale up and implement the demonstrated technology in hydrometallurgical and pyrometallurgical processes.

2. Materials and Methods
2.1. Used Materials

The article's authors used PCB waste consisting of disassembled personal computer motherboards (produced by the GIGA-BYTE Technology Co., Ltd., New Taipei City, Taiwan, from 2010 to 2015) without central processors. An operator cut a total of 6 kg PCBs into rectangular-shaped pieces with side lengths from about three up to 6 cm and subjected them to disintegration–milling experiments, as shown in Figure 1a,b.

Disassembly was conducted manually at room temperature to avoid potentially harmful emissions [39].

Figure 1. As-received computer motherboard (**a**) and after being cut into pieces before milling (**b**).

2.2. Applied Milling Procedure and Used Testing Equipment

The high-energy semi-industrial disintegration–milling system DSL-350 (Tallinn University of Technology, Tallinn, Estonia), specially designed for processing mechanically durable materials [40–42], was used to grind fragments of PCBs into finer particles. The device is grinding materials by collisions. Supplied particles collide with surfaces of grinding bodies. As a result, the intensive pressure wave propagates inside the target particles. The resulting values of stresses exceed material strength. The specification of the disintegration–milling device is demonstrated in Table 2.

Table 2. Characteristics of the high-energy disintegration–mill DSL-350.

Parameter	Value
Type of device (position of rotors)	horizontal
Grinding environment	air
Rotor system	one/two-rotor
Number of pins/blades roads	1/3
Rotation velocity of rotors, rpm	2880
Impact velocity, m/s	up to 180
Specific energy of treatment ES, $kJ \cdot kg^{-1}$	up to 13.6
Possible operating system	direct
Input (max particle size), mm	45
Productivity, $kg \cdot h^{-1}$	up to 950

The principal scheme of milling equipment—centrifugal-type disintegrator mill DSL-350 is shown in Figure 2.

Collected PCBs were preliminarily cut into smaller fragments (see Figure 1b) to feed into the disintegration–milling device. Next, the author carried out targeted mechanical cutting (slicing) to avoid damage to the main elements on the surfaces of PCBs. Therefore, only the largest contact groups were cut in half, as demonstrated in Figure 1b. Next, the operator used the obtained pieces with intact elements to investigate the effect of one and double-step disintegration–milling on target metals contents. Finally, the authors selected the disintegration–milling procedure for producing powder from the PCBs, as presented in Figure 3.

An operator milled sliced PCBs (see Figure 1a) once as raw materials. Obtained particles smaller than 2.8 mm were subjected to metal analysis and designated as X1 (see Figure 3). Subsequently, particles bigger than 2.8 mm were subjected to repeated milling and designated as X1 (>2.8) + X2 (see Figure 3).

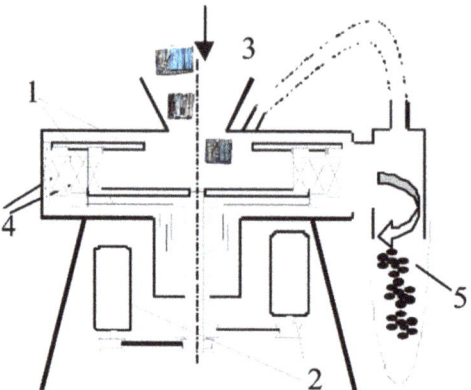

Figure 2. Schematic representation of preliminary size reduction centrifugal-type mill DSL-350. Equipment: 1—rotors; 2—electric drives; 3—material (PCBs) supply; 4—horizontally oriented grinding elements; 5—output.

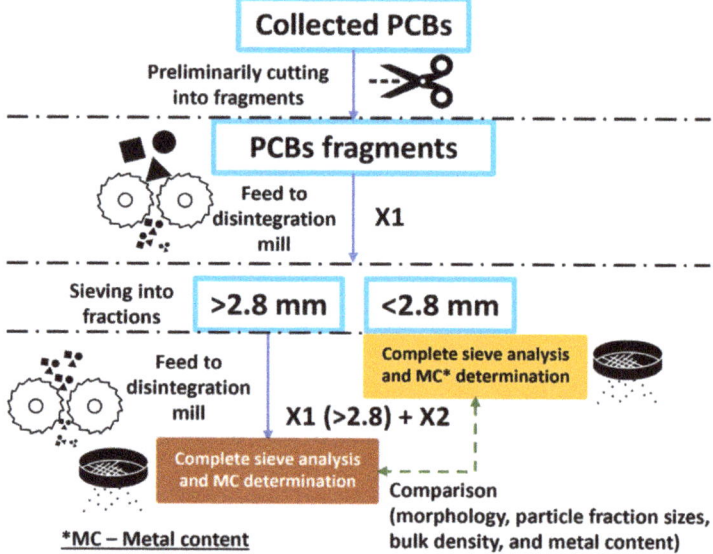

Figure 3. An applied disintegration–milling, sieving, and testing scheme show an approach for selecting fractions of powders from PCBs for metal content determination after one-step (X1) and double-step (X1 (>2.8) + X2) disintegration–milling.

Determination of particle size distribution was carried out with the help of the vibratory sieve shaker Analysette 3 PRO (FRITSCH GmbH Idar-Oberstein, Germany). Materials with particle sizes up to about 12.5 mm were fractioned using sieves with opening sizes of 0.09, 0.18, 0.35, 0.71, 1.40, 2.80, 5.60, and 11.20 mm. An operator measured the bulk density with the help of the bulk density tester (Scott volumeter, according to ASTM B 329-98, Copley, Nottingham, UK) for each fraction of sieved samples X1 and X1 (>2.8) + X2. An optical microscope (KEYENCE VHX-2000, Keyence Inc, Osaka, Japan) was used to study the morphology of the obtained fractions.

Metal content (MC) changes (in %) from one-step (X1, designated as $MC_{one-step}$) to double-step (X1 (>2.8) + X2, designated as $MC_{double-step}$) milling PCB by fractions calculated by Equation (1):

$$\frac{MC_{one-step} - MC_{double-step}}{MC_{double-step}} \cdot 100 \, (\%) \tag{1}$$

2.3. Applied Method for Metal Content Determination

Quantitative determination of MC in disintegration–milled and sieved fractions of raw material was performed with the help of inductively coupled plasma atomic emission spectroscopy (ICP-OES, Thermo scientific iCAP 7000 series, Thermo Fisher Scientific Inc., Waltham, MA, USA).

First, a representative sample was prepared using the quartering method, and chemical leaching was carried out. An aqueous solution of 6M HCl (V = 50 mL) was added to the powder sample's weight of 500 mg (±0.5 mg). The mixture was boiled until a wet residue formed. Subsequently, the same treatment approach was carried out in two portions of aqua regia (HCl:HNO$_3$ = 3:1, V = 40 mL). An excessive amount of HNO$_3$ was removed by adding concentrated HCl during the boiling process. The resulting wet residue was transferred to the filter with a 3M HCl solution and washed accurately. The resulting filtrate was brought up to a volume of 100 mL with 3M HCl solution. Afterward, the obtained solution was analyzed using ICP-OES (PerkinElmer Polska sp. z o.o., Kraków, Poland). To evaluate the metal content, it is necessary to determine the non-metallic component. In the sample under study, the filter's undissolved residue (non-metallic element) was washed with distilled water to pH ≈ 5–6, dried at 105 °C, and weighed.

3. Results and Discussion

3.1. Metal Content in Disintegration-Milled PCBs Fractions

The correlation between the determined MC and bulk densities in the milled PCB is presented in Table 3. The bulk density of specimen X1 gradually increases from 0.3 to 6.8 g·cm^{-3} by decreasing the sieved fraction size from 2.8–5.6 mm down to <0.09 mm. Measured bulk density (up to 6.8 g·cm^{-3}) correlates with an increase in MC (up to 95.2 wt.%), which indicates the positive effect of the brittleness of metals on disintegration–milling performance. The fractions from 0.09 to 0.35 mm can be attributed to high metal content with the most of ceramic impurities due to relatively high bulk densities from 4.9 to 6.8 g·cm^{-3}. At the same time, more elastic–plastic composites remain less intact and significantly reduce the densities of largest sieved fractions.

Table 3. Bulk densities and MC of one-step (X1) and double-step (X1 (>2.8) + X2) milled PCB fractions.

Fraction, mm	<0.09	0.09–0.18	0.18–0.35	0.35–0.711	0.711–1.4	1.4–2.8	2.8–5.6
Bulk density of X1, g·cm^{-3}	6.8	5.3	4.9	1.6	0.44	0.61	0.30
Bulk density of X1 (>2.8) + X2, g·cm^{-3}	2.23	0.84	0.81	0.53	0.58	0.35	
MC in X1, wt.%	95.2 ± 1.8	57.7 ± 1.5	54.6 ± 1.2	14.3 ± 0.9	8.4 ± 0.5		
MC in X1 (>2.8) + X2, wt.%	50.6 ± 1.3	26.2 ± 1.6	33 ± 0.9	8 ± 0.7	7.2 ± 0.6		

Double milling of residual fraction from specimen X1 with particle sizes above 2.8 mm significantly decreases MC by about 120% in specimen's X1 (>2.8) + X2 fraction of 0.09–0.18 mm, as compared to the same fraction's MC of the specimen X1. In addition, double milling also leads to about twice lower MC in case of each fraction from <0.09 (50.6 wt.%) to 0.711 mm (8 wt.%), as compared to the specimen X1. This result obviously demonstrates plastic and ceramic impurities, which also significantly lowers bulk densities more typical for most ceramics (2.23 g·cm^{-3}) and plastics with relatively high content of ceramic and metal impurities (0.35 to 0.84 g·cm^{-3}).

Therefore, higher concentrations of metals can be twice efficiently recovered from one-step milled fractions with sizes larger than 0.35 mm after the first disintegration–

milling procedure. In addition, it seems that the idea of combining the exact fraction sizes from samples X1 and X1 (>2.8) + X2 should be performed by preliminarily removing mechanically separated plastics from the double-step milled materials.

Therefore, most promising fractions of <0.09, 0.09–0.18, and 0.18–0.35 mm were chosen for the determination of Ag, Au, Pd, Pt, Al, Cd, Co, Cr, Cu, Fe, Mn, Mo, Ni, Pb, Sb, Sn, Ti, V, and Zn contents [43], as demonstrated in Figure 4a,b.

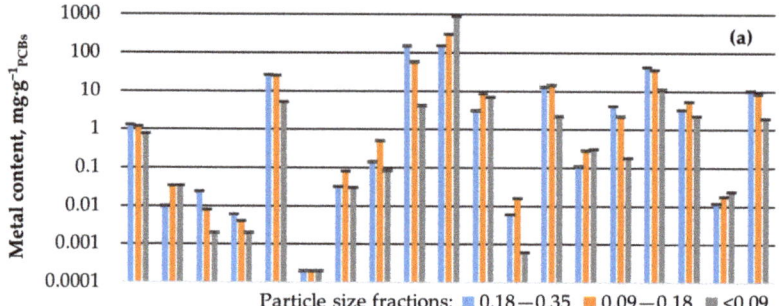

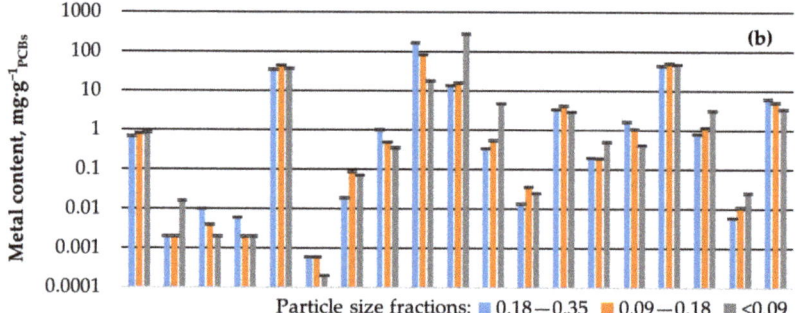

Figure 4. Dependence of element composition of milled fraction by particle size distribution for X1 (one-step) (**a**) and X1 (>2.8) + X2 (double-step for >2.5 mm fractions from first milling) (**b**) millings.

The results show that the MC varies depending on the stages of disintegration–milling and the fractional particle size. Plastic particles significantly decrease the fraction bulk density and strongly correspond to reducing the metal content in corresponded fraction.

Thus, sample X1 exhibit notably higher MC in Ag, Au, Fe, and Mn cases. Of these, recovering precious Au (from 0.01 to 0.034 mg·g^{-1}) is the most exciting and complicated process and requires chemicals that can be contaminated by side undesired side reactions. Therefore, the Au recovery from large volumes of one-step milled PCBs may provide higher economic feasibility than the recovery of the same metals from sample X1 (>2.8) + X2.

Conversely, sample X1 (>2.8) + X2 exhibits higher concentrations of Cd (from 0.0001 to 0.0006 mg·g^{-1}), Cr (from 0.35 to about 1 mg·g^{-1}), and Mo (from 0.0134 to 0.025 mg·g^{-1}), as compared to the sample X1. However, the beneficial recovery of these elements also requires large PCB volumes.

PCBs contain more than ten-fold purer precious metals compared to ore minerals, relatively rich with the same valuable metal atoms. Therefore, collecting PCBs to recover valuable metals is a crucial part of urban mining [43]. Additionally, the well-developed PCBs collection and recovery management can help avoid the leakage of environmentally harmful metals (e.g., Cr, Pb, Cd, etc.) into the soil, natural waters, and other places where oxidation into harmful compounds and absorption by living organisms may occur.

A better overview of the comparison between MC (in %) in one and double-step milled PCB samples by fractions (calculated according to Equation (1)) is demonstrated in color in Table 4. Results show the benefits of one-step milling. The total MC after one-step milling is higher in all fractions, from 46% (coarser fraction) up to 124% (finer fraction), as compared to the result of double-step milling. This result indicates the lower metal concentrations in the largest (>2.5 mm) fraction compared to tested fractions after the first milling step. Comparison of MC within factions exhibits significant benefit of the one-step milling approach in many cases of metals (gold, palladium, platinum, iron, manganese, nickel, titanium, vanadium, and zinc), indicating the increase in potentially recoverable MC by up to 1600% (precious gold) and 1770% (iron), compared to double-step milling. However, the double-step milling is more interesting (but with more minor percentage differences) for the recovery of aluminum, cadmium cobalt, chromium, copper, molybdenum, and antimony without or with the combination of disintegration milled PBC fractions after the one-step milling.

Table 4. MC changes (in %) from double (X1 (>2.8) + X2) to one-step (X1) milling PCB by fractions.

Fraction, mm	Metal Content in % after One-Step Milling in Comparison with Double-Step Milling																			Total MC *
	Ag	Au	Pd	Pt	Al	Cd	Co	Cr	Cu	Fe	Mn	Mo	Ni	Pb	Sb	Sn	Ti	V	Zn	
0.18–0.35	86	400	140	0	−25	−67	68	−86	−12	1003	798	−55	279	−45	144	−3	298	100	63	46
0.09–0.18	41	1600	100	100	−44	−67	−8	1	−31	1770	1548	−56	245	47	100	−30	357	64	70	119
<0.09	−12	113	0	0	−87	0	−58	−76	−77	207	45	−98	−26	−40	−57	−77	−31	−5	−45	124

Notes: * Total extracted MC by fraction. Green color (positive value) exhibits the benefit (higher metal content) of one-step milling in comparison with double-step milling approach. Yellow color (zero value) exhibits no effect of double-step milling in comparison to one-step milling. Red color (negative value) exhibits the benefit of double-step milling in comparison with one-step milling.

3.2. Morphology and PCBs Disintegration–Milling Dynamics

At the preliminary crushing stage by the disintegration–mill, large pieces of composite PCBs plates quickly disintegrated into parts due to fracture. Then, each separated component is crushed at its achieved linear speed. The morphology of sieved milled particles is represented in Figure 5.

Fractions larger than 2.8 mm were not analyzed in detail and were subjected to secondary milling designated as "X1 (>2.8) + X2". The fraction with particle sizes from 1.4 to 2.8 mm contained plastic particles (blue, grey, and black color) of equivalent sizes and some wire-like connections, as demonstrated in Figure 5a.

Fractions 0.711–1.40 mm (Figure 5b) and 0.355–0.711 mm (Figure 5c) also contained a significant concentration of plastic particles (blue, grey, and black color). However, the presence of visually observable metallic elements of PCBs was observed to be relatively peeled off from plastic pieces.

Fractions of 0.355–0.18 mm (Figure 5d) and 0.09–0.18 mm (Figure 5e) mainly exhibited fibers from fiberglass elements of PCB components instead of plastic chunks. Remarkably, the smaller the fraction, the lower the concentration of visually observed fibers, indicating the highest fracture resistance to the applied crushing forces inside the disintegration–mill compared to most of the other PCB components.

The X1 finer than 0.09 mm fraction consists primarily of non-plastic components with observable fiberglass, as shown in Figure 5f. Notably, the relatively large fiberglass particles can pass the narrow openings of the sieve mesh by orienting perpendicularly to the surface of the mesh.

Fractions 0.09—0.18 and <0.09 mm of X1 and X1 (>2.8) + X2 disintegration–milled samples visually (Figure 5e,f and Figure 6c,d, respectively) differ by the presence of higher fiber-like plastic particle concentration after one-step milling (X1).

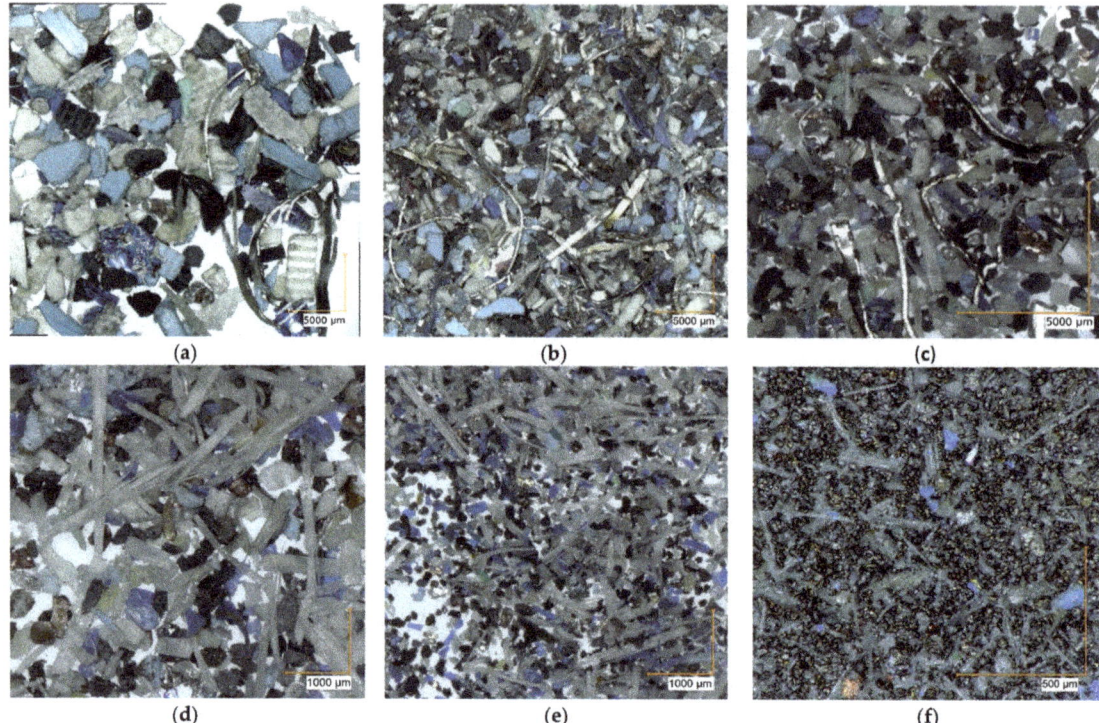

Figure 5. Optical images of the fractions 1.4–2.8 mm (**a**), 0.711–1.4 mm (**b**), 0.355–0.711 mm (**c**), 0.355–0.18 mm (**d**), 0.09–0.18 mm (**e**) and <0.09 mm with significantly lower fiberglass concentration (**f**) after one-step milling (X1).

The X1 sample had slightly higher total material amounts in sieved fractions of 2.8–5.6 (5.6 wt.%), 1.4–2.8 (0.1 wt.%), 0.711–1.4 (0.8 wt.%), 0.355–0.711 (3.9 wt.%), and <0.09 mm (0.6 wt.%), as compared to sample X1 (>2.8) + X2 (see Figure 7).

However, slightly higher total material amount in sieved X1 (>2.8) + X2 sample fractions of 0.18–0.35 (0.1 wt.%), 5.6–11.2 (11.1 wt.%), and >11.2 (1.7 wt.%) were measured. It seems more beneficial to perform total metal recovery from disintegration–milled particles with fraction sizes larger than about 1.4 mm. However, the presence of equivalent-sized plastic particles observed in Figures 5a and 6a should be removed prior to the chemical treatment process to reduce the consumption of chemically reactive and avoid harmful emissions by undesirable side reactions.

Glass fibers are poorly reactive to many applied reagents for chemical metal recovery. However, these fibers can be adhered to undesirable side reactions generating polymers. Unfortunately, many metal particles are still attached to plastic and fibrous elements; therefore, the accessibility of chemical reagents to free metals is also limited physically. Therefore, the total metal recovery feasibility of using the largest from disintegration–milled particles should be researched in detail. Particles with sizes under 1.4 mm can be chosen from specimen X1 for economically more feasible total metal recovery compared to the exact particle sizes of X1 (>2.8) + X2, especially in recycling large volumes of PCBs scraps. However, from a common point of view, it is also reasonable to combine these fractions from specimens X1 and X1 (>2.8) + X2. Additional separation based on the density or magnetic properties of the target material could be applied as a next step to remove the separated polymer from disintegration–milled powder and obtain a cleaner product in further studies.

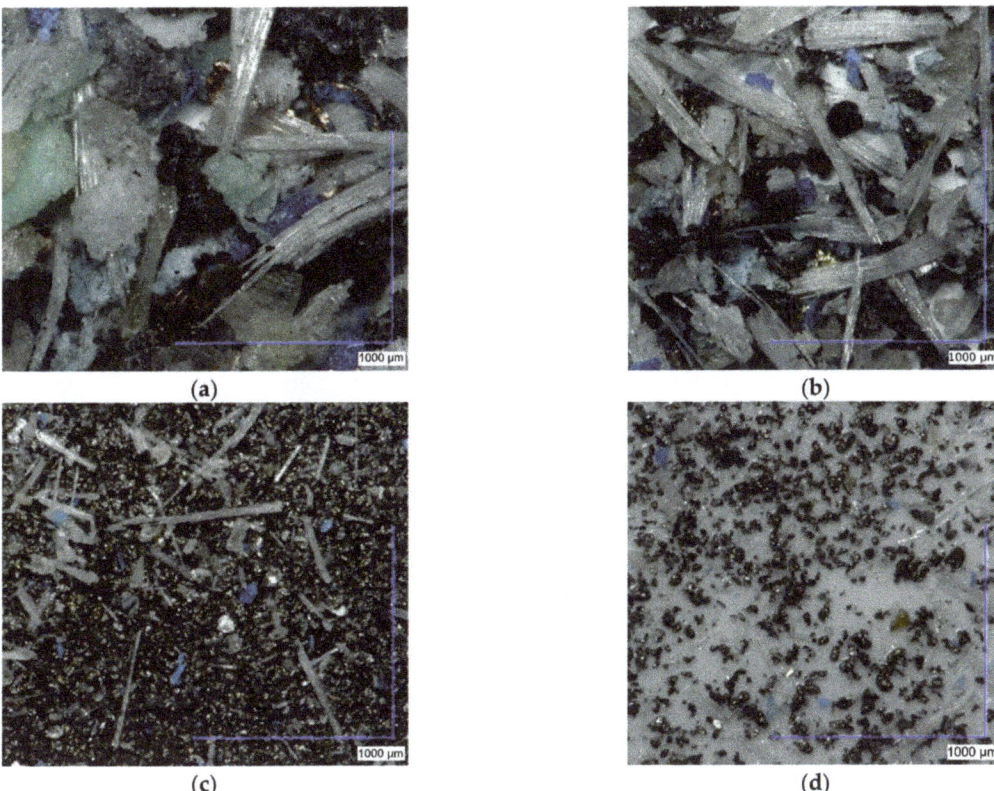

Figure 6. Optical images of the fractions 0.711–1.4 mm (**a**); 0.355–0.711 mm (**b**); 0.09–0.18 mm (**c**); and <0.09 mm (**d**) after double-step milling (X1 (>2.8) + X2) of fraction >2.8 mm collected from the one-step milling.

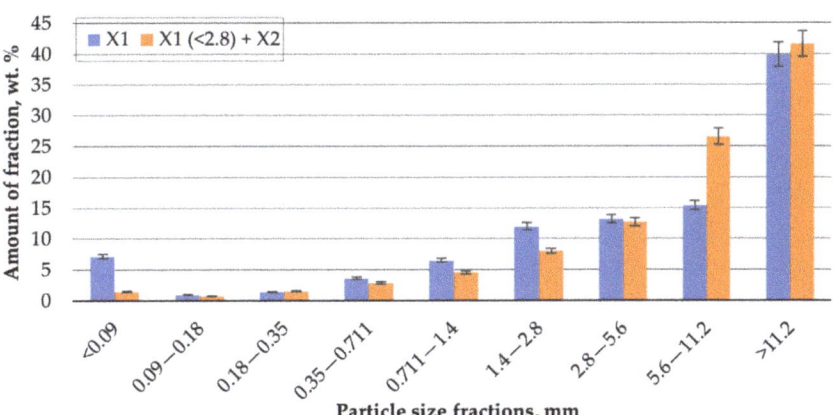

Figure 7. Particle size distribution of the disintegration–milled PCBs after first "X1" and second milling of the fraction with particle sizes larger than 2.8 mm from first milling designated as "X1 (>2.8) + X2".

The PCB is a multicomponent metal–plastic multilayer composite material with a complex structure consisting of brittle and plastic components. The mechanisms for reducing the particle size of the plastic and brittle materials are different. The collision of brittle materials with grinding elements results in a natural fracture. However, ductile metallic materials harden during every collision impact. Therefore, the fatigue fracture occurs [42] after two or more have implications inside the disintegration–milling working chamber. These impacts generate a large range of particle size PCBs fractions from <0.009 up to more than 11.2 mm during the one-step milling at 2880 rpm, as demonstrated in Figures 5 and 7.

More advanced disintegration mills have been equipped with separation systems based on aerodynamic force principles. Such systems employ a closed air or gas flow system (so-called inertial classifier) [44]. Such a system can provide ecologically clean disintegration–milling of PCBs into classified powder kinetic energy to give a material transition to collectors without support from additional mechanical transportation and gas blowing devices. Therefore, the required time for sieving into fractions and required energy consumption for plastics, ceramics, and metals separation from target metals by using liquids, gas flows, magnet and electromagnet, isostatic separators, etc., can be significantly reduced.

The performed significant plastic separation from metals with the help of disintegration–milling helps void the incineration stage. The incineration typically leads to highly toxic compounds such as volatile polybrominated dibenzo dioxins and dibenzofurans. Heavy metals can cause secondary pollution (e.g., Pb; see Table 3), and brominated flame retardants leach into groundwater. Therefore, environmentally safe recycling is an important topic for the researcher to satisfy community and local government requirements for greener urban mining.

4. Conclusions

Impact type one-step (X1) disintegration–milling is an effective way to crush PCBs into particles with high MC (from 8.4 to 95.2 wt.% by decreasing particle size fractions from 0.711–1.4 down to <0.09 mm, respectively) before chemical treatment. However, double-step (X1 (>2.8) + X2) treatment leads to a high content of polymers and glass fibers in resulting particle fractions with sizes from <0.09 to 5.6 mm with bulk densities from 2.23 down to 0.30 g·cm^{-3}, respectively. The MC also reduces to 7.2 wt.% (0.711–1.4 mm fraction) and 50.6 wt.% (<0.09 mm fraction), as compared to MC in the same fractions of X1 samples. Many visually observable (by optical microscope) metallic particles still adhere to plastic and fibrous elements after milling. These impurities require additional separation. The presence of polymer particles and fibers has no significant impact on MC compared to the result after the second milling of particles with sizes above 2.8 mm. One-step milling leads to particle fractions with sizes from <0.09 to 2.8 mm with about to six-fold higher bulk densities from 6.8 down to 0.44 g·cm^{-3}, respectively; and up to two-fold higher MC in a finer range of fractions from 0.09 to 0.35 mm, as compared to the result of double-step milling of remaining fractions with particle sizes above 2.8 mm. The approach for the removal of mechanically released particles from metallic elements should be researched in future studies.

Obtained tested particles from PCBs are relatively rich with Fe (up to 867 mg·g^{-1} in X1, <0.09 mm fraction); Cu (up to 148 mg·g^{-1} in X1, 0.18–0.35 mm fraction); Sn (up to 51.4 mg·g^{-1} in X1 (>2.8) + X2, 0.09–0.18 mm fraction); Al (up to 45.2 mg·g^{-1} in X1 (>2.8) + X2, 0.09–0.18 mm fraction); Ni (up to 14 mg·g^{-1} in X1, 0.09–0.18 mm fraction); and Zn (up to 10 mg·g^{-1} in X1, 0.18–0.35 mm fraction). The MC of other metallic elements ranges from about 9 mg·g^{-1} (Mn in X1, 0.09–0.18 mm fraction) to 0.0006 mg·g^{-1} (Cd in X1, all tested fractions). The MC of precious elements (Au, Pd, and Pt) are under 0.1 mg·g^{-1} and require large volumes of PCBs to ensure a profitable metal recovery business. The exception is Ag with MC up to about 1 mg·g^{-1} (in X1, 0.18–0.35 mm fraction).

Author Contributions: Conceptualization E.B. and A.S. (Andrei Shishkin); methodology V.S.; validation V.P., K.M. and V.Z.; formal analysis A.S. (Andrej Shishko); investigation E.B., A.S. (Andrei Shishkin), D.G. and V.S.; resources, E.B., V.S. and A.S. (Andrei Shishkin); data curation, J.B. and A.S. (Andrei Shishkin); writing—original draft preparation, A.S. (Andrej Shishko) and D.G.; writing—review and editing, J.B.; visualization, A.S. (Andrei Shishkin), A.S. (Andrej Shishko) and J.B.; supervision, A.S. (Andrej Shishko); project administration, E.B.; funding acquisition, E.B. All authors have read and agreed to the published version of the manuscript.

Funding: This research was supported by ERDF project no. 1.1.1.1/20/A/139 "Development of sustainable recycling technology of electronic scrap for precious and non-ferrous metals extraction". The project was co-financed by REACT-EU funding to mitigate the effects of the pandemic crisis. The article was published with the financial support from the Riga Technical University Research Support Fund. This research was also supported by the Institute of Solid State Physics, University of Latvia as the Center of Excellence has received funding from the European Union's Horizon 2020 Framework Program H2020-WIDESPREAD-01-2016-2017-TeamingPhase2 under grant agreement No. 739508, project CAMART2.

Data Availability Statement: Not applicable.

Acknowledgments: The authors would also like to mention the support from the "Innovation Grants for Maritime Students" performed at Latvian Maritime Academy (project no: 1.1.1.3/18/A/006, funded by the European Regional Development Fund—ERDF, Republic of Latvia).

Conflicts of Interest: The authors declare no conflict of interest. The funders had no role in the design of the study; in the collection, analyses, or interpretation of data; in the writing of the manuscript; or in the decision to publish the results.

References

1. Popov, V.V.; Kudryavtseva, E.V.; Kumar Katiyar, N.; Shishkin, A.; Stepanov, S.I.; Goel, S. Industry 4.0 and Digitalisation in Healthcare. *Materials* **2022**, *15*, 2140. [CrossRef] [PubMed]
2. Popov, V.V.; Lobanov, M.L.; Stepanov, S.I.; Qi, Y.; Muller-Kamskii, G.; Popova, E.N.; Katz-Demyanetz, A.; Popov, A.A. Texturing and Phase Evolution in Ti-6Al-4V: Effect of Electron Beam Melting Process, Powder Re-Using, and HIP Treatment. *Materials* **2021**, *14*, 4473. [CrossRef] [PubMed]
3. Kalisz, S.; Kibort, K.; Mioduska, J.; Lieder, M.; Małachowska, A. Waste Management in the Mining Industry of Metals Ores, Coal, Oil and Natural Gas—A Review. *J. Environ. Manag.* **2022**, *304*, 114239. [CrossRef]
4. Toneatti, L.; Deluca, C.; Fraleoni-Morgera, A.; Pozzetto, D. Rationalization and Optimization of Waste Management and Treatment in Modern Cruise Ships. *Waste Manag.* **2020**, *118*, 209–218. [CrossRef] [PubMed]
5. Bumanis, G.; Vitola, L.; Stipniece, L.; Locs, J.; Korjakins, A.; Bajare, D. Evaluation of Industrial By-Products as Pozzolans: A Road Map for Use in Concrete Production. *Case Stud. Constr. Mater.* **2020**, *13*, e00424. [CrossRef]
6. Sahmenko, G.; Korjakins, A.; Bajare, D. High-Performance Concrete Using Dolomite By-Products. In *Concrete Durability and Service Life Planning*; Kovler, K., Zhutovsky, S., Spatari, S., M. Jensen, O., Eds.; Springer: Berlin/Heidelberg, Germany, 2020; pp. 99–103. ISBN 978-3-030-43332-1.
7. Lapovok, R.; Popov, V.V.; Qi, Y.; Kosinova, A.; Berner, A.; Xu, C.; Rabkin, E.; Kulagin, R.; Ivanisenko, J.; Baretzky, B.; et al. Architectured Hybrid Conductors: Aluminium with Embedded Copper Helix. *Mater. Des.* **2020**, *187*, 108398. [CrossRef]
8. Rumbo, C.; Espina, C.C.; Popov, V.V.; Skokov, K.; Tamayo-Ramos, J.A. Toxicological Evaluation of MnAl Based Permanent Magnets Using Different in Vitro Models. *Chemosphere* **2021**, *263*, 128343. [CrossRef]
9. Mori de Oliveira, C.; Bellopede, R.; Tori, A.; Marini, P. Study of Metal Recovery from Printed Circuit Boards by Physical-Mechanical Treatment Processes. *Mater. Proc.* **2022**, *5*, 121.
10. Bilesan, M.R.; Makarova, I.; Wickman, B.; Repo, E. Efficient Separation of Precious Metals from Computer Waste Printed Circuit Boards by Hydrocyclone and Dilution-Gravity Methods. *J. Clean. Prod.* **2021**, *286*, 125505. [CrossRef]
11. Mandot, V.; Saraswat, V.; Jaitawat, N. Recycling Technologies of PCBs. *J. Sci. Approach* **2017**, *1*, 6–11. [CrossRef]
12. Kulu, P.; Goljandin, D. Retreatment of Polymer Wastes by Disintegrator Milling. In *Waste Material Recycling in the Circular Economy-Challenges and Developments*; Achilias, D.S., Ed.; IntechOpen: Thessaloniki, Greece, 2021; pp. 1–23. ISBN 978-1-83969-681-7.
13. Wen, X.; Zhao, Y.; Duan, C.; Zhou, X.; Jiao, H.; Song, S. Study on Metals Recovery from Discarded Printed Circuit Boards by Physical Methods. In Proceedings of the 2005 IEEE International Symposium on Electronics and the Environment, New Orleans, LA, USA, 16–19 May 2005; pp. 121–128.
14. Wang, H.; Song, X.; Wang, X.; Sun, F. Tribological Performance and Wear Mechanism of Smooth Ultrananocrystalline Diamond Films. *J. Mater. Processing Technol.* **2021**, *290*, 116993. [CrossRef]
15. Murugan, R.V.; Bharat, S.; Deshpande, A.P.; Varughese, S.; Haridoss, P. Milling and Separation of the Multi-Component Printed Circuit Board Materials and the Analysis of Elutriation Based on a Single Particle Model. *Powder Technol.* **2008**, *183*, 169–176. [CrossRef]

16. Goljandin, D.; Kulu, P. *Disintegrators and Disintegrator Treatment of Materials*; LAP LAMBERT Academic Publishing: Saarbrücken, Germany, 2015; ISBN 3659647683.
17. Oliveira, P.C.; Taborda, F.C.; Nogueira, C.A.; Margarido, F. The Effect of Shredding and Particle Size in Physical and Chemical Processing of Printed Circuit Boards Waste. *Mater. Sci. Forum* **2012**, *730–732*, 653–658. [CrossRef]
18. Kers, J.; Kulu, P.; Goljandin, D.; Kaasik, M.; Ventsel, T.; Vilsaar, K.; Mikli, V. Recycling of Electronic Wastes by Disintegrator Mills and Study of the Separation Technique of Different Materials. *Medziagotyra* **2008**, *14*, 296–300.
19. Perdigones, F.; Quero, J. Printed Circuit Boards: The Layers' Functions for Electronic and Biomedical Engineering. *Micromachines* **2022**, *13*, 460. [CrossRef]
20. Chen, Z.; Yang, M.; Shi, Q.; Kuang, X.; Qi, H.J.; Wang, T. Recycling Waste Circuit Board Efficiently and Environmentally Friendly through Small-Molecule Assisted Dissolution. *Sci. Rep.* **2019**, *9*, 17902. [CrossRef]
21. Bukhari, M.; Mohd Noor, N.; Nan, N.M.M.; Shamsul, J.B. The Application of PCB, Mounted-Components and Solder Paste in Surface Mount Technology Assembly (SMTA). In Proceedings of the 1st National Conference on Electronic DesignAt: Putra Palace Hotel, Kangar, Malaysia, 18–19 May 2005; pp. 145–151.
22. Khandpur, R.S. *Printed Circuit Boards. Design, Fabrication, Assembly, and Testing*; Tata McGraw-Hill Education: New York, NY, USA, 2006; ISBN 0070588147.
23. Tilsley, G.M.; Axon, F.J. Comparison of Dry Film and Liquid Photo-Imageable Solder Masks for Surface-Mount Assemblies. *Microelectron. Reliab.* **1988**, *28*, 659. [CrossRef]
24. PCBCart PCB Board Material, PCB Material Type. Available online: https://www.pcbcart.com/pcb-capability/pcb-materials.html (accessed on 15 June 2022).
25. Hino, T.; Agawa, R.; Moriya, Y.; Nishida, M.; Tsugita, Y.; Araki, T. Techniques to Separate Metal from Waste Printed Circuit Boards from Discarded Personal Computers. *J. Mater. Cycles Waste Manag.* **2009**, *11*, 42–54. [CrossRef]
26. Paola, M. Recycling of Printed Circuit Boards. In *Integrated Waste Management-Volume II*; InTech: Houston TX, USA, 2011; ISBN 978-953-307-447-4.
27. Evangelopoulos, P.; Arato, S.; Persson, H.; Kantarelis, E.; Yang, W. Reduction of Brominated Flame Retardants (BFRs) in Plastics from Waste Electrical and Electronic Equipment (WEEE) by Solvent Extraction and the Influence on Their Thermal Decomposition. *Waste Manag.* **2019**, *94*, 165–171. [CrossRef]
28. Jonkers, N.; Krop, H.; van Ewijk, H.; Leonards, P.E.G. Life Cycle Assessment of Flame Retardants in an Electronics Application. *Int. J. Life Cycle Assess.* **2016**, *21*, 146–161. [CrossRef]
29. Ogunniyi, I.O.; Vermaak, M.K.G.; Groot, D.R. Chemical Composition and Liberation Characterization of Printed Circuit Board Comminution Fines for Beneficiation Investigations. *Waste Manag.* **2009**, *29*, 2140–2146. [CrossRef]
30. Ribeiro, P.P.M.; dos Santos, I.D.; Dutra, A.J.B. Copper and Metals Concentration from Printed Circuit Boards Using a Zig-Zag Classifier. *J. Mater. Res. Technol.* **2019**, *8*, 513–520. [CrossRef]
31. Huang, T.; Zhu, J.; Huang, X.; Ruan, J.; Xu, Z. Assessment of Precious Metals Positioning in Waste Printed Circuit Boards and the Economic Benefits of Recycling. *Waste Manag.* **2022**, *139*, 105–115. [CrossRef] [PubMed]
32. Tatariants, M.; Yousef, S.; Denafas, G.; Bendikiene, R. Separation and Purification of Metal and Fiberglass Extracted from Waste Printed Circuit Boards Using Milling and Dissolution Techniques. *Environ. Prog. Sustain. Energy* **2018**, *37*, 2082–2092. [CrossRef]
33. Sousa, P.M.S.; Martelo, L.M.; Marques, A.T.; Bastos, M.; Soares, H. A Closed and Zero-Waste Loop Strategy to Recycle the Main Raw Materials (Gold, Copper and Fiber Glass Layers) Constitutive of Waste Printed Circuit Boards. *Chem. Eng. J.* **2022**, *434*, 134604. [CrossRef]
34. Zhou, Y.; Qiu, K. A New Technology for Recycling Materials from Waste Printed Circuit Boards. *J. Hazard. Mater.* **2010**, *175*, 823–828. [CrossRef]
35. Arslan, V. Bacterial Leaching of Copper, Zinc, Nickel and Aluminum from Discarded Printed Circuit Boards Using Acidophilic Bacteria. *J. Mater. Cycles Waste Manag.* **2021**, *23*, 2005–2015. [CrossRef]
36. Yazici, E.Y.; Deveci, H. Extraction of Metals from Waste Printed Circuit Boards (WPCBs) in H_2SO_4–$CuSO_4$–NaCl Solutions. *Hydrometallurgy* **2013**, *139*, 30–38. [CrossRef]
37. Sahin, M.; Akcil, A.; Erust, C.; Altynbek, S.; Gahan, C.S.; Tuncuk, A. A Potential Alternative for Precious Metal Recovery from E-Waste: Iodine Leaching. *Sep. Sci. Technol.* **2015**, *50*, 2587–2595. [CrossRef]
38. Van Yken, J.; Cheng, K.Y.; Boxall, N.J.; Sheedy, C.; Nikoloski, A.N.; Moheimani, N.R.; Kaksonen, A.H. A Comparison of Methods for the Characterisation of Waste-Printed Circuit Boards. *Metals* **2021**, *11*, 1935. [CrossRef]
39. Hanafi, J.; Jobiliong, E.; Christiani, A.; Soenarta, D.C.; Kurniawan, J.; Irawan, J. Material Recovery and Characterization of PCB from Electronic Waste. *Procedia-Soc. Behav. Sci.* **2012**, *57*, 331–338. [CrossRef]
40. Zimakov, S.; Goljandin, D.; Peetsalu, P.; Kulu, P. Metallic Powders Produced by the Disintegrator Technology. *Int. J. Mater. Prod. Technol.* **2007**, *28*, 226. [CrossRef]
41. Peetsalu, P.; Goljandin, D.; Kulu, P.; Mikli, V. Micropowders prozducted by disintegrator milling. *Powder Metall.* **2003**, *3*, 99–110.
42. Goljandin, D.; Sarjas, H.; Kulu, P.; Käerdi, H.; Mikli, V. Metal-Matrix Hardmetal/Cermet Reinforced Composite Powders for Thermal Spray. *Mater. Sci.* **2012**, *18*, 84–89. [CrossRef]

43. Cui, J.; Zhang, L. Metallurgical Recovery of Metals from Electronic Waste: A Review. *J. Hazard. Mater.* **2008**, *158*, 228–256. [CrossRef] [PubMed]
44. Tymanok, A.; Tamm, J.; Roes, A. Flow of Air and Particles Mixture in a Disintegrator. In Proceedings of the Estonian Academy of Sciences, Physics Mathematics, Tallinn, Estonia, 19–20 April 1994; pp. 280–292.

Review

Chelating Extractants for Metals

Pavel Yudaev and Evgeniy Chistyakov *

Department of Chemical Technology of Plastics, Mendeleev University of Chemical Technology of Russia, Moscow 125047, Russia; yudaevpavel5@gmail.com
* Correspondence: ewgenijj@rambler.ru

Abstract: In the present review, works on the classes of chelating extractants for metals, compounds with several amide and carboxyl groups, azomethines, oximes, macrocyclic compounds (crown ethers and calixarenes), phenanthroline derivatives, and others are systematized. This review focuses on the efficiency and selectivity of the extractants in the recovery of metals from industrial wastewater, soil, spent raw materials, and the separation of metal mixtures. As a result of this study, it was found that over the past seven years, the largest number of works has been devoted to the extraction of heavy metals with amino acids (16 articles), azomethines and oximes (12 articles), lanthanids with amide compounds (15 articles), lanthanides and actinides with phenanthroline derivatives (7 articles), and noble metals with calixarenes (4 articles). Analysis of the literature showed that amino acids are especially effective for extracting heavy metals from the soil; thiodiglycolamides and aminocalixarenes for extracting noble metals from industrial waste; amide compounds, azomethines, oximes, and phenanthroline derivatives for extracting actinides; amide compounds for extracting lanthanides; crown ethers for extracting radioactive strontium, rhenium and technetium. The most studied parameters of extraction processes in the reviewed articles were the distribution ratios and separation factors. Based on the reviewed articles, it follows that chelate polydentate compounds are more efficient compounds for the extraction of metals from secondary resources compared to monodentate compounds.

Keywords: extraction; diamides; amino acids; Schiff bases; oximes; crown ethers; calixarenes; phenanthroline

1. Introduction

Metals, such as lead (Pb), chromium (Cr), zinc (Zn), cadmium (Cd), copper (Cu), mercury (Hg), and nickel (Ni), and metalloids (arsenic As) entering various ecosystems as a result of anthropogenic influences can accumulate in living organisms [1], with a detrimental effect on the physiological processes of living organisms. Therefore, the removal of heavy metals and their salts from the environment is an urgent task today. To remove toxic metal ions from wastewater of industrial enterprises, extraction is used. Extraction is the most economical, technologically simple and productive process compared to other processes, e.g., electrodialysis, electrocoagulation, flotation, and ion exchange [2]. In addition, unlike membrane filtration and chemical precipitation, extraction is effective at low concentrations of metal ions, which allows it to be used for preliminary analytical concentrations of trace elements contained in water and soil samples [3].

Extraction is widely used in hydrometallurgical processes for selective extraction of the target metal from ores or wastes of enterprises—for example, for treatment of zinc oxide ores [4] and extraction of cerium (Ce) from spent catalytic converters of cars [5].

Among the most effective and selective metal extractants are chelates, i.e., compounds containing at least two donor atoms. Chelates form stable complexes with metal cations, which are poorly soluble in water and well soluble in low-polar organic solvents. This feature of chelating compounds makes them suitable for liquid and supercritical fluid extraction. Chelating extractants are, for the most part, low in toxicity, which makes them safe to work with. In addition, for a number of their mixtures, there is a cooperative effect, in which the distribution ratios of metals increase many times compared to monodentate analogues [6,7].

A number of review articles on chelate extractants are available in the literature. For example, the 2011 review considers extraction systems based on crown ethers, quinolines, fluorine-containing β-diketones, and ionic liquids used as diluents [8]. In [9], studies on the solid-phase extraction of heavy metals by polymers, chitin, chitosan, and calcium alginate are presented, and their chelating and adsorption properties are described. In [10], the synthesis methods, chelating properties and applications in analytical chemistry of azomethines are reviewed, and in [11], the application of 1-(2-pyridylazo)-2-naphthol in solid-phase extraction, microextraction and cloud point extraction is addressed. In 2017, a review on supercritical fluid extraction of heavy metals, lanthanides and actinides by some chelating extractants was published [12]. However, the presented reviews are narrowly focused, and the reference to the most recent article is dated 2015, while most of the cited literature was published in the 1980s to 1990s.

This review presents current research over the past 5 years, systematized by classes of chelating extractants that can be used to isolate and/or separate specific metals or groups of metals. The characteristics of the extractants and an evaluation of the prospects for their use are also given.

2. Extractants Containing Amide Groups

Amide extractants form chelate complexes due to the coordination of the metal ion by the oxygen atoms of the amide groups, and, additionally, by oxygen, nitrogen or sulfur atoms contained in the compound in the form of ether, amino or thioether groups, respectively. The advantages of these extractants are high affinity to a number of metal ions and good solubility in non-polar solvents.

2.1. Actinides

Dialkylamides are often used to separate actinides contained in nuclear fuel-reprocessing waste—a highly active liquid waste. For example, N,N-dihexyloctanamide **1** in n-dodecane selectively separates the pairs of neptunium Np (IV) and plutonium Pu (III), uranium U (VI) and plutonium Pu (III) with separation factors equal to 285 and 1080, respectively [13]. However, it remains unclear why these pairs of elements were chosen for separation by the authors, and how the extractant separates the overall mixture of elements. Additionally, the work did not consider the effect on the extraction and separation of the above metals of attendant impurities that are always present in nuclear fuel waste, such as Cs^+ and Sr^{2+} ions

$C_9H_{19}-C(=O)-N(C_6H_{13})-C_6H_{13}$

1

Using N,N'-dimethyl-N,N'-dioctyl-4-oxaheptanediamide **2** in the ionic liquid 1-butyl-3-methylimidazolium bis[(trifluoromethyl)sulfonyl]imide **3**, the separation of thorium ions from uranyl ions from the HNO_3 solution was studied [14].

It was found that extractant **2** has a better affinity for thorium ions than for uranyl ions. The maximum Th (IV)/U (VI) separation factor is 21.9 at pH 2.24. The mechanism of extraction of Th (IV) and U (VI) ions with the indicated extractant is cation exchange, whereby three molecules of compound **2** can be bound to one thorium ion (1) and two molecules of the extractant (2) can be bound to the uranyl ion:

$$Th^{4+}_{(aq.)} + 3 \text{ extractant 2}_{(org.)} + 4 \text{ C}_4\text{mim}^+_{(org.)} = Th(\text{extractant 2})^{4+}_{3\,(org.)} + 4 \text{ C}_4\text{mim}^+_{(aq.)} \quad (1)$$

$$UO_2^{2+}_{(aq.)} + 2 \text{ extractant 2}_{(org.)} + 2 \text{ C}_4\text{mim}^+_{(org.)} = UO_2(\text{extractant 2})^{2+}_{2\,(org.)} + 2 \text{ C}_4\text{mim}^+_{(aq.)} \quad (2)$$

where $C_4\text{mim}^+$—ionic liquid cation **3**.

In the solid-phase extraction of both plutonium Pu (IV) and americium Am (III) simultaneously using N,N,N′,N′-tetra(2-ethylhexyl) diglycolamide **4**, encapsulated in polyethersulfonic polymer granules, the authors of work [15] managed to selectively separate the above actinide mixture. Since a high extraction of both metals was observed, the strategy for their separation was to carry out two stages. First, Pu (IV) was reduced to Pu (III) and removed using ascorbic acid, leaving up to 90% of Am (III) unchanged. Then, oxalic acid was used to separate the americium.

It should be noted that in [14,15], experimental data on the separation of the target metals from the accompanying impurities that are always present in the products of nuclear fusion are also lacking.

In [16], the liquid extraction of not only Am (III), Np (IV), Pu (IV), and U (VI), but also impurities Sr (II) and Cs (I) by tripodal diglycolamides **5, 6, 7** from nitric acid media was studied.

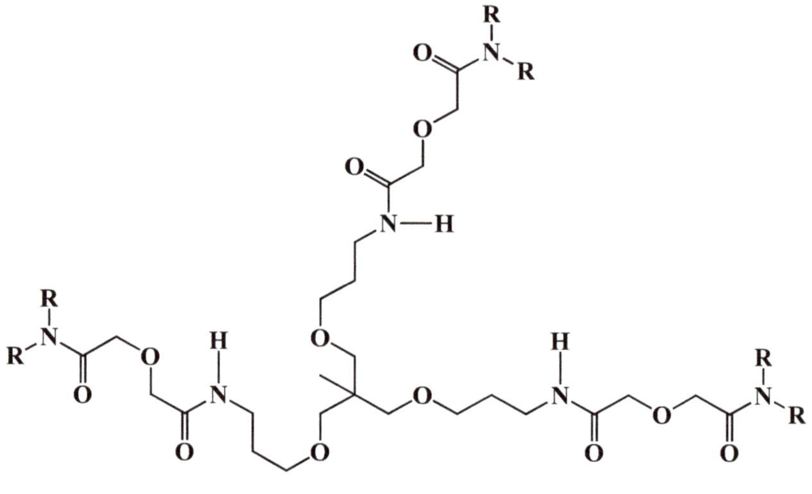

R=C$_8$H$_{17}$

5

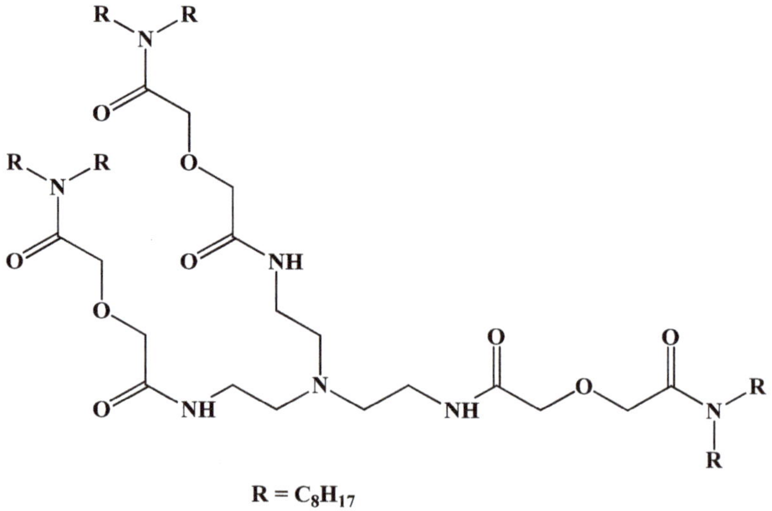

R = C$_8$H$_{17}$

6

$$R = C_8H_{17}$$

7

The following ligand extraction efficiency series was established: using **6**, the pattern of Cs(I) < Sr (II) < NpO$_2^{2+}$ ~ UO$_2^{2+}$ < Am (III) < Np (IV) < Pu (IV) was observed, which changes to Cs (I) < Sr (II) < NpO$_2^{2+}$ < UO$_2^{2+}$ < Np (IV) < Am (III) < Pu (IV) for **5** and for **7** is as follows: Cs (I) < Sr (II) < UO$_2^{2+}$ < NpO$_2^{2+}$ < Np (IV) < Pu (IV) < Am (III). The authors attribute the higher degree of extraction of plutonium compared to neptunium to the higher complexing ability of plutonium.

The compounds are arranged in a series of extraction capacities as follows: **7** > **5** > **6**. The higher extraction ability of extractant **7** is due to its higher lipophilicity (hydrophobicity) compared with **5**. The low extraction ability of extractant **6** was explained by protonation of the central nitrogen atom, which makes it difficult to extract metal ions.

The authors of [16] managed to achieve quantitative extraction of neptunium Np (IV) and plutonium Pu (IV) from nitric acid, with a detailed description of the extraction mechanism. The authors selectively separated Np (IV) and Pu (IV) from UO$_2^{2+}$, Cs (I), and Sr (II) (the values of the separation factors are given in Table 1) by extractants **5**, **6**, **7**. In addition, the obtained extractants had radiolytic stability.

Table 1. Separation factor values of Np (IV) and Pu (IV).

Ligand	UO$_2^{2+}$	Cs (I)	Sr (II)
Compound 5	435	87,000	2200
Compound 6	837	50,000	3100
Compound 7	7300	70,000	10,000

In [17], the extraction of actinides Am (III), Pu (IV), Np (IV) and U (VI) in the presence of Sr (II) using five tripodal diglycolamide ligands **8–12** with different lengths of spacers and substituents was studied. Considering the actinide distribution ratios (aqueous phase: nitric acid; organic phase: ionic liquid **3**), the most efficient ligand is compound **8** (Table 2).

Table 2. Distribution ratios and separation factor data $\beta_{Am/Me}$ (given in brackets).

Extractant	Am (III)	U (VI)	Np (IV)	Pu (IV)	Sr (II)
8	59.9	0.04 (1500)	3.11 (19.2)	3.86 (15.5)	0.06 (1000)
9	9.51	0.01 (951)	1.47 (6.47)	1.73 (5.50)	0.04 (238)
10	0.96	0.02 (48)	1.01 (0.95)	1.04 (0.92)	0.02 (48)
11	0.13	0.01 (13)	0.66 (0.20)	0.71 (0.18)	0.01 (13)
12	0.31	0.01 (31)	0.79 (0.39)	0.84 (0.37)	0.02 (16)

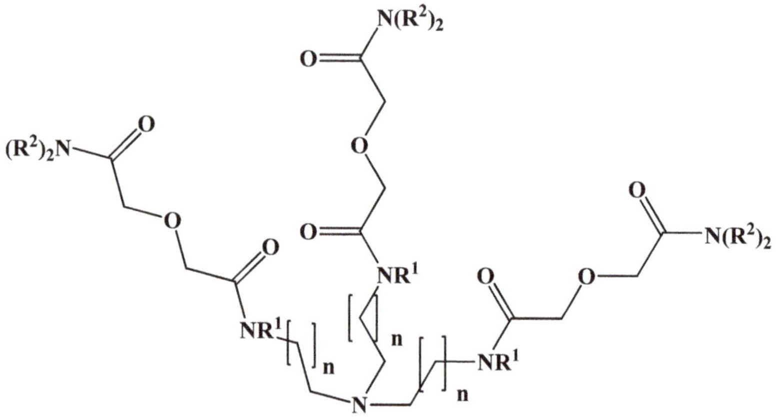

8 $n = 2$, $R^1 = H$, $R^2 = C_8H_{17}$
9 $n = 1$, $R^1 = H$, $R^2 = C_8H_{17}$
10 $n = 1$, $R^1 = CH_3$, $R^2 = C_8H_{17}$
11 $n = 1$, $R^1 = $ isopropyl, $R^2 = C_8H_{17}$
12 $n = 1$, $R^1 = H$, $R^2 = CH_3$

The following series of actinide extraction efficiency was observed: compound **8** >> compound **9** >>> compound **10** ≈ compound **11** ≈ compound **12**. According to the authors, this is due to factors such as spacer length, hydrophobicity and the branched alkyl groups at the amide nitrogen atom.

To summarize, diglycolamide compounds are excellent extractants of actinides from highly reactive wastes as well as for their separation. Therefore, the implementation of the developed formulations in various technological processes seems rather promising.

2.2. Lanthanides

The worldwide demand for rare earth elements (REEs) in 2020 was approximately 185 thousand tons, while it reached 80 thousand tons in 2000. The rapid demand for REEs used in electronics, instrumentation, aerospace, defense and engineering industries makes solving the problem of extraction and concentration of lanthanides an urgent task. Diamides containing a simple ether group are widely used for extraction of lanthanides.

A mixture of trivalent lanthanides (La, Ce, Pr, Nd, Sm, Eu, Gd, Tb, Dy, Ho, Er, Tm, Yb, and Lu) can be separated into individual elements contained in nitric acid solution by using a solution of extractants-tri-(*n*-butyl) phosphate (TBP) and *N,N,N',N'*-tetraoctyldiglycolamide (compound **13**) in ionic liquid **3**, which is more environmentally friendly than conventional molecular solvents [18]. It was found that Ho, Er, Tm, Yb, and Lu are most effectively extracted using the above system.

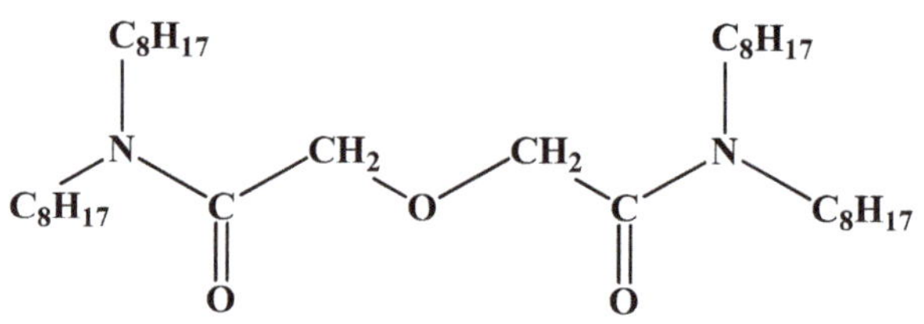

13

It is important that a synergistic effect providing excellent separation of lanthanide ions was observed when using the mentioned extraction mixture, which is explained by the formation of complex forms of lanthanides with both TBP and **13**. However, further studies are needed to establish the composition of internal and external coordination spheres of the forms formed.

It was also noted that hydrophobic anionic fragments of the ionic liquid contribute to the extraction process through the formation of lipophilic metal–ligand complexes.

Unfortunately, the authors present separation factors only for the Lu (III)/La (III), Lu (III)/Sm (III) and Lu (III)/Tb (III) pairs, equal to 1622, 45 and 4.8, respectively; for other lanthanide pairs, these factors are not presented.

Liquid high-level radioactive waste can be a source of neodymium (III), a valuable REE used in telephones and computers, and zirconium (IV), used to manufacture nuclear reactor parts.

In [19], comparative extraction of neodymium Nd (III) and zirconium Zr (IV) using compound 13 and N,N-dioctylhydroxyacetamide 14 from nitric acid medium was carried out. It was found that compound 14 forms large aggregates with neodymium ions, and 13 forms large aggregates with zirconium. The authors did not find an explanation for this fact.

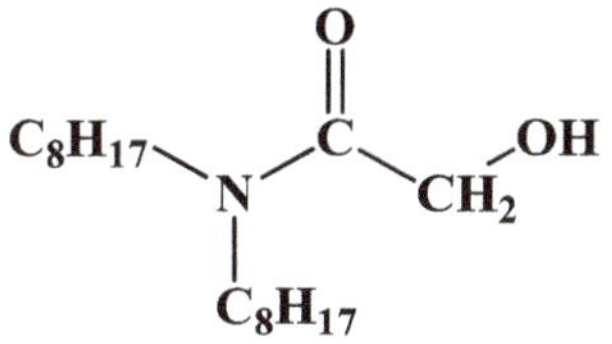

14

The extraction percentages using compound 14 were 98.8% for neodymium and 99.9% for zirconium. Extractant 14, according to the authors, can be used to extract neodymium (III) and zirconium (IV) from high-level waste generated during reprocessing of spent fuel for fast-neutron reactors (uranium-zirconium and uranium-plutonium-zirconium metal alloys). However, the separation factors of these metals and other uranium fission products are not presented in this work.

The extraction properties, extraction mechanism and third phase formation were studied [20] during extraction of neodymium (III) from nitric acid medium using asymmetric (having different substituents at the nitrogen atom) diglycolamide-N,N'-dimethyl-N,N'-dioctyl-3-oxadiglycolamide 15 in mixture of n-octanol and paraffin. A special feature of the work is that the influence of the acidity of the medium, diglycolamide concentration and temperature on the Nd (III) distribution ratio was investigated.

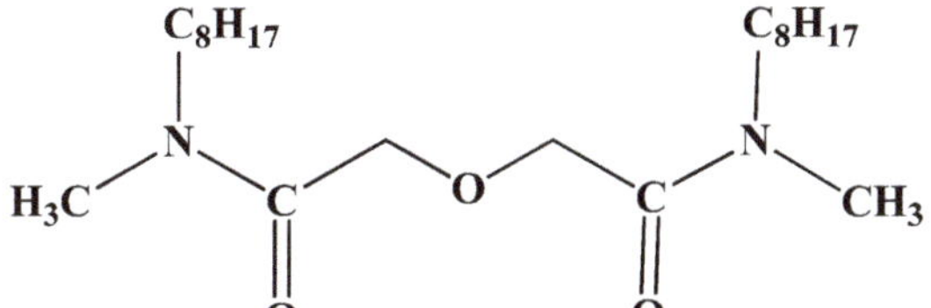

15

The authors, based on earlier work [21,22], conclude that compared to symmetrical diglycolamides, asymmetrical diglycolamides may be more promising candidates for neodymium extraction from high-activity liquid radioactive waste. This is due to the absence of third phase formation during the extraction process.

When extracting [23] neodymium (III) from nitric acid using extractant 4 in n-dodecane, the formation of aggregates 4/Nd(NO$_3$)$_3$ and 4/HNO$_3$ of large size, 30–40 nm, was observed in the organic phase, which is due to reverse micellar aggregation because the extractant molecule is amphiphilic. Under the influence of ^{60}Co (500 kGy), the formation of the third phase was minimized due to radiolytic degradation of the extractant, with the average size of the aggregates decreasing to 9 nm. The authors only studied the extraction behavior of Nd (III) under these conditions. Whether the method is applicable for other lanthanides has not yet been investigated.

To control the formation of an undesirable third phase in the extraction of neodymium from a solution of $Nd(NO_3)_3$ in nitric acid, the authors of [24] added 5 to 15 vol. % aliphatic alcohols, n-decanol, n-octanol and isodecanol to the organic phase (extractant **4** in n-dodecane). The principle is based on the fact that when alcohols with long aliphatic chains are added, they interact with polar acid solvates and metal solvates and prevent their aggregation in the organic phase. This reduced the size of the aggregates from 20 to 4 nm. The best results were achieved using n-decanol.

Extraction system 0.2 mol L^{-1} extractant **4** + 1 mol L^{-1} n-decanol/n-dodecane is proposed for the separation of trivalent lanthanides from highly reactive liquid wastes formed during the processing of uranium-plutonium carbide fuel. At the same time, the authors of works [23,24] did not study the behavior of the extractant in the presence of other elements, including radioactive ones, which can radiolytically destroy it.

When extracting Eu (III) with 4-oxaheptanediamides **16, 17** and **18**, the best results were obtained using the extractant with the longest radicals (**18**) [25]. There is some contradiction with earlier works—for example, [26], in which the study of actinide extraction with diglycolamides with different alkyl substituents showed that the increase in the length of alkyl chain of the substituent decreases the extraction ability of the extractant due to the appearance of steric hindrance.

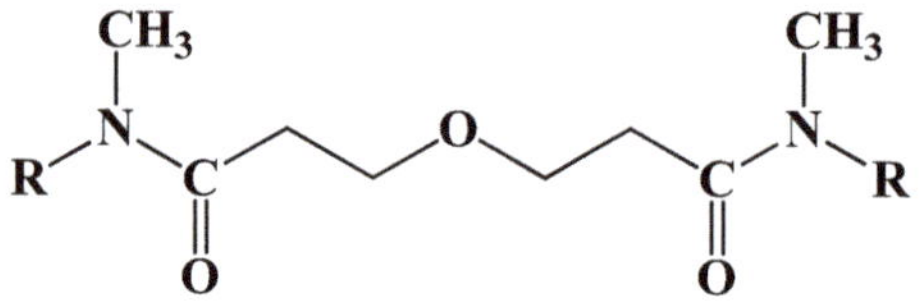

16 R = C_4H_9
17 R = C_6H_{13}
18 R = C_8H_{17}

It is proposed to use unsymmetrical heptanediamides to further investigate the possibility of extracting other lanthanides.

The authors of [27] used compound **14** as an additive to extractant **13** to prevent stratification of organic phase due to micellar aggregation during extraction of trivalent metals from nitric acid medium. The antagonistic effect of aggregation in the organic phase (n-dodecane as the diluent) when adding the above extractant was found to be advantageous in the extraction of trivalent lanthanides from high-level nuclear waste. The authors of the work explain the reduction in aggregation by the structure of extractant **14**, which is part of molecule **13**. It was found that the addition of compound **14** minimizes the average size of aggregates and increases the limit of third phase formation. However, the authors of the work, having studied the effect of extractant **14** on the recovery of Eu (III) and Nd (III), did not consider their separation from other REEs using the proposed synergistic mixture of extractants. Unfortunately, no data are presented on the effect of impurities of other REEs. The authors have not shown information regarding the separation factors of impurities of other REEs.

The authors of [28] extracted trivalent lanthanides (Ce, Eu and La) using **13** and octyl(phenyl)-N,N-diisobutylcarbonylmethylphosphinoxide **19** diluted in ionic liquid-containing imidazolium cations. A high extraction efficiency of lanthanides from dilute hydrochloric acid is noted when these extractants are used. In addition, compound **13** is more efficient and less toxic (due to its lack of phosphorous) than compound **19**.

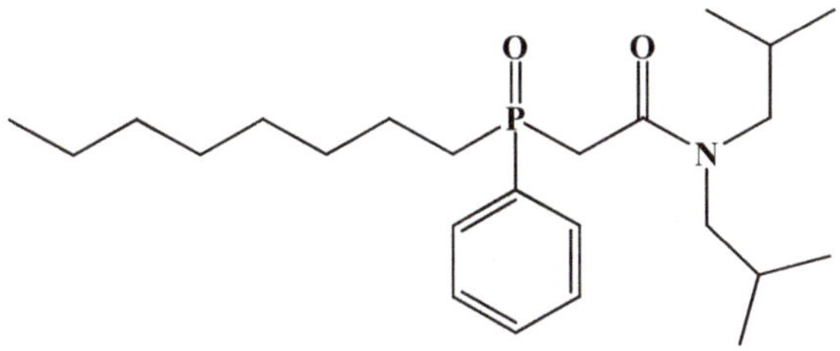

19

The authors were critical about the separation of the Eu (III)/Ce (III) pair. For the separation of this pair, both own results as well as data from scientific literature are given. A very high selectivity of extraction was achieved when 1-(2-pyridylazo)-2-naphthol was used as a synergistic agent. A marked decrease in selectivity towards 4f-ions was observed when the combination of **13** + **19** was used. The authors also point out that the choice of diluent is very important not only for extraction or synergistic enhancement but also for efficient metal separation.

Using diamides, an attempt was made to separate europium(III) and americium(III) from concentrated nitric acid solutions. This was performed using N,N'-diethyl-N,N'-bis(6-methylpyridin-2-yl)-2,2'-bipyridine-6,6'-dicarboxamide **20** dissolved in nitrobenzene [29]. However, the separation factor proved to be lower than 1. The authors attributed this result to the electron-withdrawing properties of the pyridine rings, which decrease the activity of amine groups. However, they neglected the positive mesomeric effect of these rings caused by the nitrogen atoms. In addition, the paper presents no data on the extraction of these lanthanides: the distribution ratios and effect of nitric acid and extractant concentrations on the extraction efficiency.

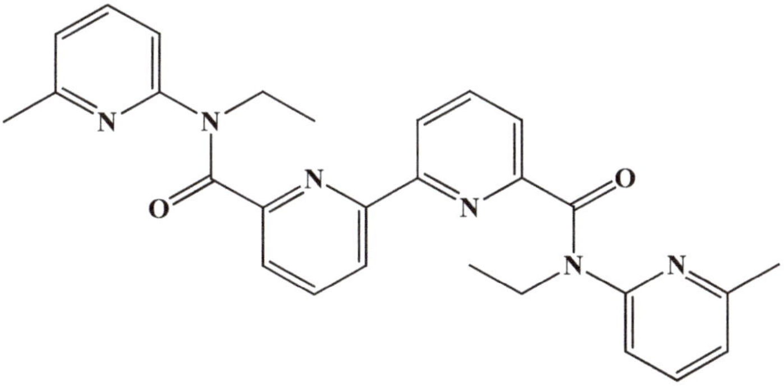

20

However, negative synergism was obtained when attempting trivalent lanthanide extraction using mixed-compound systems "**13**/**19**/ionic liquid". The challenge for successful applications of ionic liquid in synergistic extraction is to find or develop a suitable system for that particular application. Continued research to elucidate the driving force responsible for the successful application of ionic liquid in the field of liquid metal extraction will contribute to optimizing the process and introducing more efficient and environmentally friendly technologies.

Extractant **13** diluted in aviation paraffin is proposed for the extraction of 14 REEs (lanthanum La, cerium Ce, praseodymium Pr, neodymium Nd, europium Eu, gadolinium Gd, terbium Tb, dysprosium Dy, holmium Ho, erbium Er, thulium Tm, ytterbium Yb, lutetium Lu and yttrium Y) from sulfuric acid solution [30]. With **13** the authors were able to achieve a high extraction rate of 91.4–99.8%. The recovery rate of REEs increased with increasing sulfuric acid concentration from 1 to 6 mol L^{-1} and decreased with further increase. According to the authors, this is due to a decrease in activity of water molecules at high concentrations of sulfuric acid, due to which the extractant molecules **13** form complexes at high concentrations of sulfuric acid. From the description, the chemistry of the process is not very clear—the phenomenon should have been presented in more detail.

According to the authors, this extractant is more environmentally friendly than the phosphorous-containing extractants P507 (mono-2-ethylhexyl ester of 2-ethylhexyl phosphonic acid) and P204 (di(2-ethylhexyl) phosphate), which are used in industry to extract REEs.

The extraction of REEs and their complexation with tetraalkyldiglycolamides (**4, 13**, N,N,N',N'-tetrabutyldiglycolamide **21** and N,N,N',N'-tetrahexyldiglycolamide **22**) was studied [31]. Extraction ability was evaluated against light (La, Pr, Nd), middle (Eu, Gd) and heavy (Y, Er, Yb) REEs in HNO$_3$ solution. The extraction capacities of the ligands were distributed in this order: *4 < 22 < 13 < 21*.

21 R = C$_4$H$_9$
22 R = C$_6$H$_{13}$

During extraction of REEs from a hydrochloric acid medium by [32] unsymmetrical diglycolamide-N,N'-dibutyl-N,N'-di(1-methylheptyl) diglycolamide (compound **23**) diluted with kerosene/*n*-octanol mixture the following order of metal extraction ability was established: Nd (III) < Gd (III) < Er (III) < Dy (III) < Sm (III) < Yb (III), indicating that compound **23** has a better affinity for heavier lanthanides. It was noted that the distribution ratio increased with increasing concentration of the extractant. According to IR study, it was found that the extractant coordinates with ions of REEs through the oxygen of carbonyl group. The mechanism of REE extraction was established, namely, it was shown that Nd (III) can bind with two molecules **23** and other REEs with three. Consequently, the composition of the extracted REE (III) complex can be expressed as NdCl$_3$·2(**23**), MCl$_3$·3(**23**).

23

In a study [33] of liquid extraction of REEs using compound **21** in 1-octanol from hydrochloric acid medium, it was found that the use of polar 1-octanol contributed to high extraction efficiency, especially for the separation of heavy lanthanides in solutions with high chloride concentration. Experimental results showed that the distribution ratio of trivalent REE ions increased with increasing HCl concentration, metal atomic number and extractant concentration. However, the light lanthanides La and Ce were poorly extracted with **21**.

Thus, the works [32,33] failed to achieve high extraction efficiency of light lanthanides.

The polyvinylidene fluoride-based membrane containing extractant **13** enables extracting the REEs lanthanum (III), cerium (III), praseodymium (III), and neodymium (III) with purity over 95% from phosphate ore leaching solutions [34]. REE extraction was influenced by concentration **13**. The efficiency of membrane transfer increased with the increase in concentration **13**; however, when the optimum viscosity limit was reached, the transfer rate decreased. This is probably due to the fact that due to viscosity, diffusion across the membrane is impaired and becomes a limiting factor in the mass transfer process.

Nevertheless, this membrane technology makes it possible to separate more than 95% of the rare earth ions from the leaching solutions using 0.10 mol L^{-1} compound **13**. More importantly, impurity ions do not pass through the membrane, allowing high-purity REEs to be obtained. Additionally, the use of liquid membranes contributes to an environmentally friendly, highly efficient and economical method for the recovery of REEs.

Despite the available experimental studies, confirming a high degree of extraction of only La, Ce, Pr, Nd, the authors of [34] make the assumption of high selectivity of the proposed membrane in relation to all lanthanides.

It should be noted that in the above-mentioned works [18–34], separation of REEs from impurities of heavy and alkaline earth metals Fe (III), Pb (II), Ti (IV), Cu (II), Mg (II) and others as well as thorium Th (IV) contained in rare earth ores was not investigated.

For the separation of trivalent REEs (Y, La, Eu, Gd, Dy, Er, Yb, and Lu) in the presence of aluminum Al (III) and iron Fe (III), a new extraction system **24** based on *N,N*-di-(2-ethylhexyl)-diglycolamide-grafted polystyrene resin Amberlite IRA-910 was proposed to prevent formation of the third phase, especially at high acidity of water phase [35].

24

The authors established the complex-forming mechanism of REE extraction based on their interaction with carbonyl and ether groups of ligands, which was confirmed by X-ray photoelectron spectroscopy.

It is interesting to note that the data on the adsorption of metal ions show the following trend: heavy REEs (Yb, Er, and Lu) > light REEs >> Fe > Al. In addition, the adsorption of all REEs was approximately 60%, while it was less than 1% at pH 1.8 for Al and Fe. However, increasing the pH contributes to increasing the adsorption of iron and aluminum, so at pH = 3.5, the adsorption of Fe was almost 18%. In addition, at pH = 3.5, an inhibitory effect on the adsorption of REE starts to act; therefore, a pH of 3.0 is recommended for their effective purification from Al and Fe.

The method is effective in separating REEs from impurity aluminum and iron ions, which can be very high in the ore. It also involves the highly efficient extraction of heavy REEs from low-concentration leaching solutions in a single stage, providing a simpler and more environmentally friendly alternative to extraction with organic solvents. The advantage of the proposed method is durability of polymer resin and possibility of its reuse within 3–4 cycles. However, there are no data on element separation factors in the work.

The influence of phosphate ions, iron (III) and copper (II) ions on extraction of trivalent REEs (Ce, Nd, La, and Dy) from leach products of New Kankberg (Sweden) and Covas (Portugal) deposits was studied [36]. The extraction was carried out with solvate (**13**) and acidic extractant (di-(2-ethylhexyl) phosphoric acid **25**) in different media in the presence of the mentioned impurities. As a result, the advantage of the acid extractant over the solvate extractant was shown. The selectivity with acid was significantly higher, allowing easier extraction of individual REEs. Additionally, compound **13** showed satisfactory results in nitric acid medium, while it was much worse in other acids. Di-(2-ethylhexyl) phosphoric acid is able to extract REEs from all acids, which is a positive economic and technological factor.

25

The separation of rare earth metals from recycled raw materials is a very topical issue. For example, some REEs, such as Nd (III), Pr (III), and Dy (III), can be extracted from neodymium magnets that are present in various process wastes, such as hard drives, electrical generators for wind turbines, and electric motors. Separation of REEs from other magnet components, such as iron, which is the main part of the alloy, and further processing of REEs was the main task of the authors of the paper [37]. The magnetic powder was leached using harmless and cheap organic substances: maleic, glycolic and ascorbic acids, with the most efficient extraction of REEs from the maleic filtrate. When comparing the extraction ability of compound **13** with the organophosphorous extractants TBP, compound **25**, Cyanex 272 and Cyanex 923 compared to REEs, compound **13** showed the best REE extraction and selectivity between elements in the maleic leachate. As a result, Nd (distribution ratio $D_{Nd} \approx 200$), Pr ($D_{Pr} \approx 90$) and Dy ($D_{Dy} \approx 250$) were efficiently extracted from it. It can be concluded that compound **13** could potentially be used in the future on a large scale for the selective separation of REEs from impurities of other metals.

In [38], the process of leaching and solvent extraction for REE extraction from magnetocaloric materials including Ce, Fe, La, Mn, and Si was described. Leaching was carried out using solutions of nitric, hydrochloric and sulfuric acids with the selection of optimum temperature, acid concentrations and solid–liquid ratios. Extraction of REEs from leach products of nitric, hydrochloric and sulfuric acids was carried out using **13** in paraffin. High distribution ratios of REEs, expressed in terms of concentrations of REEs in the aqueous phase before and after extraction, and good selectivity for Fe and Mn were achieved. The selectivity was better in nitric acid media. Extractant **13** showed very good selectivity between REEs compared to impurities.

Imidophosphoric acid esters, which are polydentate ligands that form strong complexes with these elements and can be used for recovery of REEs from ores can be used for recovery of REEs. In [39], 2-ethylhexyl derivatives of imidophosphoric acid (**26** and **27**) were described and showed high efficiency for collective extraction of trivalent REEs Y, Pr, La, Ce, Nd, Sm, Gd, Dy, Yb and Ho from nitric acid water solutions. A mixture of C_9–C_{13} hydrocarbons was used as a diluent.

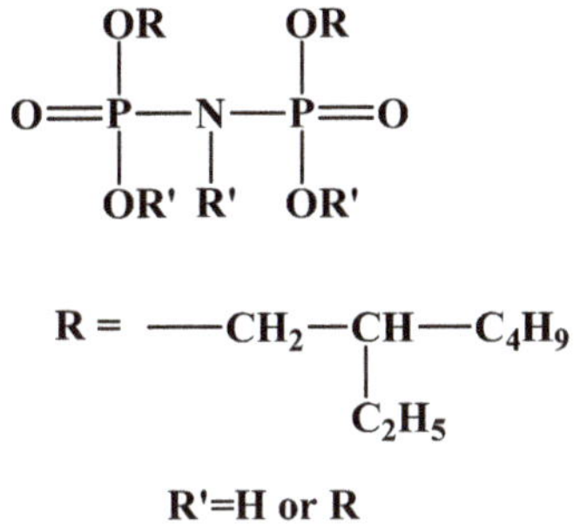

$$O=\underset{OR'}{\underset{|}{P}}-\underset{R'}{\underset{|}{N}}-\underset{OR'}{\underset{|}{P}}=O$$
(with OR substituents on each P)

$$R = -CH_2-\underset{C_2H_5}{\underset{|}{CH}}-C_4H_9$$

R' = H or R

26

$$OR'-\left(\underset{O}{\overset{OR}{\underset{\|}{P}}}-\underset{R'}{\underset{|}{N}}\right)_{4-7}-\underset{OR'}{\overset{OR}{\underset{|}{P}}}=O$$

$$R = -CH_2-\underset{C_2H_5}{\underset{|}{CH}}-C_4H_9$$

R' = H or R

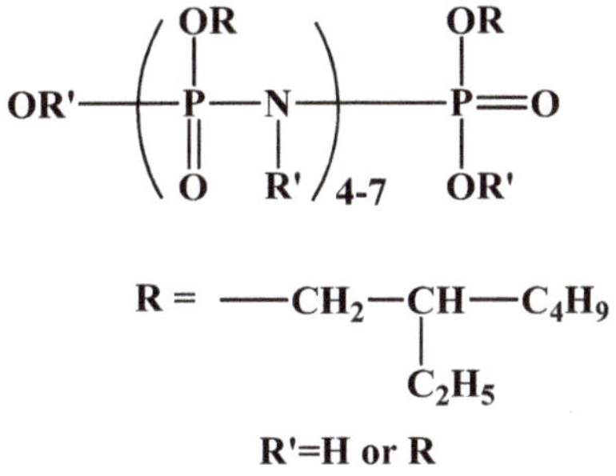

27

The results obtained by the authors were compared with the data for commercial extractant polyalkylphosphonitrile acid, whose structure is not fully established. Higher extraction capacity of synthesized compounds but lower selectivity compared to polyalkylphosphonitrile acid were noted.

The authors proposed a cation-exchange extraction mechanism (3):

$$Ln^{3+} + 3HA = [Ln(A)_3] + 3H^+ \quad (3)$$

The authors explain the low selectivity of the extractants by the fact that the properties of the resulting complex $[Ln(A)_3]$ depend to a greater extent on the -P(O)(OR)-NR'- fragments in the outer sphere of the complex and depend less on the nature of the lanthanide.

It is also reported that increasing the chain length of the ester leads to a better extraction capacity towards the lighter lanthanides. However, the authors of the present work do not explain this fact. The obtained extractants are proposed to be used for the extraction of REEs from ores.

It should be noted that unlike industrial phosphorous-containing extractants having low separation factors of REEs and requiring a large number of stages to obtain one REE of high purity, amide compounds enable separation of REEs in one stage.

2.3. Heavy and Noble Metals

Thioglycolamides are widely used as extractants for noble metals.

For example, in a study [40] of palladium Pd (II), platinum Pt (IV), and rhodium Rh (III) extraction from chloride solutions, N,N'-dimethyl-N,N'-dibutylthiodiglycolamide **28** dissolved in toluene was used.

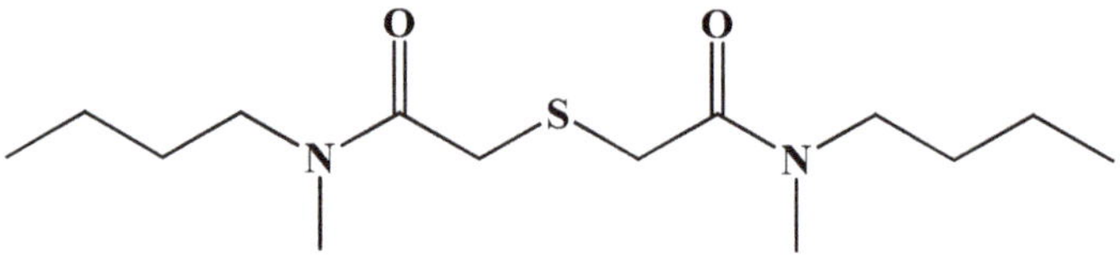

28

It was reported that this compound efficiently extracts palladium, but poorly extracts platinum and rhodium. The palladium/platinum and palladium/rhodium separation factors (at 100 mg L^{-1} concentration of each metal in solution) were 356 and 2880, respectively. It was also shown that at high concentrations of hydrochloric acid (more than 5.5 mol L^{-1}), palladium extraction was accompanied by the loss of extractant to the aqueous phase. However, no interpretation for this fact was given by the authors. Thus, compound **28** is promising for the selective extraction of palladium from low-concentration chloride solutions.

A tridentate ligand, N,N,N',N'-tetraoctylthiodiglycolamide (compound **29**), is able to selectively extract silver Ag (I), palladium Pd (II), gold Au (III) and mercury Hg (II) from nitric, sulfuric, hydrochloric and chloric acid solutions [41]. Table 3 gives data on extraction of **29** different metals. They can be divided into three groups: hardly extracted metals (D < 0.1), extractable metals (D = 0.1–10) and well-extractable metals (D >10).

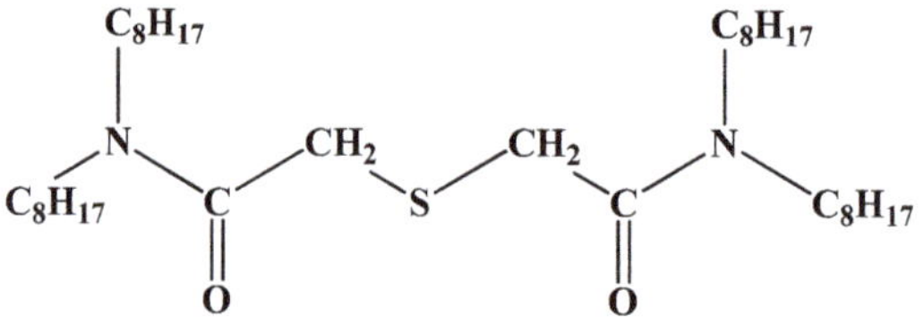

29

Table 3. Distribution ratio values obtained by extraction with compound **29** in *n*-dodecane (condition: 3 mol L^{-1} HNO_3 and 0.2 mol L^{-1} compound **29**).

Hardly Extracted Metals		Extractable Metals		Well-Extractable Metals	
Metal	D	Metal	D	Metal	D
W	0.080	Pu	8.260	Pd	470
Ca	0.045	Re	1.990	Ag	240
Co	0.035	Bi	1.870	Hg	110
Ni	0.027	U	1.610	Au	65
Cu	0.026	Zr	1.430		
Fe	0.016	Sb	0.810		
Zn	0.024	Tc	0.761		
Cd	0.017	Ta	0.740		
Nd	0.015	Ir	0.710		
Mg	0.010	Ru	0.500		
Pt	0.005	Mo	0.270		
V	0.003	Pb	0.120		

As can be seen from Table 3, compound **29** extracts platinum (IV) and iron (III) poorly and excels in extracting palladium (II), silver (I), mercury (II) and gold (III).

The authors of [41] suggested that Pd, Ag, Hg and Au are well extracted by compound **29** because they have different stable oxidation degrees: Ag (I), Au (III), Pd (II) and Hg (II); and ionic radii: Ag 67 pm [coordination number (CN) = 2], Au 85 pm [CN = 6], Pd 64 pm [CN = 4], Hg 96 pm [CN = 6]. It can be concluded that compound **29**, extracts predominantly soft acid metals, i.e., the interaction is based on the hard and soft acids and bases principle (HSAB). Sulfur as a donor gives preference to soft acid metals that promotes high values of the distribution ratios of compound **29** for Pd (II), Hg (II), Au (III) and Ag (I). However, the mechanism of extraction of these metals by compound **29** is not described in the work, which is a significant drawback. In addition, the authors of [41] did not give values for the separation factors of soft and hard metals.

A comparison of extraction of metals by compound **29** with extractants containing other donor atoms (compound **13** and methylimino-bis-N,N'-dioctylacetamide **30**) established that compound **13** shows high extraction ability in relation to hard acid metals U (VI), Pu (IV), Nd (III), and Zr (IV) due to the central oxygen donor atom [41].

Compound **30** has a high extraction capacity for soft acid metals Pd (II), Au (III), Hg (II), and Ag (I) and oxonium anions Tc (VII), Re (VII), Mo (VI), and W (VI) due to the central donor nitrogen atom.

The authors of [42] successfully extracted silver (I), palladium (II) and platinum (II) from aqueous solutions using 2,6-bis(4-methoxybenzoyl)-diaminopyridine **31** in chloroform with a more than 99% efficiency. In addition, polymer membranes based on poly(vinyl chloride) impregnated with compound **31** were used for extraction. The percentage of gold and silver extraction with the membranes was more than 96% after 48 h. However, the percentage of desorption of platinum and palladium was low (less than 70%).

Copper (II) was extracted from nitric acid solutions using compound **31** dissolved in chloroform [43]. This provided extraction of 99.13% of copper at a ligand concentration of 0.001 mol dm^{-3} and a copper (II) concentration of 0.001 mol dm^{-3}. In the opinion of the authors, extraction occurred owing to the copper coordination to the nitrogen atom of the pyridinium ring and to the oxygen atom of the amide group in the ionized enol form, thus giving complex **32**.

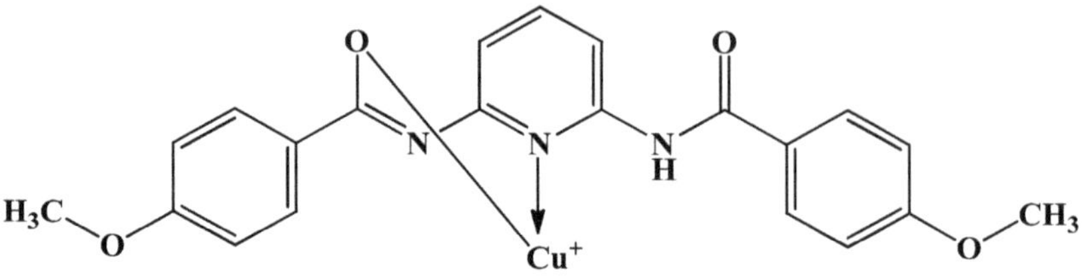

32

We can conclude that during the extraction of heavy metals by diamides, the determining factor is the nature of the central donor atom in the molecule, which, according to the HSAB principle, allows the extractant to form strong complexes with some metals and not to form them with others. Due to this fundamental approach, diglycolamides hold great promise for the isolation and separation of toxic, heavy and precious metals in the presence of various impurities.

3. Amino Acids

Amino acids, such as ethylenediaminetetraacetic acid (EDTA, compound **33**), diethylenetriaminepentaacetic acid (compound **34**), nitrilotriacetic acid (compound **35**), and (S,S)-ethylenediamine-N,N′-disuccinic acid (compound **36**), are widely used for the extraction of heavy metals from soil, since they cause less destruction of the mineral base of the soil compared to inorganic acids.

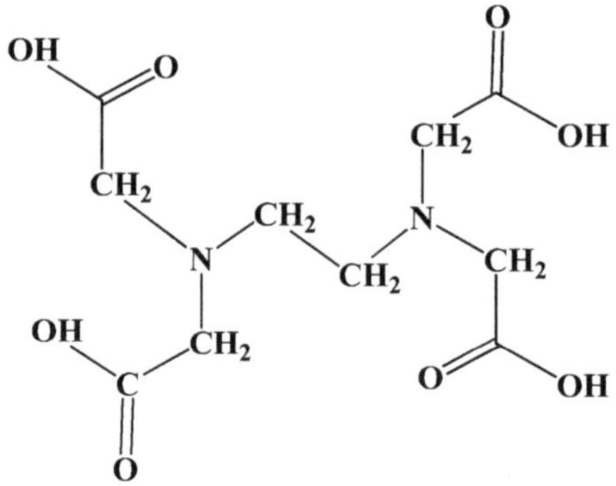

33

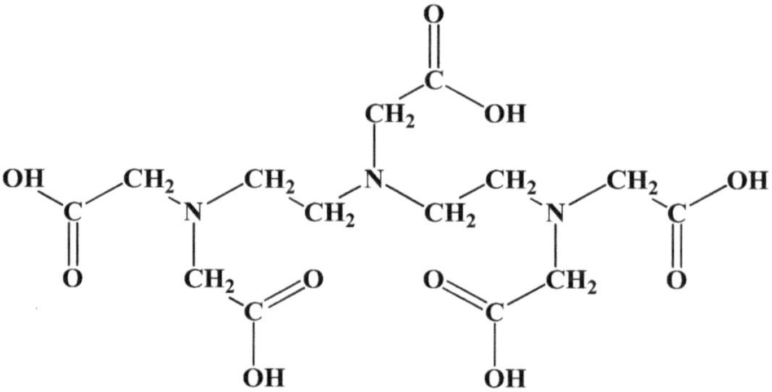

34

Heavy Metals

It was shown in [44] that an equal volume of 0.05 mol L^{-1} EDTA and 0.2 mol L^{-1} organic acids (citric acid, oxalic acid, and tartaric acid) makes it possible to extract more than 80% of heavy metals (copper, nickel, zinc) from the soil, which is higher compared to pure EDTA. However, the authors do not explain this fact. In addition, the use of mixed extractants reduces cost and secondary environmental pollution (pressure on the environment). The above is supposed to be used for the remediation of contaminated industrial soils. However, the presented technology is quite time consuming. The heavy metals extraction process takes 6 h.

EDTA can be used to extract cadmium, copper, nickel, zinc, lead and calcium from soil. The authors of [45] compared the efficiency of extraction of these metals by EDTA and proteinogenic natural amino acids, and tried to identify the factors that determine the efficiency of metal extraction. The research results show that hydrophobic, nucleophilic and steric properties do not impact the process. However, the functional groups of the side chains play an important role in the extraction of heavy metals. In particular, the effect of hydroxy groups on the side chain was clearly manifested. The advantage of tridentate amino acids over similar bidentate ones has also been shown.

The selection of the optimal conditions for the extraction of lead Pb (II) from the soil was carried out by washing the surface soil of the laterite soil with EDTA [46]. A high extraction efficiency was achieved—89.6%, which is lower than the theoretically calculated 90.8% due to the presence of iron, which also forms complexes with the extractant **33**. Nevertheless, the authors of the work [46] were unable to establish whether the extraction of lead is selective compared to other heavy metals present in the soil—zinc, cadmium, and copper, which are also capable of complexation with EDTA. The proposed extraction process is laborious and involves multiple stages.

When using EDTA to separate cadmium, lead, zinc, copper and nickel contained in soil [47], the following sequence of selectivity was established: Cd (II) > Pb (II) ≥ Zn (II) ≥ Cu (II) ≥ Ni (II). The authors propose to isolate heavy metals from lake sediments.

Sludge treatment is a rather long process and the average degree of heavy metals extraction reached only 46.4–78.8% after 21 days.

It was found in [48] that EDTA is more effective in removing lead and copper from the soil than cadmium and zinc, which the authors explain by the higher complexing ability of the extractant compared to lead and copper (higher stability constants of complexes with lead and copper). Using 0.07 mol L^{-1} EDTA, it was possible to remove 94.8% Cu (II), 99.4% Pb (II), and 77.8% Zn (II) from the soil. At the same time, the work did not consider the effect on the extraction compounds Ca (II), Mg (II), Al (III), and Fe (II and III), in the soil which contains much more than heavy metals, on the removal efficiency of target metals.

EDTA can be used to extract molybdenum, nickel, and cobalt from spent hydrodesulfurization catalyst of the oil refining industry [49]. Having established the optimal extraction conditions, the authors of this work extracted 90.22% Mo (IV), 96.71% Co (II), 95.31% Ni (III), and 19.98% Al (III). The extraction of metals was carried out in two stages: the chelation reaction and chemical precipitation.

EDTA was used to extract copper from a spent low-temperature shift catalyst (CuO, ZnO, and Al$_2$O$_3$) [50]. Under optimal process conditions (EDTA concentration 0.5 mol L^{-1}, temperature 100 °C, solid-to-liquid ratio 1:25 (g mL^{-1}), particle size 120 µm and reaction time 4 h), 95% of copper was extracted.

This study demonstrates the possibility to isolate copper by EDTA from a spent catalyst for its recovery and reuse. It remains unclear whether the zinc and aluminum contained in this catalyst can be extracted in this way. In addition, the structure of the copper EDTA complex is presented incorrectly.

The ability of compounds **33**, **34** and ethylenediamine-N,N'-bis(2-hydroxyphenylacetic acid) **37** was studied as a solution of iron (III) and aluminum (III) contained in the soil [51]. It is known that copper (II), manganese (II), and aluminum (III) compete with iron (III). It was found that extractants **33** and **34** form low-stability complexes with iron, and **37** forms more stable complexes with it due to the binding of the metal through the phenolate. Compound **37** was most efficiently extracted for iron and aluminum, and compound **33** for copper, manganese and zinc. It was also found that pH has a stronger effect on the extraction of compounds **33** and **34**, compared with compound **37**. Selectivity for iron for **37** is higher than for **33** and **34**, as evidenced by the highest value of the ratio of iron concentration to the sum of the concentration of all metals in the soil.

37

Currently, there is considerable interest in the use of magnetic nanoparticles functionalized with chelating agents for the extraction of metals from solutions. Thus, the possibility of functionalization of nanoparticles of magnetite coated with silicon dioxide with compound **34** was investigated [52] to remove potentially hazardous (cadmium, cobalt, and copper) and "non-toxic" (calcium Ca and manganese Mn) metals from solutions. It was found that extractant **33** is capable of extracting metal ions with a fairly high efficiency (more than 70%).

Superparamagnetic nanoparticles based on a magnetite core and a silica shell with an immobilized EDTA derivative [53] were used to quantitatively remove heavy metals Cd (II), Cu (II), Cr (III), and Pb (II) from contaminated media for 15 min, as well as analytical determination of metal ions in wastewater samples. The presented method is accurate and reliable and can be recommended for use in industry.

The work [53] is also interesting because the authors take into account both organic compounds and inorganic ions, which are present in the form of macrocomponents in wastewater and are of decisive importance in the evaluation of adsorbents. Organic substances can form complexes with metals in solution, which reduces the adsorption capacity. Adsorption can be influenced by humic acid containing phenolic (-OH) and carboxyl (-COOH) groups, which can bind heavy metals by complexation or chelation. Nevertheless, it was found that the adsorption of humic acid by nanoparticles was insignificant and did not affect the adsorption of heavy metal ions.

The effect of Ca (II), Mg (II), Fe (III), Na (I), Co (II) and a number of anions on the adsorption and determination of target heavy metal ions in multicomponent solutions has been studied. The results are shown in Table 4.

Table 4. The effect of coexisting ions on recoveries of determined metal ions (pH = 5.5).

Ions	Concentration (mg L^{-1})	Recovery (%)			
		Cu^{2+}	Cd^{2+}	Pb^{2+}	Cr^{3+}
Ca^{2+}	5000	95	92	96	99
Na^{+}	1000	95	98	99	98
Fe^{3+}	500	89	87	95	91
Mg^{2+}	500	84	90	94	87
Co^{2+}	300	82	89	93	85
NO$_3^{-}$	5000	95	98	99	98
SO$_4^{2-}$	5000	95	92	98	99
PO$_4^{3-}$	5000	99	98	99	97
F^{-}	1000	96	96	97	99

It follows from Table 4 that the cations Na$^+$, Ca^{2+}, Fe^{3+}, and Mg^{2+} do not have a significant effect on the adsorption of target ions, even though iron and magnesium are capable of forming highly stable complexes with EDTA. The authors explain this phenomenon by the co-deposition of small amounts of Fe^{3+} and Mg^{2+} ions with ions of extracted heavy metals. However, the explanation given is not convincing.

To isolate cadmium (II) and lead (II) from agricultural soils using extractants **33** and **34** in 30 min, at pH = 5, the ratio of soil mass/volume of extractant 1:10, 59% Cd and 63% Pb (compound **35**), 52% Cd, and 51% Pb (compound **36**) were extracted [54]. The authors of the work do not provide explanations of what caused the low efficiency of metal extraction.

In addition, compounds **33** and **34** are ineffective for the simultaneous extraction of heavy metal cations and arsenic metalloid anions present in the soil. This was shown by Swedish scientists [55], who extracted Pb (II), Cd (II), As (V), and Zn (II) using **33**, **34** and **35** from a mixture of soil and waste glass taken from a landfilling of the wastes from the glass daily production process. Unfortunately, the extraction efficiency of these metals turned out to be low (no more than 41%). At the same time, the authors of this work do not explain the mechanism of the effect of arsenic on the extraction efficiency.

The problem associated with the presence of arsenic was solved by E.J. Kim et al. [56]. They showed that the addition of reducing agents such as sodium oxalate, ascorbic acid, and sodium dithionite significantly enhances the extraction of both heavy metals Cu (II), Pb (II), Zn (II) and As (V) itself when using the same extractant **33**. The efficiency of extraction of As, Cu, Zn and Pb when using a combination of sodium dithionite and compound **33** was approximately 90% in a wide pH range, which is quite satisfactory. The authors of [56] explain the improvement of extraction in the presence of reducing agents by the weakening of the bond strength of metals with soil minerals due to the reduction of metals to lower oxidation states.

Compound **33** can be used for the extraction of heavy metals from green liquor dregs generated during the production of kraft pulp. For example, in [57], using **33**, it was possible to achieve 59 wt% Cd (II), 13 wt% Co (II), 62 wt% Cu (II), 3 wt% Mn (II), 12 wt% Ni (II), 43 wt% Pb (II), 16 wt% Zn (II) and less than 1 wt% Ca (II) at ratios of 0.035 g EDTA per 1 g liquor, liquid phase to solid phase 6.25 mL g^{-1}. Unfortunately, the authors of this work do not explain why compound **33** fails to quantitatively extract heavy metals from green liquor dregs.

With all the advantages of compounds **33** and **34**, they have a significant drawback—low biodegradability in soil, which is due to the high stability of metal complexes. Therefore, amino acids are often used for the extraction of heavy metals. For example, to remove copper (II), zinc (II) and lead (II) from soil, an equimolar mixture of chelating extractants **33** and **36** can be used [58]. It was found that extractant **36** reduces the stability of metal complexes based on **33** (for example, for complex **33** and Pb (II) logK is 17.9, and for **36** and Pb (II) logK is 12.7) and increases its biodegradability.

In [59], the extraction of chromium (VI) from soil was accomplished using biodegradable and environmentally safe amino acid, N-acetyl-L-cysteine **38**, containing thiol chelating groups.

In comparison with EDTA, a lower concentration of extractant **38** in water is required for chromium (VI) extraction. In particular, 4128 mg kg^{-1} of EDTA extract and 14.3% of chromium, while in the case of 300 mg kg^{-1} of compound **38**, this value is 65.7% for soil with a pH of 5.5. However, compound **38** is 4-fold more expensive than EDTA.

4. Shiff Bases (Azomethines) and Oximes

In contrast to amino acid-based extractants, Schiff bases and oximes are easily biodegradable, can form stable complexes with almost all metal ions due to the π-acceptor properties of the azomethine nitrogen atom, and have high values of the distribution ratios, which makes this class of compounds promising for creating highly effective chelates [60].

4.1. Actinides and Lanthanides

The introduction of water-soluble Schiff bases into the aqueous phase increases the selectivity of separation of actinides and lanthanides contained in spent nuclear fuel and high-level radioactive waste by di-(2-ethylhexyl) phosphoric acid. During the extraction process, the Schiff base predominantly forms a complex with actinides, and di-(2-ethylhexyl) phosphoric acid with lanthanides. For example, the authors of [61] compared the extraction of U (VI), Eu (III), and Np (IV) with a solution of di-(2-ethylhexyl) phosphoric acid in toluene with and without the addition of N,N′-bis (5-sulfonatosalicylidene) ethylenediamine (compound **39**) in the aqueous phase.

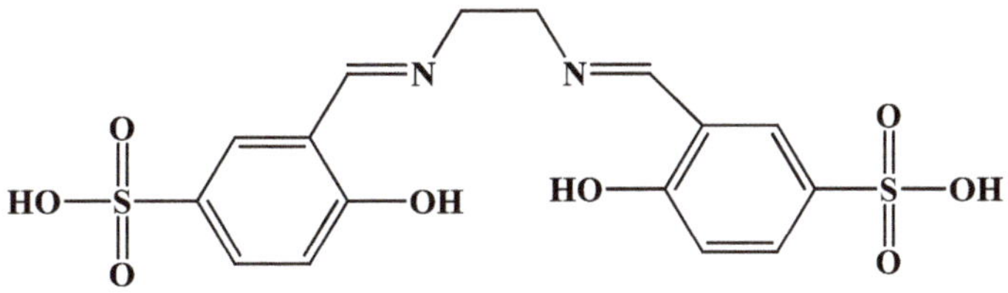

39

In the absence of the Schiff base, the lowest values of separation factors Eu (III) and U (VI) (0.2 after 30 min of extraction), Eu (III) and Np (IV) (4 after 30 min of extraction) are observed. Upon the addition of 0.01 mol L^{-1} of compound **39** to the aqueous phase after 30 min of extraction, the separation factor for Eu/U and Eu/Np was increased to 30 and 230, and after 1.5 h it reaches 210 and 1800, respectively.

Uranium extraction from alkaline leachate of uranium ores is a promising area of research, as uranium is the main raw material of the nuclear industry. The authors of [62] investigated the process of uranium (VI) sorption from aqueous solutions of uranyl sulfate using amidoxime-grafted chitosan magnetic microparticles **40**.

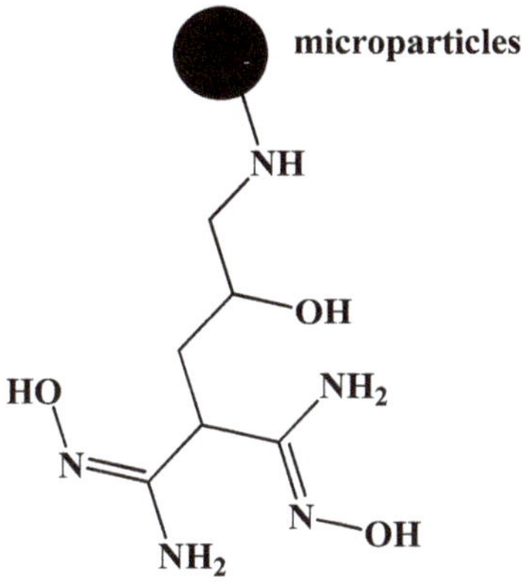

40

The obtained sorbent makes it possible to separate uranium and Eu (III), which is also contained in uranium ores. The selectivity coefficient for uranium (VI) in relation to Eu (III) was high and equal to 14 at pH = 4.9. The sorption and desorption efficiencies of europium with 0.5 mol L^{-1} HCl were 76% and 100% in one cycle, and for uranium in both cases were close to 99%.

The authors of works [61,62] consider the compounds obtained by them as extractants for the isolation of separate metals from multimetal raw materials. However, the studies carried out are limited to only a few elements; under real conditions, impurities of other ions can significantly affect the extraction results.

Thorium is an indispensable element for various technologies, especially for the peaceful uses of atomic energy. However, thorium occurs in nature in low concentrations, so it is necessary to extract it from nuclear waste.

The authors of [63] studied the extraction of thorium (IV) from Th(NO$_3$)$_4$ aqueous solution using Amberlite XAD-4 resin loaded with a tetradentate Schiff base—bis (2-hydroxybenzaldehyde)-1,2-ethylenediimine diaminoethane **41**.

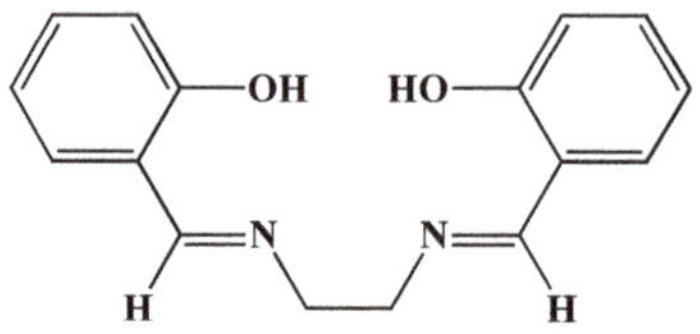

41

It was possible to achieve a quantitative release of thorium at an initial concentration of 20 mg L^{-1}, under optimal conditions, pH = 5, amount of adsorbent 0.1 g, contact time 45 min, and temperature 25 °C. In addition, it was found in the work that the REEs lanthanum (III), cerium (III), samarium (III), europium (III), dysprosium (III), holmium (III) and erbium (III), which are contained in radioactive waste, have no significant effect on thorium adsorption. However, the authors of the work do not give an explanation for this fact.

4.2. Heavy Metals

As mentioned above, the recovery of heavy metals from wastewater is an urgent task. A number of studies were devoted to its solution with the help of chelating azomethines and oximes.

In [64], by condensation of azoaldehyde compounds with hydroxylamine in new ethanol diazo-containing phenolic oximes **42–45**, which are capable of extracting copper (II) from sulfate medium at low pH values, were obtained.

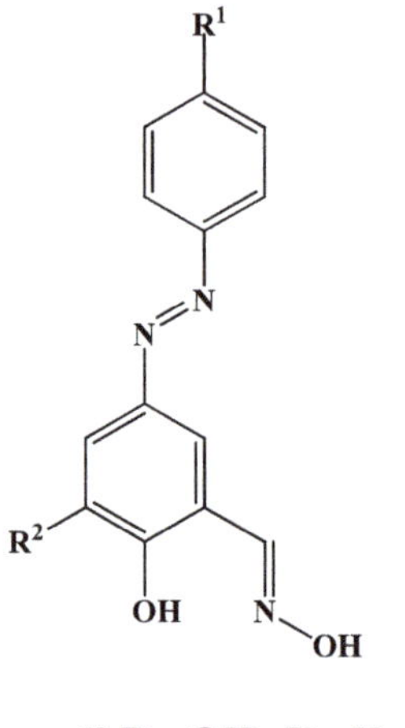

42 R^1 = C$_2$H$_5$, R^2 = H
43 R^1 = C$_2$H$_5$, R^2 = C(CH$_3$)$_3$
44 R^1 = Cl, R^2 = OCH$_3$
45 R^1 = C(CH$_3$)$_3$, R^2 = CH$_3$

A higher extraction efficiency of Cu (II) compared to Ni (II) and Zn (II) was noted. It was found that the efficiency of copper extraction with compound **42** at pH = 1 is approximately 55% and only 30% for compounds **43**, **45**. Copper extraction efficiency reached 75% only at pH = 3 with compound **45**. However, the authors of [64] did not answer the question, whether the mentioned extractants have high selectivity of copper (II) extraction in the presence of zinc (II), cadmium (II), nickel (II), iron (III), cobalt (II), manganese (II), arsenic (V), mercury (II), lead (II) and chromium (III), which, along with copper, can be contained in industrial waste, as well as in ash from municipal solid waste incineration.

During extraction of cobalt (II) from chloride/sulfate solutions by oxime 1-(2-pyridyl)tridecan-1-one (compound **46**) in toluene-decanol-1 mixture (9:1 vol./vol.), it was found that the amount of extracted cobalt decreases with increasing acidity of water phase, increases with increasing concentration of extractant **46** and cobalt (II) ions, but does not depend on concentration of chloride ions [65]. At the same time, the explanation of the obtained dependences is not given in the work. The cobalt extraction rate was 90–92% at a chloride ion concentration of more than 1 mol L^{-1}. The paper describes the possibility of selective separation of Co (II) from Ni (II), Zn (II) and Cu (II). However, there are no data on the extraction of Co (II) in the presence of Fe (III) and Mn (II), also contained in laterite ores.

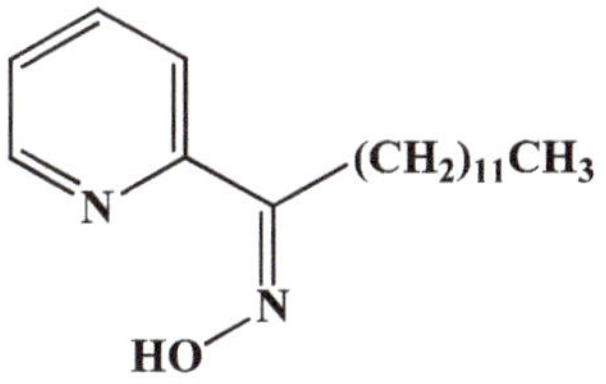

46

The extraction of nickel (II), zinc (II) and copper (II) from nitric acid solutions with N,N′-bis(salicylidene)ethylenediamine (salen) **47** in chloroform was studied [66].

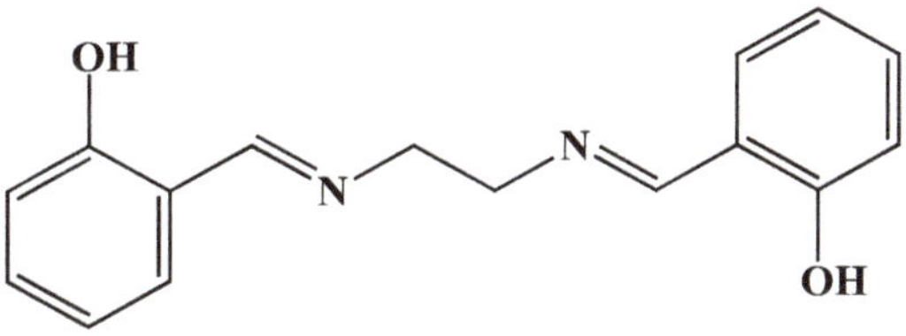

47

The efficiency of copper extraction was 99.79%, while the zinc and nickel extraction efficiencies were 87.68 and 60.98%, respectively. The authors did not give an explanation for lower extraction efficiency of zinc and nickel compared with copper.

The adsorption capacity of a membrane based on poly(vinyl chloride) containing 20 and 40 wt% compound **47** with respect to these metals was also studied. The adsorption capacity was shown to rapidly increase during the first hour of extraction. In the authors' opinion, this was due to the large number of accessible active sites for adsorption. It was also noted that an increase in the content of salen by 20% leads to a 10-fold increase in the percentage of sorption. However, the percentage of sorption by a membrane containing **47** was below 50%.

During the extraction of trace amounts of copper, chromium and mercury from the aqueous phase (0.1 mol L^{-1} KCl), using compound **48** in chloroform as an extractant, it was found that the extraction efficiency of Cr (III) is higher than Cu (II). The extraction of chromium was achieved—more than 90% at pH = 6–7—but Hg (II) could not be isolated under experimental conditions (extraction efficiency less than 1%) [67].

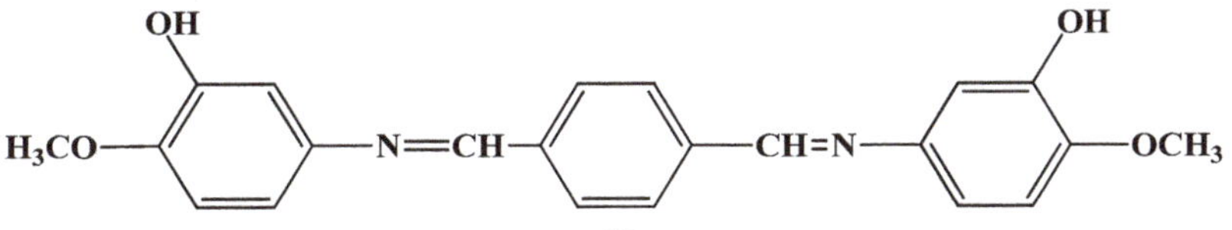

48

The use of extractant **48** is expected to extract low concentrations ($1.5 \cdot 10^{-3}$ mol L^{-1}) of toxic chromium (Cr) and copper (Cu) from industrial wastewater. However, the authors of this work did not consider the effect of trace amounts of arsenic that interfere with extraction.

In [68], 2-pyridylketoximes **46**, **49** and **50** were used for cadmium (II) extraction, and the chelate complexes of extractants **46** and **49** with cadmium salts were characterized. In addition, a negative effect of increasing chloride concentration in the aqueous phase on metal extraction was reported.

49 R = C$_{14}$H$_{29}$
50 R = C$_6$H$_5$

The authors established the structural formula of the complex between the cadmium ion and the extractant, without aiming at the selection of suitable extraction and separation conditions for the different metals. That said, the optimization of the extraction process is the basis of applied research, so it can be concluded that the work is purely fundamental.

Synergistic extraction can be used to extract cobalt (II) from wastewater using Schiff bases. For example, using a mixture of *N*-(2-hydroxybenzylidene)-aniline **51** with 1-octanol dissolved in chloroform increases from 15% to 55% extraction efficiency of Co (II) from sulfate medium, while the process occurs without emulsification and third phase formation, compared to pure compound **51** [69].

The authors of this work explain the synergistic effect by the solvation of the central Co (II) ion in the extracted complex by 1-octanol, and the low degree of extraction by the strong interaction of 1-octanol with compound **51**.

51

A synergistic effect was also observed for mixtures of extractants based on oximes and organophosphoric acids. When copper (II), zinc (II), nickel (II) and cadmium (II) were extracted and separated from calcium (II) and magnesium (II) by a Mextral 84H extractant (2-hydroxy-5-nonylacetophenone oxime, compound **52**) solution with the addition of Cyanex 272 (di-(2,4,4-trimethylpentyl) phosphinic acid, compound **53**) in the aliphatic diluent Mextral DT-100, it was found that the addition of Cyanex 272 extractant to Mextral 84H causes synergistic effects for zinc and cadmium and an antagonistic effect for nickel [70].

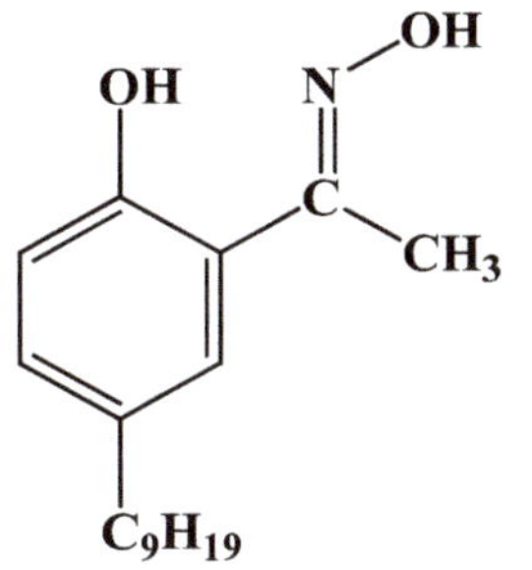

52

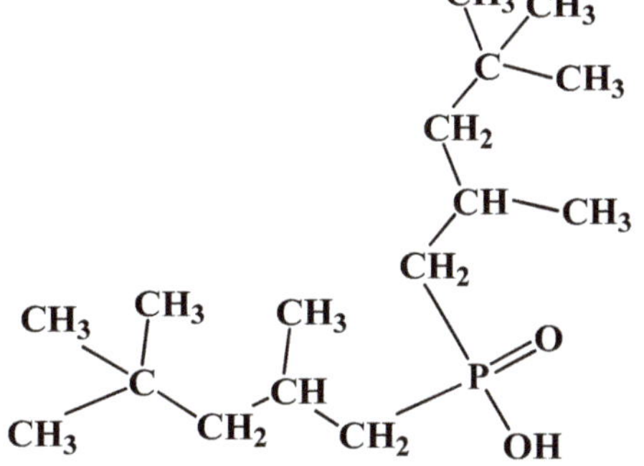

53

The extraction mechanism, which is a cationic exchange, is expressed by the authors of this paper with the following Equation (4):

$$M_{aq.}^{2+} + nH_xR_{x\ org} + mH_yL_{y\ org} = MR_{nx}L_{my}H_{(my+nx-2)\ org.} + 2H_{aq.}^+ \quad (4)$$

where $M_{aq.}^{2+}$—metal cation; H_xR_x and H_yL_y—Mextral 84H and Cyanex 272, respectively; $MR_{nx}L_{my}H_{(my+nx-2)}$—the molecular formula of the extractable complex. However, this mechanism does not explain why there is an antagonistic effect for nickel.

Quantitative metal re-extraction was achieved with dilute sulfuric acid at a ratio of organic/water phase = 1:1 and a temperature of 40 °C in 1 min. The extraction of copper and nickel reached 100%, zinc 99.5% and cadmium 98.6%, while extraction of calcium and magnesium was only 17.2% and 1.7%, respectively. This process, in which all heavy metals were extracted simultaneously and selectively separated under optimal conditions, is fully applicable to the separation of copper, zinc, nickel and cadmium from calcium and magnesium in concentrated wastewater [70].

A significant synergistic effect was observed in the extraction of nickel (II) with compound **52** mixed with di-(2-ethylhexyl) phosphoric acid [71]. The synergistic coefficient reached 3.4, which is explained by the interaction of the phosphorous-containing extractant with nickel and oxime to form the octahedral $Ni(H_2A_2L_2)$ complex **54**.

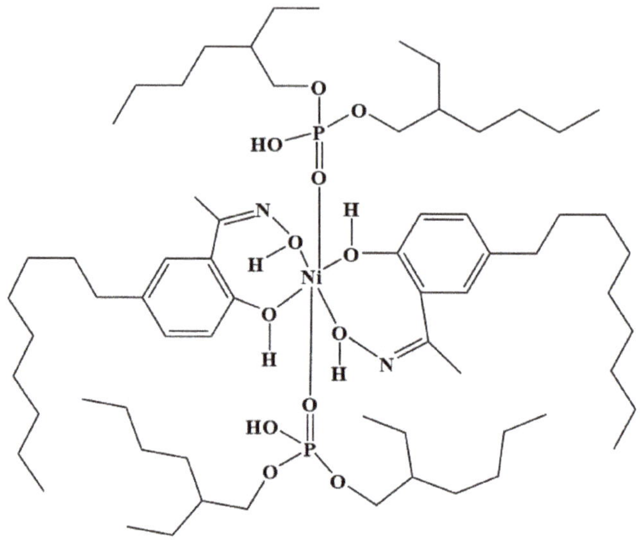

54

The mechanism of nickel extraction with a mixture of extractants is described by the following Equation (5):

$$Ni_{aq.}^{2+} + y/2\ H_2A_{2\ org.} + z\ HL_{org.} = Ni(H_{y+z-2}A_yL_z)_{org.} + 2H_{aq.}^+ \quad (5)$$

where H_2A_2—di-(2-ethylhexyl) phosphoric acid and HL—oxime **52**.

The authors explain the synergistic effect by the formation of a more stable and hydrophobic complex compared to compound **52**.

Pyridine oxime-ethers can be used for the separation of heavy metals. It is known, that to extract zinc (II) from spent pickling liquor, which is formed at a steel pickling line in steel rolling mills, it is necessary to separate zinc and iron. For example, *N*-decoxy-1-(pyridin-3-yl)ethaneimine **55** can selectively separate Zn (II) and Fe (III) in chloride media. The authors of [72] found that the separation factor $\beta_{Zn/Fe}$ for extractant **55** is 90-fold higher than that of TBP.

To evaluate the selectivity, the authors performed Zn (II) extraction in the presence of Fe (III). The separation factor under different conditions was between 58 and 80. Thus, the authors describe extractant **55** as satisfactory for the separation of zinc and iron. One can agree with this conclusion, because TBP, used in the metallurgical industry for the separation of zinc and iron, has a much lower separation factor.

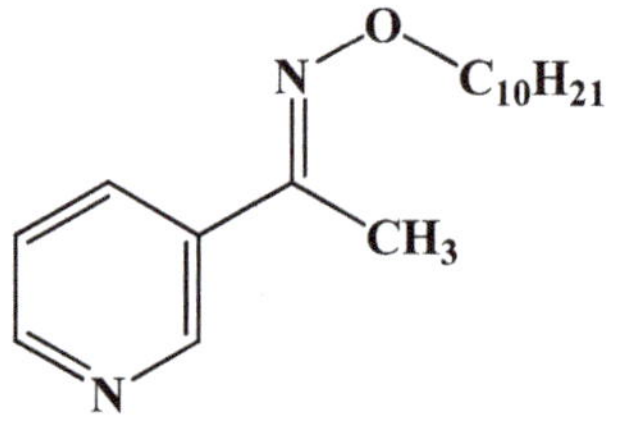

55

Oximes are also suitable for microfluidic extraction. For example, oxime LIX 984N, a mixture of 5-nonylsalicylaldoxime **56** and compound **52**, was used for the separation of Cu^{2+} and Co^{2+} ions. It was found that at optimum operating parameters (initial pH 2.5, volume flow rate 0.035 mL min^{-1} and extractant concentration 17.36%), the copper extraction rate could reach 96.73% with the low cobalt extraction rate (2.41%) [73]. The result shows that the microfluidic method can provide a higher copper extraction rate, a shorter equilibrium setting time, fewer extraction stages and higher copper and cobalt separation factors compared to other methods such as conventional extraction and emulsion liquid membranes.

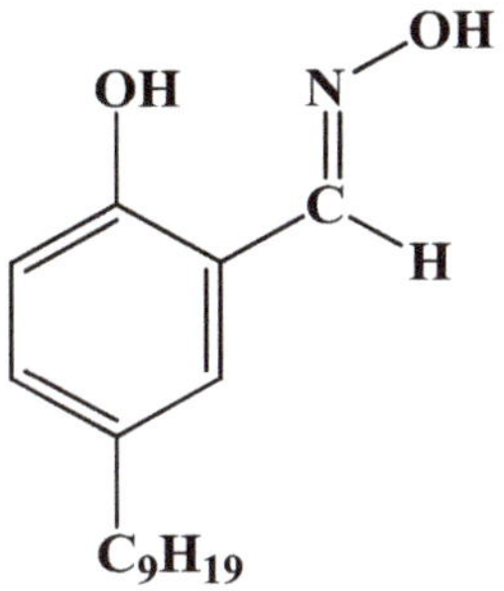

56

Microfluidic extraction of copper from sulfate solution containing Cu (II), Fe (III) and Zn (II) was also carried out using DZ988N extractant, which is a mixture of **56** and 5-dodecylsalicylaldoxime **57** (volume ratio of 1:1). The copper extraction efficiency was 99% in one step. The separation factors, $\beta_{Cu/Fe}$ and $\beta_{Cu/Zn}$, reached maximum values of 644 and 7417 at pH 1.96, extractant concentration 15 vol.% in aliphatic diluent Mextral DT-100, at flow rate 15 mL min^{-1} [74].

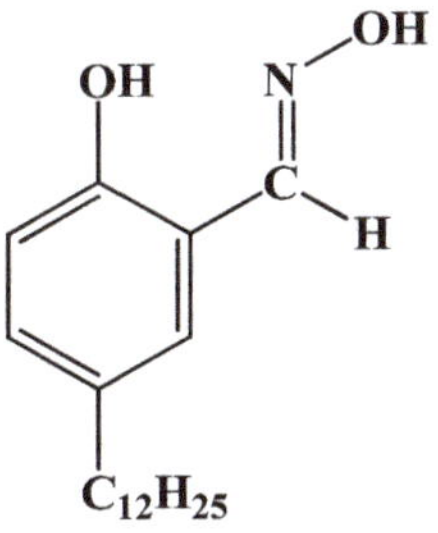

57

A drawback of the works [73,74] is the lack of explanation for the high selectivity of copper extraction.

The oxime Mextral 984H, which is a mixture of compounds **56** and **52**, selectively extracts vanadium (V) from vanadium-bearing shale leachate in the presence of iron Fe (III), aluminum Al (III), magnesium Mg (II), potassium K (I) and calcium Ca (II) impurities. In [75], the efficiency of vanadium (V) extraction by Mextral 984H dissolved in sulfonated kerosene was 90%, while the extraction efficiency for impurities was less than 4% at a temperature of 25 °C, extraction time of 12 min, filtrate pH of 0.53, extractant concentration of 20 vol. % and A/O ratio of 2:1. The authors also studied the extraction mechanism and found that vanadium is coordinated to the hydroxyl oxygen atom and oxime nitrogen atom [75].

4.3. Noble Metals

Compound **47** dissolved in chloroform was used to extract noble metals such as palladium (II), silver (I), platinum (II) and gold (III) from aqueous solutions [76]. Compound **47** was shown to be a promising extractant for platinum (extraction efficiency above 94%), palladium, silver (extraction efficiency above 96%) and gold (above 99%) from polymetallic solutions. However, when polymer membranes containing compound **46** were used, the percentage of sorption of noble metals decreased after 24 h (down to 93% for palladium, 84% for gold, 81% for silver and 48% for platinum). The authors gave no explanation for this fact.

Commercial-grade extractants LIX 84 (compound **52**) and LIX 860-I (compound **57**), dissolved in toluene, can extract up to 99% Pd (II) from nitric acid media in the presence of Cu (II), which is due to the stability of PdL_2 complexes in the organic phase [77]. The quantitative extraction of palladium requires a higher concentration of extractant **57** (3.4 mmol L^{-1}) compared to **52** (1.4 mmol L^{-1}). Additionally, the extraction constants for oxime **52** are higher than for **57**; therefore, LIX 84 (**52**) represents a significant prospect for the extraction of palladium from spent automotive catalysts.

It can be concluded that Schiff bases and oximes are promising compounds for the extraction of thorium from nuclear waste and uranium from uranium ores in the presence of lanthanide impurities, as well as low concentrations of heavy metals and noble metals from industrial wastewater. The separation of heavy metals can also be achieved.

The disadvantage of this class of extractants is the difficulty of quantitative extraction of heavy metals from industrial wastewater. Mixtures of oximes with organophosphorous compounds can be used for this purpose.

5. Crown Ethers

Macrocyclic polyethers, crown ethers, due to their defined ring size and ion–dipole interactions, are able to efficiently extract specific metals, such as lithium, strontium, palladium, as well as seventh group elements present in high-level waste, such as rhenium and technetium.

5.1. Lithium

A team of researchers from South Korea has developed a simple and environmentally friendly scheme for the liquid extraction of lithium cations (Li^+) from seawater in the presence of sodium, potassium and magnesium ions [78]. The extractant used was a dibenzo-14-crown-4 ether with a long lipophilic C_{18}-alkyl chain and a side carboxylic group (compound **58**), diluted with ionic liquid CYPHOSIL 109 $[(C_6H_{13})_3(C_{14}H_{29})P]^+[(CF_3SO_2)_2N]^-$. The quantitative separation of lithium was found to be quite fast, within 10 min, and the extractant could be easily regenerated with dilute hydrochloric acid. Therefore, compound **58** has great potential for the extraction of lithium ions from brine or seawater. However, the need to use expensive ionic liquids as diluents is likely to limit the industrial application of the proposed method.

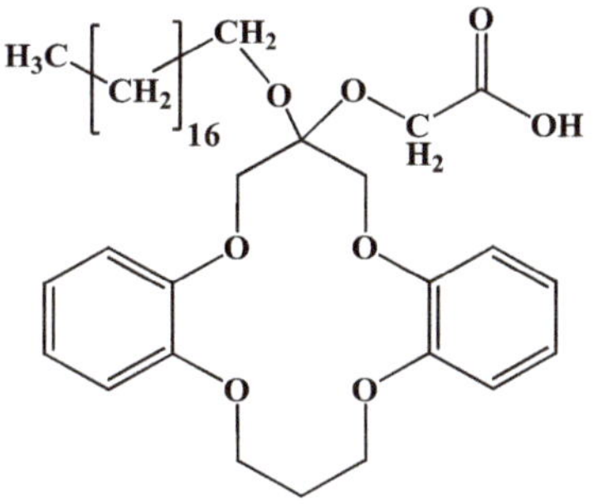

58

Benzo-15-crown-5 ether **59** in dichloromethane (concentration 0.05 mol L^{-1}) can be used for the selective extraction of Li$^+$ from spent lithium-ion batteries. At pH = 6.0, a temperature of 30 °C and an extraction time of 2 h, the extraction degree of Li$^+$ was 37%, which is significantly higher than for impurity ions Ni (II) (extraction degree 5.18%), Co (II) and Mn (II) (extraction degree close to zero) [79]. Unfortunately, high-purity lithium could not be obtained by this method.

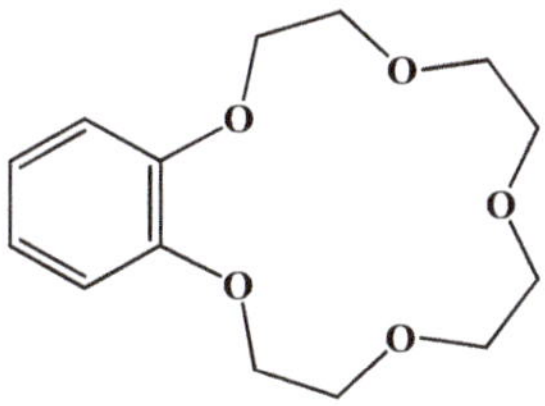

59

5.2. Heavy Metals and Strontium

The extraction of lead (II) and cadmium (II) from nitric acid solutions with benzo-18-crown-6 ether **60** in benzene was studied [80]. The extraction properties of compound **60** were evaluated by determining the extraction constants of these metals. The logK_{ex} values were more than 7 for various Cd/Pb ratios. When the Cd/Pb ratio increased, logK_{ex} decreased. In particular, when Cd/Pb was 1.06, logK_{ex} was 7.143, while an increase in the Cd/Pb ratio to 178 resulted in a logK_{ex} of 7.06.

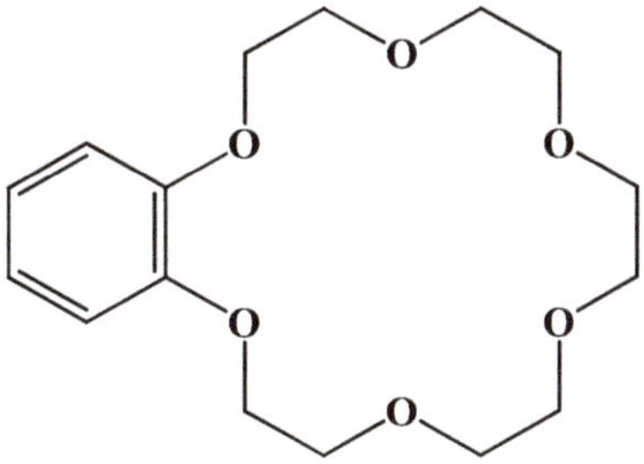

60

However, the authors of [80] did not present distribution ratios or separation factors, which are the key characteristics of extraction processes.

The isotope ^{90}Sr is among the most dangerous radionuclides, as it can accumulate in bone tissue on a par with calcium and lead to radiation damage of surrounding tissues. Therefore, an important goal is to develop an effective method for the extraction of strontium from high-level waste.

In [81], 4,4'(5')-di-tret-butylcyclohexano-18-crown-6 ether (compound **61**) was used for Sr (II) extraction from nitric acid solution. The authors proposed a method consisting in passing the nitric acid solution through a chromatographic column (compound **61** in 1-octanol on an inert polymeric carrier) is simple, economical and allows the analysis of a large volume of samples, but slower compared to extraction with di-(2-ethylhexyl) phosphoric acid in toluene in a liquid–liquid system. The extraction rate of strontium by the developed method was more than 60%.

This method can be used for the extraction of the long-lived radioactive strontium isotope ^{90}Sr from high-level waste, as well as for the separation of parent ^{90}Sr and its daughter isotope ^{90}Y (used in radiation therapy).

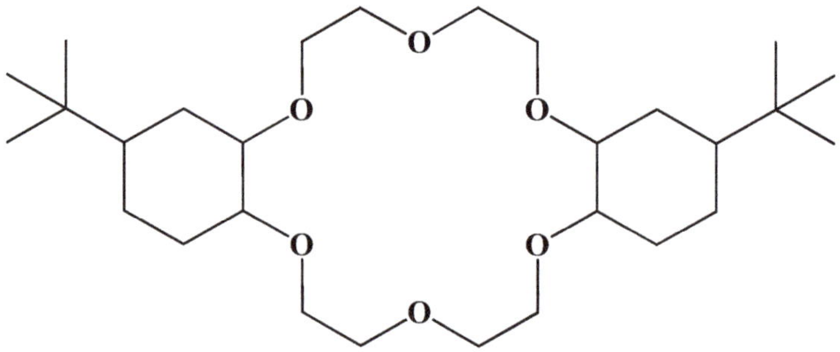

61

The extraction of Sr (II) from nitric acid media using dicyclohexano-18-crown-6 ether **62** dissolved both in common organic solvents and in ionic liquids **3, 63, 64** was studied [82]. The dependence of Sr (II) extraction degree on concentrations of nitric acid, crown ether and initial concentration of strontium was determined. It was found that a shorter carbon chain in the cations of ionic liquids contributed to an increase in the efficiency of metal extraction (the distribution ratio increased from 0.3 for ionic liquid **63** to 2.5 for ionic liquid **64** at a concentration of nitric acid 1 mol L^{-1}). The highest distribution ratio of 3.9 was achieved using a diluent mixture of n-octanol and acetylene tetrachloride (50:50 vol.%) at a nitric acid concentration of 1 mol L^{-1}. However, the authors of the present work provide no explanation for this fact.

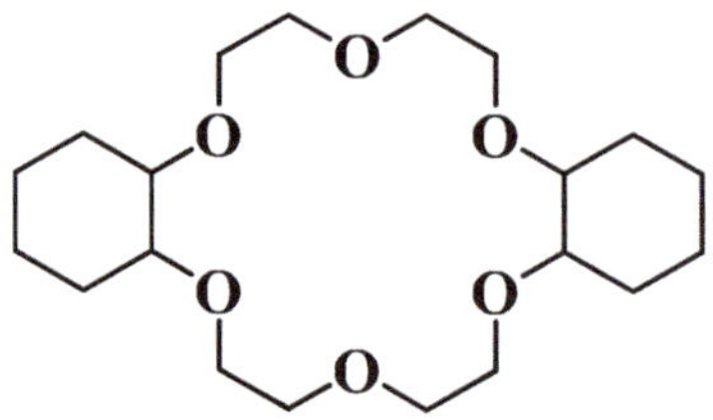

62

63 R = C₂H₅   **64** R = C₆H₁₃

The use of **62** in a mixture of acetylene tetrachloride with n-octanol has great potential for industrial applications for the recovery of radioactive strontium from spent nuclear fuel, in contrast to the much more expensive ionic liquids.

A new bis (crown ether) **65** was synthesized from carboxyl-containing crown ethers and 4,4′-dihydroxy azobenzene by the authors of [83]. Its ability to extract alkali, alkaline earth and transition metal ions was considered. Interestingly, the rate of extraction of metal ions by compound **65** after light irradiation was slightly higher than under natural light. The authors attribute this to the formation of a more compact complex due to the transition from the trans-configuration to the cis-configuration.

For example, when exposed to UV light, the extractant is highly selective for Sr (II) as a result of the formation of a sandwich complex with metal ions, which indicates its promising potential application for the selective extraction of strontium ions.

65

However, a common disadvantage of the works [80–82] is the lack of data on the influence of other radionuclide impurities on ^{90}Sr extraction.

5.3. Elements of 7th Group of the Periodic Table of Elements

Significant amounts of rhenium (Re) are present in waste materials from the petrochemical, aerospace and nuclear fuel industries, so the extraction of rhenium from secondary raw materials is promising for technologies to produce this rare element.

Using dicyclohexano-18-crown-6 ether **62** in chloroform in the presence of high concentrations of uranium (VI), rhenium (VII) was extracted from nitric acid solution in the form of perrhenate ions ReO_4^- [84].

The highest extraction efficiency of ReO_4^- (58%) was observed at a nitric acid concentration of 3 mol L^{-1} and then decreased. This is explained by the extraction mechanism, which is an anion exchange involving the protonated form of the crown ether (6):

$$ReO_4^-{}_{aq.} + [NO_3^- (H_3O^+ \cdot \mathbf{62})]_{org.} = [ReO_4^- (H_3O^+ \cdot \mathbf{62})]_{org.} + NO_3^-{}_{aq.} \quad (6)$$

However, the extraction efficiency of the perrenate ions was practically independent of the uranium concentration when the nitric acid concentration was approximately 3 mol L^{-1} [84]. The authors of the present work provide no explanation for this fact. Separation of rhenium (VII) from impurities of other radionuclides was also not considered.

The treatment of high-level waste produces low-level waste that is treated with alkali for storage in tanks. Extraction of TcO$_4$$^-$ from alkaline waste is difficult due to the presence of competing NO$_3$$^-$ and OH$^-$ ions [85]. Therefore, it is necessary to develop an effective method of extraction of long-lived radionuclide ^{99}Tc from alkaline waste. This is due to the fact that ^{99}Tc has a high mobility in soil and is easily transferred from it to living organisms.

At a high-level waste recycling plant, di-tret-butyl-dibenzo-18-crown-6 ether **66** dissolved in a mixture of isodecyl alcohol and n-dodecane was used to selectively isolate technetium in the form of pertechnetate ions TcO$_4$$^-$ from wastewater [86]. Extractant **66** is capable of forming Na$^+$·crown-ether·TcO$_4$$^-$, which was established experimentally by the increase in distribution ratio D$_{Tc}$ with increasing concentration of sodium ion. Therefore, according to the authors, there is a preferential extraction by crown ether TcO$_4$$^-$ in the presence of competing anions.

66

The separation factor for the mentioned extraction system was more than 22 for technetium with respect to Cs^{2+} (22.5), Sr^{2+} (112.5), RuO$_4$$^-$ (450). The distribution ratio for TcO$_4$$^-$ (4.5) is much higher than the distribution ratios for Cs^{2+} (0.2), Sr^{2+} (0.04) and RuO$_4$$^-$ (0.01).

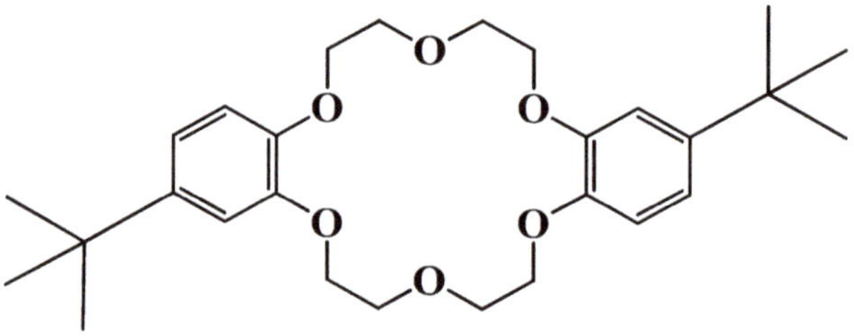

Thus, crown ether **66** is a suitable compound for the recovery of technetium from alkaline low-level waste.

5.4. Noble Metals

It was found in [87] that the logarithms of silver Ag (I) extraction constants from nitric acid, perchlorate and picrate solutions with benzo-18-crown-6 ether **60** dissolved in 1,2-dichloroethane were 1.04, 2.99 and 5.31, respectively. Hence, it is appropriate to extract silver with crown ether **60** from picrate media. However, the authors did not report data on the selectivity of the extraction of silver, i.e., separation factors of silver from other noble metals.

To extract palladium (II) and platinum (II) from spent automobile catalyst leachate and separate these metals, dioxa-dithiacrown ether derivatives **67–69** in toluene were tested [88].

High separation factors of palladium and platinum were observed (on the order of 10^4–10^5), with the highest separation ratio observed for compound **67**. High Pd (II) purity and 99.5% extraction efficiency were achieved using crown ether **67** in toluene.

67

68

69

At the same time, competing metal ions (cerium (III), aluminum (III), iron (III), nickel (II), manganese (II), and chromium (III)) did not impede palladium extraction, as predominantly palladium binds to "soft" ligands containing the S-heteroatom.

Crown ethers are therefore of interest for the extraction and separation of radionuclides from spent nuclear fuel. Crown ethers make it possible to separate similar elements, such as noble metals. However, crown ethers also have a number of drawbacks. These include difficult synthesis, very high cost, difficulty in regenerating the extractant, and low solubility in organic solvents.

6. Calixarenes

Along with crown ethers, another group of macrocyclic compounds, calixarenes, obtained by condensation of p-alkylphenol and formaldehyde, have recently attracted considerable research attention. This class of compounds is known for its excellent performance properties, such as low toxicity, resistance to heating and various types of radiation.

6.1. Lanthanides and Actinides

The process of europium (III) and neodymium (III) extraction by carboxyl-containing calix[6] arene **70** was studied by a group of Chinese scientists led by X. Lu [89].

70

Carboxyl groups, due to cation exchange, increased the extraction efficiency of metal ions from aqueous solutions. The maximum extraction of europium (III) and neodymium (III) was 95% and 70%, respectively, at an extractant concentration of $0.1 \cdot 10^{-3}$ mol L^{-1}. A higher extraction of europium and neodymium could not be achieved. Unfortunately, the authors do not provide data regarding the study of the influence of Y (III), Dy (III), Tb (III) and other REEs on europium and neodymium extraction. In addition, the authors do not give an explanation for the lower degree of extraction of neodymium compared to europium.

In [90], the extraction ability of functionalized calix[4] arenes **71–75** and their non-macrocyclic analogues was compared on the example of Eu (III) and Am (III), which are part of highly active liquid wastes. Extraction was carried out both from alkaline and nitric acid solutions using extractants dissolved in 1-nitro-3-(trifluoromethyl)benzene. As a result, it was concluded that calixarenes extract americium and europium more efficiently than their non-macrocyclic counterparts. For example, distribution ratio $D_{Am, max}$ for extractant **74** is 2 and for extractant **76** is 0.067. Distribution ratio $D_{Eu, max}$ for extractant **74** is 1.7 and for extractant **76** is 0.0335.

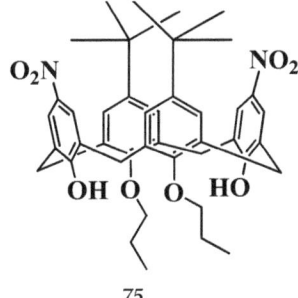

71

72 R^1 and R^2 = OC_3H_7
73 R^1 and R^2 = OC_4H_9
74 R^1 = C_4H_9, R^2 = OC_3H_7

75

76

It is also reported that the functionalization of *p-tret*-butylcalix[4]arene by phosphonate groups increases the distribution ratios of the Am and Eu (extraction efficiency), but has little effect on the separation factors β$_{Am/Eu}$ (extraction selectivity). The extraction efficiency for phosphorylated calixarenes 72 and 73 was higher from nitric acid solutions, while the selectivity was higher in alkaline solutions. Calix[4]arenes with pyridine ring 71 showed the highest selectivity with a separation factor β$_{Am/Eu}$ exceeding 3 at pH = 12–13.

At the same time, it should be noted that the presented calixarenes are inferior in efficiency and selectivity to the extraction of americium from alkaline solutions with the known extractant–2-hydroxy-5-alkylbenzyldiethanolamine 77.

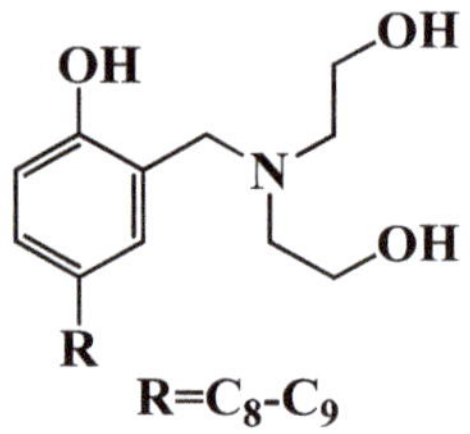

R=C$_8$-C$_9$

77

The extraction behavior of lanthanides, except for promethium Pm (III), from nitric acid medium with calix[4]arene tetraphosphonic acid 78 in chloroform has been considered [91]. It has been observed that the extraction of REEs decreases with increasing nitric acid concentration. For example, the extraction rate of dysprosium (III) decreased from 99.9% at 0.5 mol L^{-1} HNO$_3$ to 17.5% at 1.1 mol L^{-1} HNO$_3$. It was also found that heavy rare earth ions are better extracted than light rare earth ions and are arranged in the following order: Lu^{3+} > Yb^{3+} > Tm^{3+} > Er^{3+} > Ho^{3+} > Dy^{3+} > Tb^{3+} > Gd^{3+} > Eu^{3+} > Sm^{3+} > Nd^{3+} > Pr^{3+} > Ce^{3+} > La^{3+}.

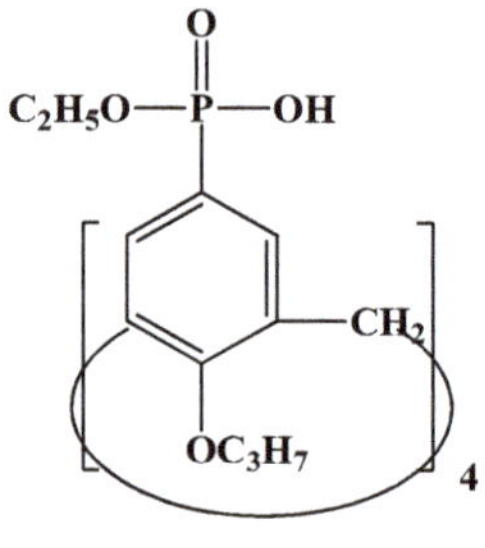

78

The authors explain this extraction sequence by HSAB theory, namely, the stronger bond between the extractant and heavy REEs compared to light REEs.

However, the authors of [91] did not explain why they extracted all lanthanides except Pm (III).

6.2. Noble Metals

Calixarenes can be used for fast and selective extraction of noble and heavy metal ions from ore leaching products and industrial wastewater by microfluidic and liquid extraction.

The extraction of silver (I) from aqueous solutions by thiacalix[4]monocrown ethers 79–87 was reported [92].

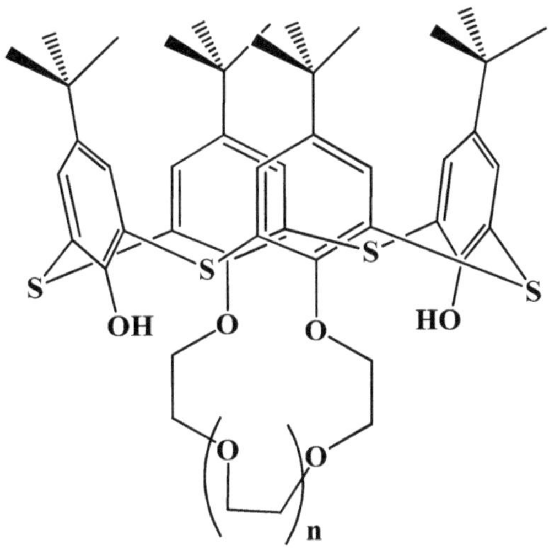

79 n = 4
80 n = 5
81 n = 6

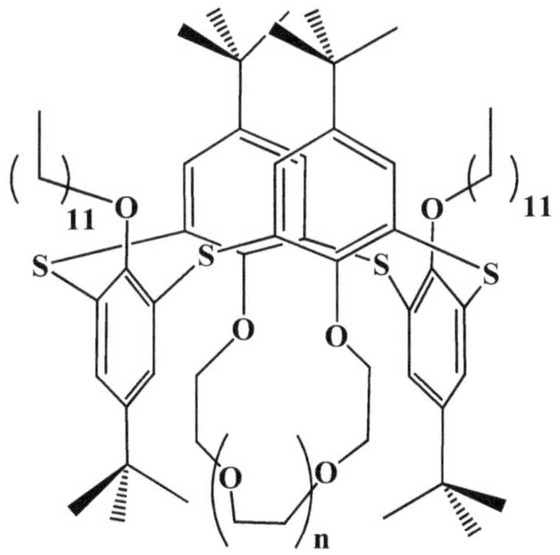

82 n = 4
83 n = 5
84 n = 6

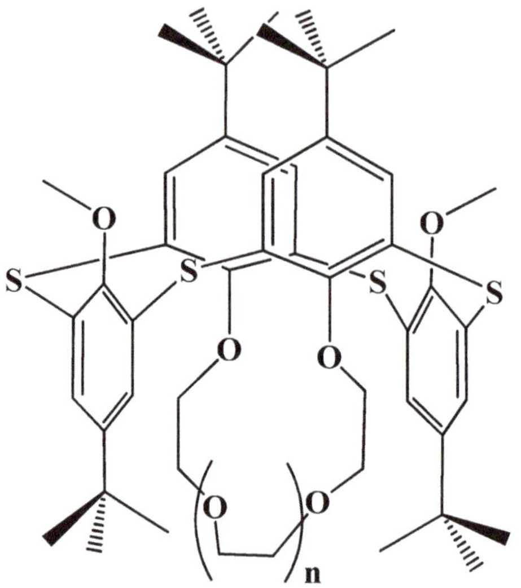

85 n = 4
86 n = 5
87 n = 6

The best efficiency of silver extraction (61%) was found for thiacalix[4]monocrown ether 81. For other compounds, the efficiency of metal extraction was less than 40%. Thus, thiacalix[4]monocrown ethers 79–87 are poor extractants for silver (I).

Nevertheless, the authors of [92] studied the extraction mechanism and found that the silver coordination to thiacalix[4]monocrown ethers 79–81 involves the sulfur atom, the phenolic oxygen atom and two crown ether oxygen atoms.

The authors of [93] reviewed extraction of silver Ag (I) from nitric acid with the help of a number of *p-tret*-octylcalix[4]arenes 88–91 diluted with chloroform. The difference of extraction ability of keto-containing and amide-containing *p-tret*-octylcalix[4]arenes was revealed to be due to the difference in interfacial activity at the chloroform–nitric acid interface. In particular, the interfacial surface tension of amide derivatives 90 and 91 sharply decreased with increasing extractant concentration and the extraction efficiency of Ag (I) was extremely high compared to that of other extractants. This fact is explained by the ability of the extractant to adsorb at the liquid–liquid interface due to the high polarity of the amide group. Unfortunately, the authors do not provide data on the separation of silver (I), gold (III), palladium (II), and platinum (IV) contained in secondary waste from the production of electronic devices, catalysts and jewelry.

It is not clear from the work why chloroform was used as the solvent. From a technological point of view, it is not very convenient as it is very volatile, toxic and can react with silver salts.

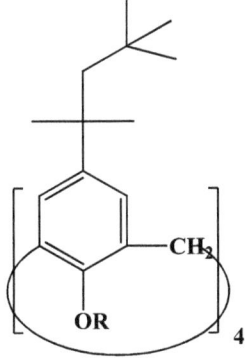

R= —CH₂—C(=O)—CH₃

88

R= —CH₂—C(=O)—C₆H₅

89

R= —CH₂—C(=O)—N(CH₂C₆H₅)₂

90

R= —CH₂—C(=O)—N(C₂H₅)₂

91

High extraction ability with respect to Pd (II) and Pt (IV) ions was found during the extraction of metals from leaching solutions of car catalysts in dialkylaminomethyl-calix[4]arenes **92**, **93** and **94** dissolved in saturated hydrocarbons [94,95].

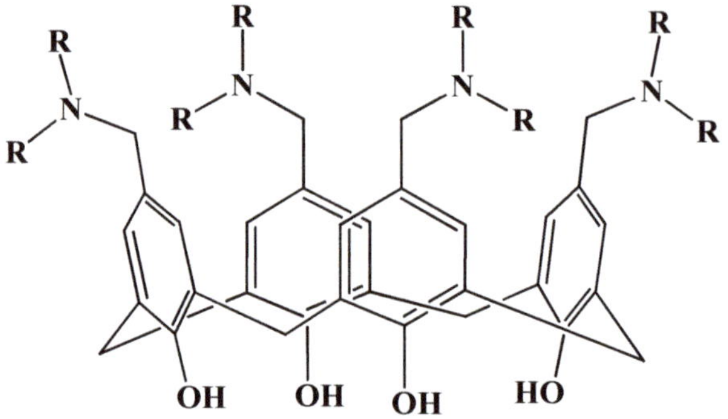

92 R = C_6H_{13}
93 R = C_7H_{15}
94 R = C_8H_{17}

The extraction efficiencies for these compounds were 94.2%, 93.0%, and 97.7% in the case of palladium and 91.7%, 94.0%, and 92.5% for platinum [94]. As a result, palladium and platinum were extracted with 93.4% and 97.3% efficiency, respectively, in five extraction/re-extraction cycles using *p*-di-*n*-octylaminomethylcalix[4]arene **94** in kerosene/1-decanol diluent mixture [95].

The mechanism of metal extraction using compounds **92–94** was described in [94]. The mechanism involves partial extraction of HCl molecules by amino fragments of calixarenes (7) followed by exchange of Cl$^-$ ions for [PdCl$_4$]$^{2-}$ (8) or [PtCl$_6$]$^{2-}$ (9).

$$\text{Extractant} + 2\,\text{HCl} = (\text{Extractant} \cdot 2\text{H})^{2+} + 2\text{Cl}^- \tag{7}$$

$$(\text{Extractant} \cdot 2\text{H})^{2+} \cdot 2\text{Cl}^- + [\text{PdCl}_4]^{2-} = [(\text{Extractant} \cdot 2\text{H})^{2+} \cdot \text{PdCl}_4{}^{2-}] + 2\text{Cl}^- \tag{8}$$

$$(\text{Extractant} \cdot 2\text{H})^{2+} \cdot 2\text{Cl}^- + [\text{PtCl}_6]^{2-} = [(\text{Extractant} \cdot 2\text{H})^{2+} \cdot \text{PtCl}_6{}^{2-}] + 2\text{Cl}^- \tag{9}$$

The impurities in the leaching solution were Rh (III), Zr (IV), Ce (III), Ba (II), Al (III), La (III), and Y (III). The efficiency of impurity extraction was less than 0.01%. Consequently, amino calixarenes with long hydrocarbon chains at the nitrogen atom are promising compounds for the extraction of platinum group metals from secondary raw materials.

6.3. Heavy Metals

The extraction of lead (II) with 25,26,27,28-tetrakis(N,N-diethylamino- carbonylmethoxy)-5,11,17,23-tetrakis(1,1,3,3-tetramethylbutyl)calix[4]arene (compound **95**) in chloroform from nitric acid media in a microfluidic reactor was studied [96]. The extractant showed very high selectivity for Pb (II) in the presence of Fe (III), Zn (II), Cu (II), Ni (II) and Co (II) with concentrations hundreds of times higher than lead. The authors suggest that this fact is due to the formation of complex **82** in the organic phase, but do not explain why the complex does not form with other metals. At the same time, complex **96** itself has not been isolated and studied.

The use of a microfluidic system reduced the extraction and re-extraction times to 2 s, which were initially 1.5 and 24 h, respectively. It is intended to use this development for the rapid removal of lead from industrial wastewater by constructing a large-scale reactor.

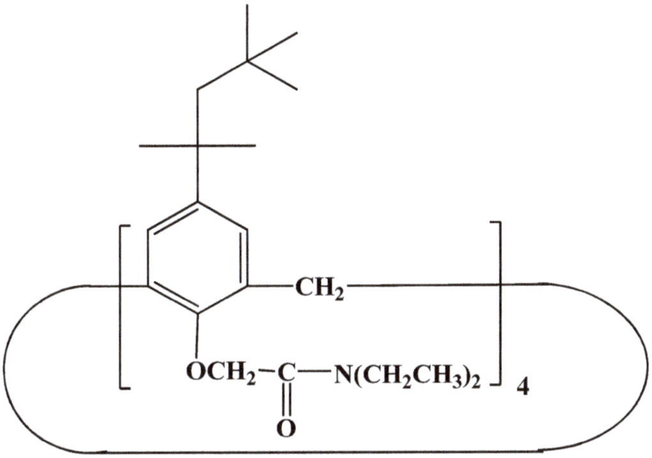

95

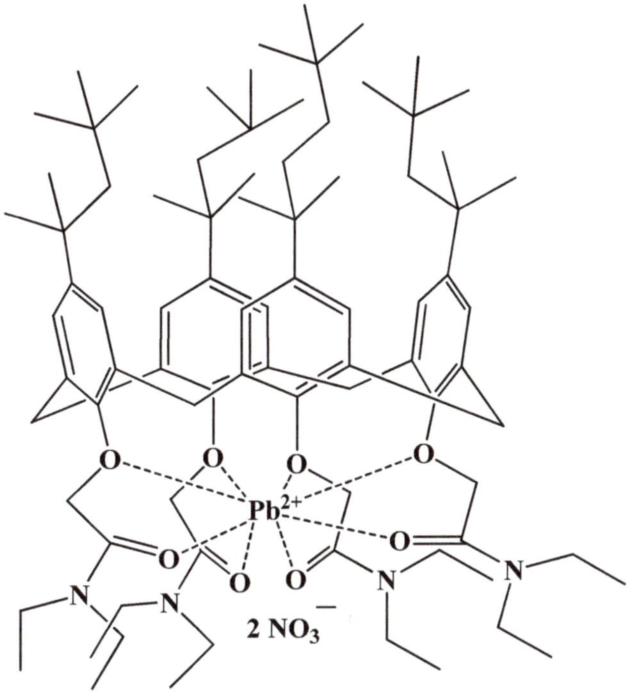

96

The use of calix[n]arenes with chitosan fragments **97–99** (R = -CS or -OH) dissolved in chloroform allows the extraction of Hg (II) from industrial wastewater with efficiencies of 89.46% (extractant **97**), 98.96% (extractant **98**) и 97.85% (extractant **99**) [97].

97 $n = 4$ **98** $n = 6$ **99** $n = 8$

However, in comparison with Hg (II), the extraction efficiency of these rings for other metals was low: 51% for Fe (III), 83% for Cd (II), 56% for Pb (II), 49% for La (III), 54% for Ce (III), and 52% for Eu (III). However, the authors do not explain why this is the case.

6.4. Strontium

Of particular interest are calix[4]arenes containing crown ethers in their structure, which allows for the efficient isolation of certain metals. For example, by using calix[4]arene-crown-6 ether **100**, it was possible to isolate strontium (Sr) from nitrate medium with high efficiency (the value of the distribution ratio D_{Sr} reached 120).

When the process was carried out using ionic liquid **101** and in *n*-dodecane, it was found that the "anion exchange" mechanism prevails in the ionic liquid and the "solvation" mechanism prevails in *n*-dodecane [98].

100

101

This is due to the formation of a neutral metal–extractant complex in *n*-dodecane (10) and an ionic–cationic (11) or anionic (12) complex in the ionic liquid due to its ionic nature:

$$Sr^{2+}_{aq.} + 2NO_3^-{}_{aq.} + m100_{\text{n-dodecane}} = Sr(NO_3)_2 \cdot m100_{org.} \quad (10)$$

$$Sr^{2+}_{aq.} + nNO_3^-{}_{aq.} = m100_{IL} + (2-n)Py^+{}_{IL} = [Sr(NO_3)_n \cdot m100]^{2-n}{}_{IL} + (2-n)Py^+{}_{aq.} \quad (11)$$
$$(n = 2)$$

$$Sr^{2+}_{aq.} + nNO_3^-{}_{aq.} = m100_{IL} + (n-2)NTf^-{}_{2IL} = [Sr(NO_3)_n \cdot m100]^{n-2}{}_{IL} + (n-2)NTf^-{}_{2aq} \quad (12)$$
$$(n = 2)$$

In addition, it was found that the maximum value of the distribution ratio of Sr (II) in the ionic liquid is 5-fold higher than in *n*-dodecane. This is explained by the polar nature of the ionic liquid. It was noted that with changing the concentration of 1-ethyl-2-methylpyridinium cation the values of the strontium distribution ratio did not change, which evidences that the cation was not involved in the complex formation. Consequently, the cation exchange mechanism is not predominant. A decrease in the distribution ratio with increasing concentration of NTf_2^- anion was observed; therefore, the anion is transferred to the aqueous phase during the extraction process and the anion exchange mechanism prevails. The authors propose to use the **100 + 101** extraction system for the extraction of long-lived radionuclides, which are products of ^{235}U, nuclear fission, from high-level nuclear waste. However, the authors of the work studied the extraction of the radionuclide ^{90}Sr in the absence of impurities.

The disadvantages of calixarenes are their limited solubility in organic solvents, which does not allow achieving a high concentration of extractant and high extraction efficiency. Although it is possible to increase the solubility with various additives, this leads to additional material costs.

7. Phenanthroline Derivatives

Phenanthroline is a fused heterocyclic compound capable of coordinating metals owing to the lone pairs of electrons at nitrogen atoms. However, the coordinating action of nitrogen atoms is insufficient for metal extraction. In addition, the selectivity of phenanthroline to various ions is moderate. Therefore, additional functional groups are introduced into the phenanthroline molecule, thus providing for efficient extraction and separation of lanthanides, actinides, and some precious metals.

7.1. Actinides and Lanthanides

The extraction behaviors of phenanthroline (**102**), bipyridine (**103**), and pyridine (**104**) derivatives towards Am (III) and Eu (III) dissolved in highly concentrated nitric acid solutions were compared [99]. 3-Nitrobenzotrifluoride was used as the diluent.

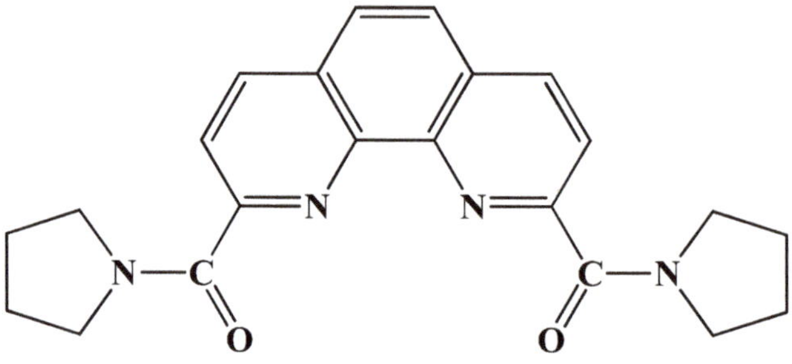

102

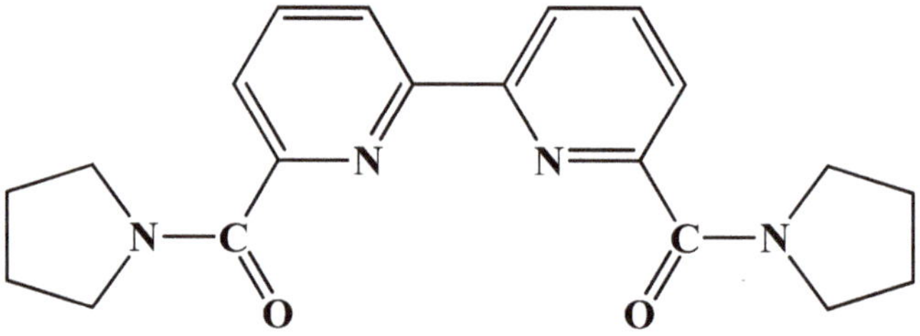

103

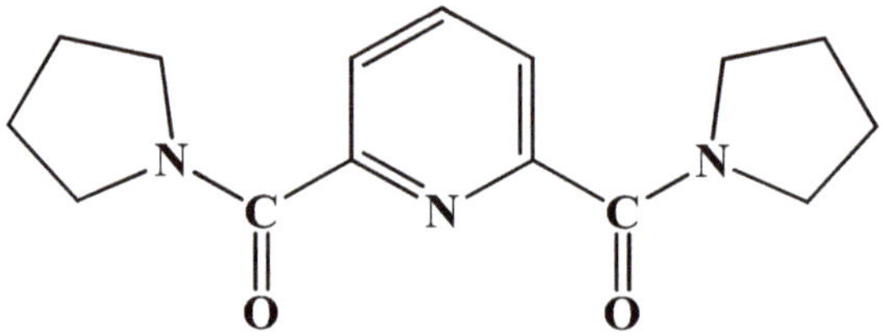

104

Among the tested ligands, compound *102* showed the highest selectivity to americium, with the Am (III) and Eu (III) separation factors being 12, while for compounds *103* and *104*, these factors were 1.1 and 0.5, respectively. This result was attributed to the higher affinity of trivalent actinides for the rigid 1,10-phenanthroline core.

The other two phenanthroline derivatives, tetrabutyl (1,10-phenanthroline-2,9-diyl) phosphonate *105* and tetraethyl (1,10-phenanthroline-2,9-diyl)phosphonate *106*, selectively extract Am (III) from a highly acidic 3 mol L^{-1} HNO_3 solution [100,101]. The distribution ratio of Am (III) was an order of magnitude higher than that of Eu (III): more than 100 for extractant *104* and less than 20 for extractant *106*, respectively.

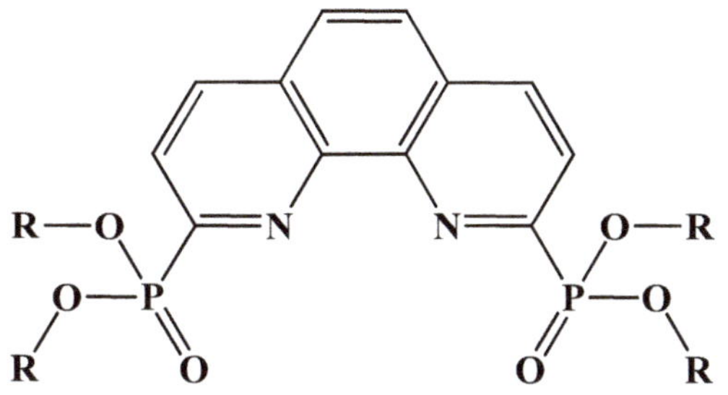

105 R = C₄H₉
106 R = C₂H₅

The higher distribution ratio obtained for extractant **105** was attributed to the fact that compound **105** contains stronger electron-donating groups than **106**.

Additionally, using the dependence $D_{Am(III)}$ = f(log[extractant]), the authors established that for extractant **105** in the loaded organic phase, the 2:1 metal–ligand complex predominates, while in the case of extractant **106**, the predominant species is the 1:1 metal–ligand complex. Hence, the mechanisms of americium(III) extraction with structurally related compounds **105** and **106** are different.

In [102], a number of tetradentate ligands, 4,7-dichloro-1,10-phenanthroline-2,9-dicarboxylic acid diamides (compounds **107–114**), were prepared. It was found that these compounds as solutions in 3-nitrobenzotrifluoride show high selectivity in the separation of americium (III) and lanthanum (III), or americium (III) and europium (III). According to HSAB, these ligands contain two types of coordination sites, hard carbonyl oxygen atoms and soft phenanthroline nitrogen atoms. Being moderately strong Brønsted bases and strong Lewis bases, compounds **107–114** can efficiently bind to metal cations in highly acidic media and form stable metal complexes soluble in polar organic solvents.

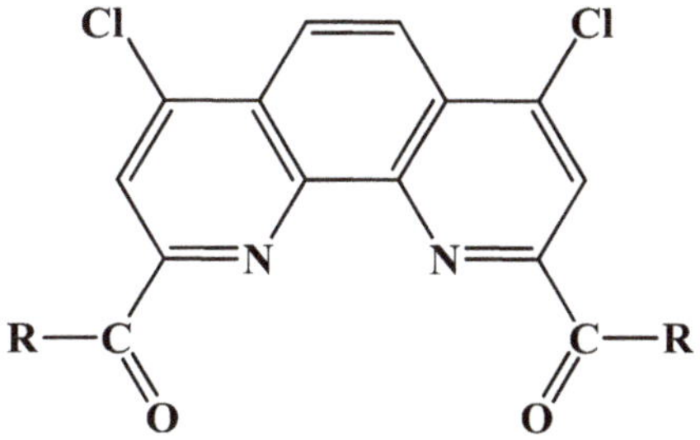

107: R= pyrrolidine
108: R= piperidine
109: R= azepane
110: R= morpholine
111: R= N-methyl-piperazine
112: R= indoline
113: R= 1,2,3,4-tetra-hydroquinoline
114: R= 9H-carbazole

The extraction capacity and selectivity of ligands towards americium (III) were found to decrease on going from **107** to **114**. The decrease in the extraction efficiency for compound **112** in comparison with **107** or **108** was attributed to the fact that the amide groups in the crystals of **112** are located in the *anti*-periplanar position, while in the crystals of **107** and **108**, they are in the periplanar position. Additionally, in the authors' opinion, high barriers for the rotation around the phenanthroline-CO bonds in **112** and **113** contributed to the decrease in the extraction capacity of diamides **112** and **113** in comparison with diamides **107** and **108**.

The separation of americium Am (III) and curium Cm (III) from europium Eu (III) can be accomplished using 2,9-bis(1-(2-ethylhexyl)-*1H*-1,2,3-triazol-4-yl)-1,10-phenanthroline **115** diluted with *n*-octanol or with the [A336][NO$_3$] ionic liquid [103].

115

It was found that in *n*-octanol, selective extraction with **115** is possible only in the presence of 2-bromohexanoic acid as a synergistic agent; the separation factor of Am (III) and Eu (III) is thus more than 200. For extraction in the [A336][NO$_3$] ionic liquid, there is no need to use a synergistic agent, but the americium/europium and americium/curium separation factors are markedly lower, 70 and 1.9–2.2, respectively.

Drawbacks of extractant **115** are the slow extraction rate and formation of a precipitate upon contact between the organic phase and the highly acidic aqueous phase; therefore, extraction can be carried out only from aqueous phases with low acidity (pH = 2–3). In the authors' opinion, this is due to protonation of the triazole ring nitrogen atoms followed by formation of intermolecular hydrogen bonds. Thus, compound **115** cannot be used in continuous extraction processes.

Three tetradentate extractants **116–118** showed high efficiency in the extraction of uranium (VI) (as uranyl ions UO$_2^{2+}$) from nitric acid solutions; the compounds were found to be applicable for the recovery of actinides from nuclear waste [104]. The UO$_2^{2+}$ distribution ratios for compounds **116**, **117**, and **118** were 118, 92, and 90, respectively.

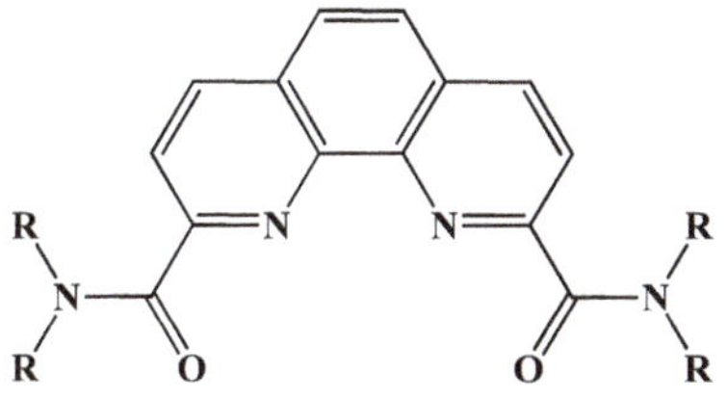

116 R = C₄H₉
117 R = C₆H₁₃
118 R = C₈H₁₇

However, compound *119* with an ethyl group at the amide nitrogen showed a poor extraction capacity. This result was attributed [104] to the fact that the short alkyl chain hampers the ligand dissolution in the organic solvent (1-(trifluoromethyl)-3-nitrobenzene). It was also noted that compound *116* is the most promising extractant for uranium (VI). The authors attributed this result to the steric effect of long alkyl chains, which weaken the coordination of the uranyl ion and decrease the stability constant of the uranyl ion complex.

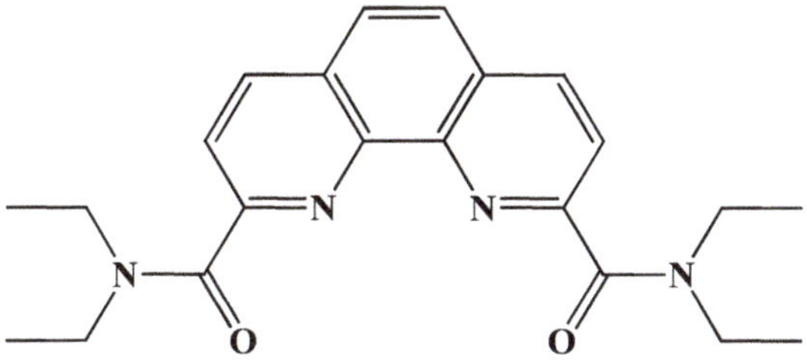

119

Further, the authors indicated that one ligand binds one uranyl ion via two amide oxygen atoms and two phenanthroline nitrogen atoms. Subsequently, the authors plan to study the radiation stability of ligands *116–118*.

N,N'-Diethyl-N,N'-ditolyl-2,9-diamide-1,10-phenanthroline *120* dissolved in 1-(trifluoromethyl)-3-nitrobenzene was used to separate actinides and lanthanides in a nitric acid solution [105]. It was found that compound *120* selectively extracts UO_2^{2+} over a broad range of aqueous-phase acidities.

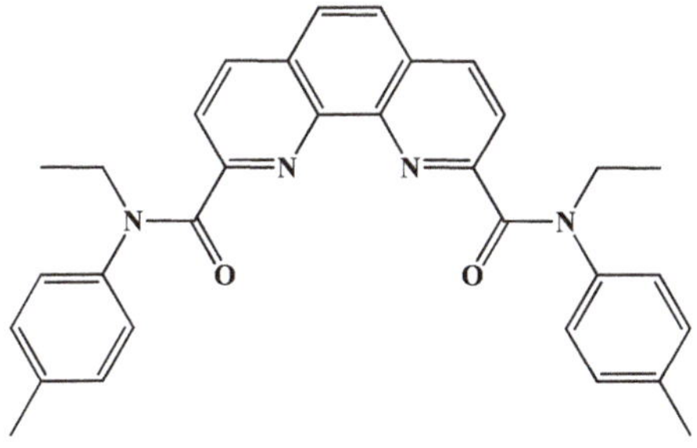

120

The distribution ratio for UO_2^{2+} was more than 300, and the recovery amounted to 99.4%. Meanwhile, trivalent lanthanides (La, Pr, Yb, Nd, Sm, Ce, Gd, Dy, Eu, and Tb) were not extracted from nitric acid solutions. In the opinion of the authors, the nitrogen atoms of the phenanthroline moiety provide for fast and thermodynamically favorable complex formation with the uranyl ion UO_2^{2+}, which is due to the higher effective charge and ionic potential of the uranyl ion in comparison with the lanthanide ions. Quantitative (100%) back extraction of UO_2^{2+} was performed in one stage with a 5% aqueous solution of Na_2CO_3.

7.2. Palladium

Phenanthroline-based compounds form stable complexes with palladium ions and can selectively extract palladium from aqueous solutions. The extraction of palladium from nitric acid solutions with compound **120** dissolved in 3-nitrobenzotrifluoride was studied in [106]. In the opinion of the authors, compound **120** would be suitable for the extraction of Pd (II) from high-level liquid waste.

The extraction equilibrium was attained in 30 min, being indicative of the fast extraction rate. The highest distribution ratio was approximately 10^2.

Additionally, the extractant was found to be reusable for 10 extraction–back extraction cycles, which is a substantial advantage over the other extractants.

Furthermore, unlike dialkylamides, extractants based on phenanthroline have a higher affinity for actinides owing to the presence of additional coordination sites (nitrogen atoms).

However, the need to often use specific expensive solvents, laborious synthesis, and high cost of the phenanthroline derivatives prevent their use on an industrial scale.

8. Other Chelating Extractants

8.1. Lanthanides and Actinides

Polydentate neutral and acidic phosphorus-containing compounds are promising reagents for the extraction of neodymium (III) from nitric acid solutions and for separation of neodymium from light and heavy lanthanides.

The authors of [107] described extraction of neodymium (III) from a nitric acid solution using a polymeric sorbent based on styrene-divinylbenzene copolymer impregnated with bidentate phosphine oxide **121** and ionic liquid **122**.

121

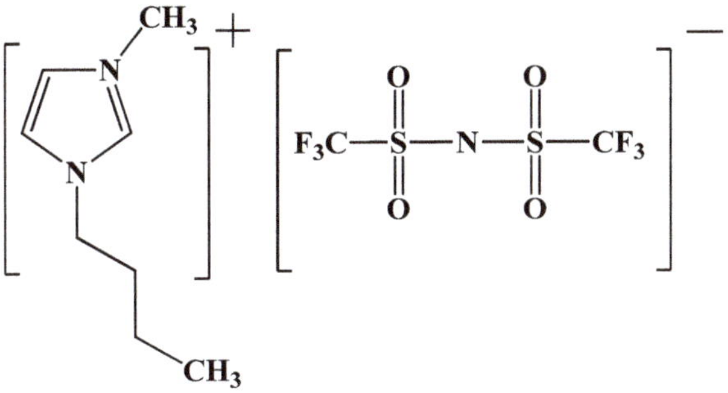

122

The sorbent containing 40% of a **121** + **122** mixture at a **121:122** molar ratio of 2:1 (nitric acid concentration not higher than 0.01 mol L^{-1}) was found to be most efficient. The extraction efficiency with this sorbent was higher than that attained with the sorbent impregnated with only compound **121** or only ionic liquid **122**. The synergistic action of ionic liquid **122** was attributed to the replacement of nitrate anion by more hydrophobic Tf$_2$N$^-$ anion in the extracted neodymium complex.

It was also noted that neodymium is more easily separated from heavy than from light lanthanides. In particular, the β(La/Nd), β(Dy/Nd) and β(Tm/Nd) separation factors were 1.03, 1.15 and 1.30, respectively. The authors explained this fact by higher stability of complexes with heavy lanthanides than with light ones.

In the opinion of the authors, the mechanism of lanthanide extraction with a **121** + **122** mixture can be described by the Equation (13):

$$Ln^{3+}{}_{(a)} + s\mathbf{121}_{(o)} + [C_nmim][Tf^+{}_2N]^-{}_{(o)} = Ln\mathbf{121}_s(Tf_2N)_{3(o)} + 3[C_nmim]^+{}_{(a)} \quad (13)$$

where s is the solvate number, Ln is lanthanide, and (a) and (o) are the aqueous and organic phases, respectively.

Polyamines are promising extractants for lanthanides from scraps of electrical and electronic equipment. For example, lanthanum (III) and neodymium (III) extraction with a sorbent based on montmorillonite clay impregnated with pentaethylenehexamine **123** has been studied [108].

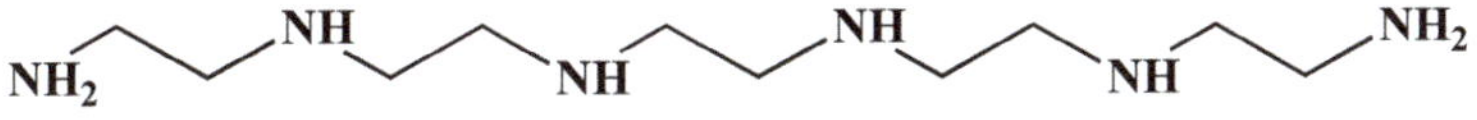

123

The authors proposed three sorption mechanisms: ion exchange, surface adsorption and coordination to the amino groups of compound **123**. It was also noted that the amount of captured ions was higher for lanthanum than for neodymium, with the starting ion content being 0.5–2.0 mmol g^{-1}. When the content of ions was 0.5 mmol g^{-1}, no statistically significant difference between the uptakes of these metals was observed. However, no explanation for this fact is given in the publication [108].

Spent nuclear fuel contains radioactive actinides, in particular radioactive plutonium; therefore, it is necessary to develop efficient extractants and fluorescent sensors for Pu extraction and detection in spent fuel.

In [109], the ability to bind Pu (IV) and PuO$_2$ (VI) in water was investigated for six tridentate N-donor ligands **124–129**.

124 R^1 and R^2 = [2-methylpyridine]

125 R^1 and R^2 = CH$_2$NH$_2$

126 R^1 = CH$_2$NH$_2$, R^2 = [2-methylpyridine]

127 [2,2':6',2''-terpyridine]

128 [2-(2-pyridyl)-1,10-phenanthroline]

129

It was ascertained that the nitrogen-PuO$_2$ (VI) bonds are longer than the nitrogen-Pu (IV) bonds. The authors attributed this to weaker interaction between plutonium and nitrogen caused by the slight positive charge on the central ion of PuO$_2^{2+}$.

It was noted that among ligands **124–129**, the most stable complex with Pu (IV) and PuO$_2$ (VI) is formed with **129**; this complex is characterized by the highest bond energy, indicating that coordination to the pyridine nitrogen atom ensures high stability of complexes. Thus, ligand **129** is a promising compound for plutonium (IV) extraction from aqueous solutions.

Unfortunately, the authors of [109] give no explanation for the lower stability of Pu (IV) and PuO$_2$ (VI) complexes with ligands **127** and **128**, which are structurally similar to compound **129**.

8.2. Heavy Metals

Direct extraction of heavy metals (mercury, lead, cadmium, nickel, etc.) from water samples is a challenging task, which is due to low concentration of the metals in water; therefore, instrumental determination of heavy metals should be preceded by extraction and preconcentration.

Extraction of trace amounts of mercury (II) from wastewater with a hybrid organic-inorganic material, polyaniline-modified molybdenum disulfide (MoS$_2$) nanosheets, was investigated [110]. The adsorbent (0.25 g) provided extraction of 100% of mercury (at a mercury concentration of 100 mg L^{-1}). The adsorption capacity of the adsorbent was 240 mg g^{-1}, which is comparable with the previously studied adsorbents, for example, Fe$_3$O$_4$@SiO$_2$SH.

The mechanism of adsorption at pH = 6.0, associated with chelation of mercury with sulfur ions on the surface of the hybrid material, was described.

Thus, the developed material can be used to monitor the content of mercury in water samples.

The extraction of heavy metals, that is, lead (II), copper (II) and cadmium (II), was performed using cellulose nanofibers covalently functionalized with diethylenetriamine penta(methylene phosphonic acid) **130** [111]. The highest adsorption capacity (180.3 mg g^{-1} for lead, 76.2 mg g^{-1} for copper and 103.4 mg g^{-1} for cadmium) were observed at pH = 6; this was attributed to favorable interaction between the indicated metal ions (soft acids) and groups of phosphonic acids (soft bases). The lower adsorption capacities observed for low pH were attributed to protonation of functional groups and positive surface charges. This material showed low limits of detection of copper, lead and cadmium in trace amounts (0.03–0.05 µg L^{-1}) and can be used to determine these metals in environmental water samples.

130

Cadmium magnetic ion-imprinted polymer MIIP was developed for the extraction of cadmium from water samples [112]. The polymer was synthesized by non-covalent imprinting using two functional monomers: methacrylic acid and acrylamide. $CdCl_2 \cdot 5/2$ H_2O was used as the template and a mixture of acetonitrile and water served as the solvent. The synthesis of MIIP is shown in the scheme below, where EGDMA is ethylene glycol dimethacrylate (cross-linking monomer), AIBN is azobisisobutyronitrile (initiator) and TEOS is tetraethyl orthosilicate:

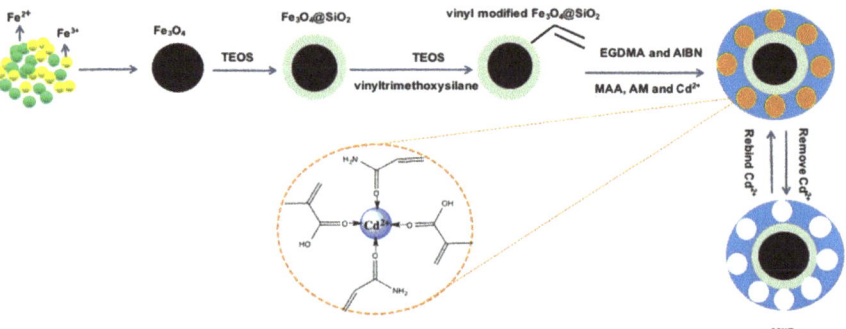

The adsorption capacity of the obtained polymer increased as the content of cadmium in the sample increased from 30 to 80 µg mL^{-1}. The highest adsorption capacity was 46.8 mg g^{-1} (pH = 6). The adsorption capacity remained invariable for six adsorption-desorption cycles. The selectivity factors for lead, copper and nickel were greater than 1.

Thus, the proposed polymer is a promising material for selective extraction of traces of cadmium from water samples in the presence of copper and nickel.

Thenoyltrifluoroacetone *131* was used to extract nickel Ni (II) from nitric acid solutions and to detect traces of nickel (II) in seawater (collected from the Toyama Bay, Toyama, Japan) [113].

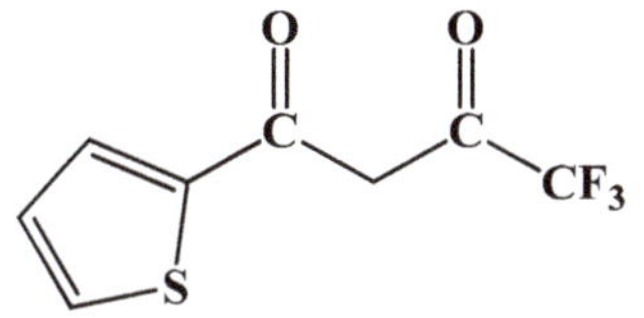

131

Acetone was used as the solvent, because among polar solvents such as methanol, ethanol, and 2-methoxyethanol, quantitative extraction of nickel was attained only in acetone.

The authors proposed the following equations to describe the extraction of nickel at pH = 8 with the enol form of *131* (TTA$^-$):

$$Ni^{2+} + 2\ TTA^- = Ni(TTA)_2 \quad (14)$$

$$Ni^{2+} + 3\ TTA^- = [Ni(TTA)_3]^- \quad (15)$$

It was found that nickel can be extracted from sea water samples when the initial nickel concentration is 0.125 and 0.25 µg L^{-1} with efficiency of 99 and 95%, respectively.

8.3. Platinum-Group Metals

Rhodium is widely used in the production of catalysts; therefore, low concentrations of rhodium salts get into the environment. Hence, it is necessary to develop an efficient method for detection of rhodium in water samples.

In [114], 2-(5-iodo-2-pyridylazo)-5-dimethylaminoaniline *132* was used as the chelating extractant for the recovery of rhodium (III) from water samples.

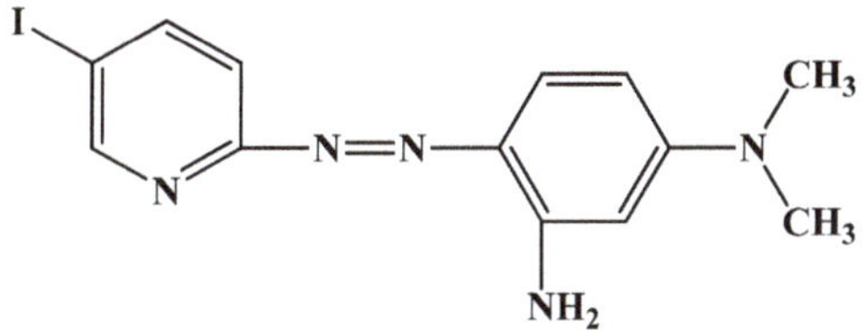

132

The authors assumed that compound *132* coordinates rhodium (III) to give complex *133*.

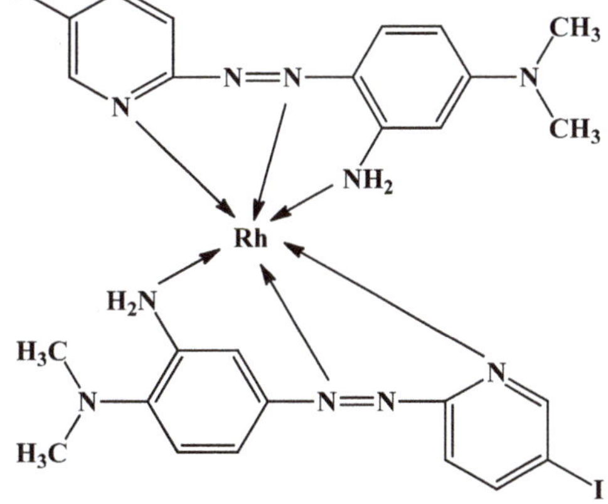

133

The authors established that compound **132** extracts rhodium from tap, well, spring and river water samples with an efficiency of more than 98%. However, no comparative data for extraction with nitrogen-containing agents studied previously or experimental evidence for the mechanism of formation of this complex are given in the paper.

1-Alkyltriazoles are promising compounds—soft ligands (in terms of the HSAB theory) for the separation of soft platinum group metal cations (Pt, Pd, etc.) from hard heavy metal cations (Co, Zn, Ni, etc.).

For example, 1-alkyltriazole **134** (alkyl = pentyl, hexyl, heptyl, octyl, nonyl, decyl, dodecyl, tetradecyl or hexadecyl) dissolved in dichloromethane was used to extract palladium (II) and to separate palladium (II) from zinc (II) and nickel (II) [115].

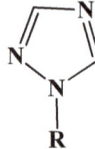

134

The highest percentage of palladium extraction was attained for 1-hexadecyltriazole. This fact is attributable to the increase in the stability constants of complexes with increasing alkyl chain length. However, quantitative extraction of palladium could not be attained.

Conversely, the highest selectivity was observed for a compound with fewer carbon atoms, 1-pentyltriazole. The authors gave no interpretation for this fact.

A similar trend was observed for polymer membranes based on cellulose triacetate impregnated with 1-alkyltriazole **133**. These membranes can be used at low pH (below 2).

9. Conclusions

Dialkylamide compounds are non-toxic, easy to synthesize and dispose after extraction, cheap and practical. According to the literature, dialkylamide compounds are promising compounds to separate actinides contained in high-level waste, to separate them from impurities in uranium fission products and to extract REEs from natural ores and secondary raw materials. Thereby, dialkylamide compounds are a suitable alternative to toxic phosphorous-containing extractants.

However, the extraction capacity of this class of extractants is strictly dependent on the oxidation state of the element to be extracted. In addition, dialkylamide extractants have a low capacity and a tendency to form a third phase.

In turn, amino acids are cheap, form stable complexes with heavy metal ions, but take a long time to establish extraction equilibrium. In addition, amino acids are toxic, accumulate in the environment and are poorly biodegradable. In most of the studied works, the authors do not give an explanation for the low (less than 70%) extraction rate of heavy metals from soil.

Oximes and azomethines form stable chelate complexes with heavy metals and can be used for their extraction (e.g., cobalt, copper, nickel, and chromium) from primary (ores) and secondary (industrial waste) sources. Oximes and azomethines are promising compounds to separate actinides from lanthanide impurities found in radioactive ores. The main disadvantage of this class of extractants is the problem of achieving a complete extraction of the heavy metals, especially those contained in multi-metal solutions.

Crown ethers are promising compounds for reprocessing spent nuclear fuel because they are resistant to radiolysis and hydrolysis. They can also be used for the separation of platinum group metals. The disadvantage of crown ethers is their high cost compared to phosphorous-containing extractants.

Calixarenes are effective for separation of silver, palladium, platinum, lead, mercury, heavy REEs and also products of uranium fission. However, calixarenes are mostly soluble in halogenated or aromatic hydrocarbons, which creates potential problems for the environment and human health. This necessitates the use of expensive green solvents such as ionic liquids.

Compounds containing phenanthroline moieties are promising extractants for the selective separation of actinides and lanthanides during the reprocessing of spent nuclear fuel, as well as for the extraction of palladium from high-level liquid waste. Phenanthroline derivatives have a high extraction rate, high extraction capacity, excellent selectivity towards actinides in a wide range of acidities of the aqueous phase, and high resistance to chemical and radiation corrosion.

According to the literature, it can be concluded that chelating compounds, for the most part, are much more efficient and selective metal extractants compared to monodentate ligands. They are also safer to use than common industrial organophosphorous extractants. However, the high cost, complexity of synthesis and regeneration of most chelating extractants limits their wide practical application. Therefore, the current line of research on this topic is to optimize a number of characteristics such as cost, efficiency, safety, processability and the available raw material base.

For example, it is possible to reduce the cost of extraction processes in general through automation, the use of cheaper natural raw materials, and the development of technologies for the regeneration of components used in extraction. To increase the safety of the process, extraction is carried out in the absence of toxic and flammable solvents, for example, using environmentally friendly and non-flammable ionic liquids. The use of natural amino acids such as glycine and sarcosine for the synthesis of extractants makes the extraction process environmentally friendly. To optimize the extraction efficiency, it is necessary to select the optimal values of pH of the medium, extractant concentration, contact time and temperature, which can be achieved through computer simulation.

Author Contributions: Conceptualization, E.C. and P.Y.; methodology, E.C.; validation, E.C. and P.Y.; writing—original draft preparation, P.Y.; writing—review and editing, E.C.; visualization, P.Y.; supervision, E.C.; project administration, E.C.; funding acquisition, E.C. All authors have read and agreed to the published version of the manuscript.

Funding: This research received no external funding.

Institutional Review Board Statement: Not applicable.

Informed Consent Statement: Informed consent was obtained from all subjects involved in the study.

Data Availability Statement: Not applicable.

Conflicts of Interest: The authors declare no conflict of interest.

References

1. Zhao, M.; Xu, Y.; Zhang, C.; Rong, H.; Zeng, G. New trends in removing heavy metals from wastewater. *Biotechnology* **2016**, *100*, 6509–6518. [CrossRef] [PubMed]
2. Crini, G.; Lichtfouse, E. Advantages and disadvantages of techniques used for wastewater treatment. *Environ. Chem. Lett.* **2018**, *17*, 145–155. [CrossRef]
3. Davidson, C.M. Methods for the Determination of Heavy Metals and Metalloids in Soils. In *Heavy Metals in Soils. Environmental Pollution*; Alloway, B.J., Ed.; Springer: Dordrecht, The Netherlands, 2013; Volume 22, pp. 97–140. [CrossRef]
4. Soltani, F.; Darabi, H.; Aram, R.; Ghadiri, M. Leaching and solvent extraction purification of zinc from Mehdiabad complex oxide ore. *Sci. Rep.* **2021**, *11*, 1566. [CrossRef] [PubMed]
5. Ilyas, S.; Kim, H.; Srivastava, R.R. Hydrometallurgical Recycling of Rare Earth Metal–Cerium from Bio-processed Residual Waste of Exhausted Automobile Catalysts. *JOM* **2021**, *73*, 19–26. [CrossRef]
6. Li, X.Z.; Zhou, L.P.; Yan, L.L.; Dong, Y.M.; Bai, Z.L.; Sun, X.Q.; Diwu, J.; Wang, S.; Bunzli, J.C.; Sun, Q.F. A supramolecular lanthanide separation approach based on multivalent cooperative enhancement of metal ion selectivity. *Nat. Commun.* **2018**, *9*, 547. [CrossRef] [PubMed]
7. Yudaev, P.A.; Kolpinskaya, N.A.; Chistyakov, E.M. Organophosphorous extractants for metals. *Hydrometallurgy* **2021**, *201*, 105558. [CrossRef]
8. Hirayama, N. Chelate extraction of Metals into Ionic Liquids. *Solvent Ext. Res. Dev.* **2011**, *18*, 1–14. [CrossRef]
9. Zalloum, H.M.; Mubarak, M.S. Antioxidant Polymers: Metal Chelating Agents. In *Antioxidant Polymers: Synthesis, Properties, and Applications*; Cirilo, G., Iemma, F., Eds.; John Wiley & Sons Inc.: Hoboken, NJ, USA, 2012; pp. 87–114. [CrossRef]

10. Memon, S.Q.; Memon, N.; Mallah, A.; Soomro, R.; Khuhawar, M.Y. Schiff Bases as Chelating Reagents for Metal Ions Analysis. *Curr. Anal. Chem.* **2014**, *10*, 393–417. [CrossRef]
11. Laskar, M.A.; Siddiqui, S.; Islam, A. Reflection of the physiochemical characteristics of 1-(2-pyridylazo)-2-naphthol on the preconcentration of trace heavy metals. *Crit. Rev. Anal. Chem.* **2016**, *46*, 413–423. [CrossRef]
12. Ding, X.; Liu, Q.; Hou, X.; Fang, T. Supercritical Fluid Extraction of Metal Chelate: A Review. *Crit. Rev. Anal. Chem.* **2016**, *47*, 99–118. [CrossRef]
13. Mahanty, B.; Kanekar, A.S.; Ansari, S.A.; Bhattacharyya, A.; Mohapatra, P.K. Separation of neptunium from actinides by monoamides: A solvent extraction study. *Radiochim. Acta* **2018**, *107*, 369–376. [CrossRef]
14. Ren, P.; Li, Y.; Wang, Z.; Geng, Y.; Yu, T.; Hua, R. Extraction and separation of thorium(IV) and uranium(VI) with 4-oxaheptanediamide into ionic liquid system from aqueous solution. *Chem. Pap.* **2020**, *74*, 2049–2057. [CrossRef]
15. Kumar, S.S.; Rao, A.; Yadav, K.K.; Lenka, R.K.; Singh, D.K.; Tomar, B.S. Selective removal of Am(III) and Pu(IV) from analytical waste solutions of quality control operations using extractant encapsulated polymeric beads. *J. Radioanal. Nucl. Chem.* **2020**, *324*, 375–384. [CrossRef]
16. Gujar, R.B.; Mohapatra, P.; Verboom, W. Extraction of Np4+ and Pu4+ from nitric acid feeds using three types of tripodal diglycolamide ligands. *Sep. Purif. Technol.* **2020**, *247*, 116986. [CrossRef]
17. Ansari, S.A.; Mohapatra, P.K.; Verma, P.K.; Leoncini, A.; Yadav, A.K.; Jha, S.N.; Bhattacharyya, D.; Verboom, W. Highly Efficient Extraction of Trivalent f-Cations Using Several N -Pivot Tripodal Diglycolamide Ligands in an Ionic Liquid: The Role of Ligand Structure on Metal Ion Complexation. *Eur. J. Inorg. Chem.* **2020**, *2*, 191–199. [CrossRef]
18. Turanov, A.N.; Karandashev, V.K.; Boltoeva, M. Solvent extracvation of intra-lanthanides using a mixture of TBP and TODGA in ionic liquid. *Hydrometallurgy* **2020**, *195*, 105367. [CrossRef]
19. Prathibha, T.; Swami, K.R.; Sriram, S.; Venkatesan, K.A. Interference of Zr(IV) during the extraction of trivalent Nd(III) from the aqueous waste generated from metallic fuel reprocessing. *Radiochim. Acta* **2020**, *108*, 543–554. [CrossRef]
20. Liu, Y.; Gao, Y.; Wei, Z.; Zhou, Y.; Zhang, M.; Hou, H.; Tian, G.; He, H. Extraction behavior and third phase formation of neodymium (III) from nitric acid medium in N, N′-dimethyl-N, N′-dioctyl-3-oxadiglcolamide. *J. Radional. Nucl. Chem.* **2018**, *318*, 2087–2096. [CrossRef]
21. Ravi, J.; Venkatesan, K.A.; Antony, M.P.; Srinivasan, T.G.; Vasudeva Rao, P.R. Tuning the diglycolamides for modifier-free minor actinide partitioning. *J. Radioanal. Nucl. Chem.* **2013**, *295*, 1283–1292. [CrossRef]
22. Venkatesan, K.A.; Antony, M.P.; Srinivasan, T.G.; Vasudeva Rao, P.R. New unsymmetrical digycolamide ligands for trivalent actinide separation. *Radiochim. Acta* **2014**, *102*, 609–617. [CrossRef]
23. Swami, K.R.; Venkatesan, K.A.; Selvan, B.R. Studies on the aggregation behaviour of radiolytically degraded tetra (2-ethyhexyl) diglycolamide in n-dodecane medium during the extraction of trivalent metal ions. *J. Radioanal. Nucl. Chem.* **2020**, *325*, 283–291. [CrossRef]
24. Swami, K.R.; Venkatesan, K.A. Unraveling the role of phase modifiers in the extraction of Nd(III) from nitric acid medium in tetra-bis(2-ethylhexyl)diglycolamide in n-dodecane containing long chain aliphatic alcohols. *J. Mol. Liq.* **2019**, *296*, 111741. [CrossRef]
25. Niu, Y.-N.; Ren, P.; Zhang, F.; Yan, Z.-Y. Solvent extraction of Eu3+ with 4-oxaheptanediamide into ionic liquid system. *Sep. Sci. Technol.* **2018**, *53*, 2750–2755. [CrossRef]
26. Chen, J.; Wang, S.; Wang, X. Studies on the extraction of actinides, europium and technetium by diamide derivcatives. In Proceeding of the International Conference Global 2003, New Orleans, LA, USA, 16–20 September 2003; Volume 2, pp. 1915–1919.
27. Parvathy, N.; Swami, K.R.; Prathibha, T.; Venkatesan, K.A. Antagonism in the aggregation behaviour of N,N,N′,N′-tetraoctyldiglycolamide in n-dodecane upon adding N, N-dioctylhydroxyacetamide during trivalent metal extraction. *J. Mol. Liq.* **2020**, *317*, 113940. [CrossRef]
28. Atanassova, M. Investigation of synergism and selectivity using mixture of two neutral extractants in IL media for lanthanoids extraction. *Sep. Purif. Technol.* **2016**, *169*, 253–261. [CrossRef]
29. Borisova, N.E.; Ivanov, A.V.; Kharcheva, A.V.; Sumyanova, T.B.; Surkova, U.V.; Matveev, P.I.; Patsaeva, S.V. Effect of Heterocyclic Ring on Ln III Coordination, Luminescence and Extraction of Diamides of 2,2′-Bipyridyl-6,6′-Dicarboxylic Acid. *Molecules* **2020**, *25*, 62. [CrossRef]
30. Yuan, H.; Hong, W.; Zhou, Y.; Pu, B.; Gong, A.; Xu, T.; Yang, Q.; Li, F.; Qiu, L.; Zhang, W.; et al. Extraction and back-extraction behaviors of 14 rare earth elements from sulfuric acid medium by TODGA. *J. Rare Earth* **2018**, *36*, 642–647. [CrossRef]
31. Chen, Z.; Yang, X.; Song, L.; Wang, X.; Xiao, Q.; Xu, H.; Feng, Q.; Ding, S. Extraction and complexation of trivalent rare earth elements with tetraalkyl diglycolamides. *Inorg. Chim. Acta* **2020**, *513*, 119928. [CrossRef]
32. Sun, G.J.; Yang, J.H.; Yang, H.X.; Sun, G.X.; Cui, Y. Extraction study of rare earth elements with N,N′-dibutyl–N,N′-di (1-methylheptyl)-diglycolamide from hydrochloric acid. *Nucl. Sci. Tech.* **2016**, *27*, 75. [CrossRef]
33. Case, M.E.; Fox, R.V.; Baek, D.L.; Mincher, B.J.; Wai, C.M. Extraction behavior of selected rare earth metals from acidic chloride media using tetrabutyl diglycolamide. *Solvent Extr. Ion Exch.* **2017**, *35*, 496–506. [CrossRef]
34. Xu, D.; Shah, Z.; Sun, G.; Peng, X.; Cui, Y. Recovery of rare earths from phosphate ores through supported liquid membrane using N,N,N′,N′-tetraoctyl diglycol amide. *Miner. Eng.* **2019**, *139*, 105861. [CrossRef]

35. Cui, H.; Feng, X.; Shi, J.; Liu, W.; Yan, N.; Rao, G.; Wang, W. A facile process for enhanced rare earth elements separation from dilute solutions using N,N-di (2-ethylhexyl)-diglycolamide grafted polymer resin. *Sep. Purif. Technol.* **2020**, *234*, 116096. [CrossRef]
36. Tunsu, C.; Menard, Y.; Eriksen, D.Ø.; Ekberg, C.; Petranikova, M. Recovery of critical materials from mine tailings: A comparative study of the solvent extraction of rare earths using acidic, solvating and mixed extractant systems. *J. Clean. Prod.* **2019**, *218*, 425–437. [CrossRef]
37. Gergoric, M.; Barrier, A.; Retegan, T. Recovery of rare-earth elements from neodymium magnet waste using glycolic, maleic, and ascorbic acids followed by solvent extraction. *J. Sustain. Metall.* **2018**, *5*, 85–96. [CrossRef]
38. Tunsu, C.; Petranikova, M. Perspectives for the recovery of critical elements from future energy-efficient refrigeration materials. *J. Clean. Prod.* **2018**, *197*, 232–242. [CrossRef]
39. Bredov, N.S.; Gorlov, M.V.; Esin, A.S.; Bykovskaya, A.A.; Kireev, V.V.; Sinegribova, O.A.; Ryabochenko, M.D. Linear 2-Ethylhexyl Imidophosphoric Esters as Effective Rare-Earth Element Extractants. *Appl. Sci.* **2020**, *10*, 1229. [CrossRef]
40. Paiva, A.P.; Martins, M.E.; Ortet, O. Palladium (II) recovery from hydrochloric acid solutions by N,N'- dimethyl-N,N'-dibutylthiodiglycolamide. *Metals* **2015**, *5*, 2303–2315. [CrossRef]
41. Sasaki, Y.; Morita, K.; Saeki, M.; Hisamatsu, S.; Yoshizuka, K. Precious metal extraction by N,N,N',N'-tetraoctyl-thiodiglycolamide and its comparison with N,N,N',N'-tetraoctyl-diglycolamide and methylimino-N,N'- dioctyacetamide. *Hydrometallurgy* **2017**, *169*, 576–584. [CrossRef]
42. Bożejewicz, D.; Witt, K.; Kaczorowska, M.A.; Urbaniak, W.; Ośmiałowski, B. The Application of 2,6-Bis (4-Methoxybenzoyl)-Diaminopyridine in Solvent Extraction and Polymer Membrane Separation for the Recovery of Au (III), Ag (I), Pd (II) and Pt (II) Ions from Aqueous Solutions. *Int. J. Mol. Sci.* **2021**, *22*, 9123. [CrossRef]
43. Bożejewicz, D.; Witt, K.; Kaczorowska, M.A.; Ośmiałowski, B. The copper (II) ions solvent extraction with a new compound: 2,6-bis (4-methoxybenzoyl)-diaminopyridine. *Processes* **2019**, *7*, 954. [CrossRef]
44. Cheng, S.; Lin, Q.; Wang, Y.; Luo, H.; Huang, Z.; Fu, H.; Chen, H.; Xiao, R. The removal of Cu, Ni, and Zn in industrial soil by washing with EDTA-organic acids. *Arab. J. Chem.* **2020**, *13*, 5160. [CrossRef]
45. Dolev, N.; Katz, Z.; Ludmer, Z.; Ullmann, A.; Brauner, N.; Goikhman, R. Natural amino acids as potential chelators for soil remediation. *Environ. Res.* **2020**, *183*, 109140. [CrossRef]
46. Nair, N.M.; Varghese, G.K. Optimization of parameters for the extraction of Pb from lateritic soil using EDTA. *SN Appl. Sci.* **2020**, *2*, 1344. [CrossRef]
47. Ayyanar, A.; Thatikonda, S. Enhanced Electrokinetic Removal of Heavy Metals from a Contaminated Lake Sediment for Ecological Risk Reduction. *Soil Sediment Contam.* **2021**, *30*, 12–34. [CrossRef]
48. Sun, T.; Beiyuan, J.; Gielen, G.; Mao, X.; Song, Z.; Xu, S.; Ok, Y.S.; Rinklebe, J.; Liu, D.; Hou, D.; et al. Optimizing extraction procedures for better removal of potentially toxic elements during EDTA-assisted soil washing. *J. Soils Sediments* **2020**, *20*, 3417–3426. [CrossRef]
49. Alpaslan, O.; Yaras, A.; Arslanoglu, H. A Kinetic Model for Chelating Extraction of Metals from Spent Hydrodesulphurization Catalyst by Complexing Agent. *Trans. Indian Inst. Met.* **2020**, *73*, 1925–1937. [CrossRef]
50. Sharma, S.; Gautam, A.; Gautam, S. A Greener Approach to Extract Copper from Fertilizer Industry Spent Catalyst. *Arab. J. Sci. Eng.* **2020**, *45*, 7529–7538. [CrossRef]
51. Orr, R.; Hocking, R.K.; Pattison, A.; Nelson, P.N. Extraction of metals from mildly acidic tropical soils: Interactions between chelating ligand, pH and soil type. *Chemosphere* **2020**, *248*, 126060. [CrossRef]
52. Hughes, D.L.; Afsar, A.; Laventine, D.M.; Shaw, E.J.; Harwood, L.M.; Hodson, M.E. Metal removal from soil leachates using DTPA-functionalised maghemite nanoparticles, a potential soil washing technology. *Chemosphere* **2018**, *209*, 480–488. [CrossRef]
53. Kobylinska, N.; Kostenko, L.; Khainakov, S.; Garcia-Granda, S. Advanced core-shell EDTA-functionalized magnetite nanoparticles for rapid and efficient magnetic solid phase extraction of heavy metals from water samples prior to the multi-element determination by ICP-OES. *Microchim. Acta* **2020**, *187*, 289. [CrossRef]
54. Liang, F.; Guo, Z.H.; Men, S.H.; Xiao, X.Y.; Peng, C.; Wu, L.H.; Christie, P. Extraction of Cd and Pb from contaminated-paddy soil with EDTA, DTPA, citric acid and FeCl$_3$ and effects on soil fertility. *J. Cent. S. Univ.* **2019**, *26*, 2987. [CrossRef]
55. Jani, Y.; Hogland, W. Chemical extraction of trace elements from hazardous fine fraction at an old glasswork dump. *Chemosphere* **2018**, *195*, 825–830. [CrossRef] [PubMed]
56. Kim, E.J.; Jeon, E.-K.; Kitae, B. Role of reducing agent in extraction of arsenic and heavy metals from soils by use of EDTA. *Chemosphere* **2016**, *152*, 274–283. [CrossRef] [PubMed]
57. Golmaei, M.; Kinnarinen, T.; Jernström, E.; Häkkinen, A. Extraction of hazardous metals from green liquor dregs by ethylenediaminetetraacetic acid. *J. Environ. Manag.* **2018**, *212*, 219–227. [CrossRef]
58. Beiyuan, J.; Tsang, D.C.W.; Valix, M.; Baek, K.; Ok, Y.S.; Zhang, W.; Bolan, N.S.; Rinklebe, J.; Li, X.-D. Combined application of EDDS and EDTA for removal of potentially toxic elements under multiple soil washing schemes. *Chemosphere* **2018**, *205*, 178–187. [CrossRef]
59. Shukla, M.; Baksi, B.; Mohanty, S.P.; Mahanty, B.; Mansi, A.; Rene, E.R.; Behera, S.K. Remediation of chromium contaminated soil by soil washing using EDTA and N-acetyl-L-cysteine as the chelating agents. *Prog. Org. Coat.* **2022**, *165*, 106704. [CrossRef]
60. Al Zoubi, W. Solvent extraction of metal ions by use of Schiff bases. *J. Coord. Chem.* **2013**, *66*, 2264. [CrossRef]

61. Hawkins, C.A.; Bustillos, C.G.; May, I.; Copping, R.; Nilsson, M. Water-soluble Schiff base-actinyl complexes and their effect on the solvent extraction of f-elements. *Dalton Trans.* **2016**, *45*, 15415–15426. [CrossRef] [PubMed]
62. Hamza, M.F.; Roux, J.C.; Guibal, E. Uranium and europium sorption on amidoxime-functionalized magnetic chitosan microparticles. *Chem. Eng. J.* **2018**, *344*, 124–137. [CrossRef]
63. Shiri-Yekta, Z. Removal of Th(IV) ion from wastewater using a proper Schiff base impregnated onto Amberlite XAD-4. *Particul. Sci. Technol.* **2020**, *38*, 495–504. [CrossRef]
64. Karabörk, M.; Kirpik, H.; Sayin, K.; Köse, M. New diazo-containing phenolic oximes: Structural characterization, computational studies, and solvent extraction of Cu(II), Ni(II), and Zn(II) ions. *Turk. J. Chem.* **2019**, *43*, 197–212. [CrossRef]
65. Wieszczycka, K.; Krupa, M.; Wojciechowska, A.; Wojciechowska, I.; Olszanowski, A. Equilibrium studies of cobalt (II) extraction with 2-pyridineketoxime from mixed sulphate/chloride solution. *J. Radioanal. Nucl. Chem.* **2016**, *307*, 1155–1164. [CrossRef]
66. Witt, K.; Bożejewicz, D.; Kaczorowska, M.A. N,N′-Bis (salicylidene) ethylenediamine (Salen) as an Active Compound for the Recovery of Ni (II), Cu (II), and Zn (II) Ions from Aqueous Solutions. *Membranes* **2020**, *10*, 60. [CrossRef]
67. Al Zoubi, W.; Kandil, F.; Chebani, M.K. Solvent extraction of chromium and copper using Schiff base derived from terephthaldialdehyde and 5-amino-2-methoxy-phenol. *Arab. J. Chem.* **2016**, *9*, 526–531. [CrossRef]
68. Mazarakioti, E.C.; Beobide, A.S.; Angelidou, V.; Efthymiou, C.G.; Terzis, A.; Psycharis, V.; Voyiatzis, G.A.; Perlepes, S.P. Modeling the Solvent Extraction of Cadmium (II) from Aqueous Chloride Solutions by 2-pyridyl Ketoximes: A Coordination Chemistry Approach. *Molecules* **2019**, *24*, 2219. [CrossRef] [PubMed]
69. Almi, S.; Bouzgou, M.; Adjal, F.; Barkat, D. Methyl-isobutyl ketone and 1-octanol as synergistic agents and phase modifiers in solvent extraction of cobalt (II) by the N-(2-hydroxybenzylidene) aniline from sulfate medium. *Inorg. Nano-Met. Chem.* **2020**, *50*, 8–15. [CrossRef]
70. Sun, Q.; Wang, W.; Yang, L.; Huang, S.; Xu, Z.; Ji, Z.; Li, Y.; Hu, Y. Separation and recovery of heavy metals from concentrated smelting wastewater by synergistic solvent extraction using a mixture of 2-hydroxy-5- nonylacetophenone oxime and bis (2,4,4-trimethylpentyl)-phosphinic acid. *Solvent Extr. Ion Exch.* **2018**, *36*, 175–190. [CrossRef]
71. Sun, Q.; Yang, L.; Huang, S.; Xu, Z.; Li, Y.; Wang, W. Synergistic solvent extraction of nickel by 2-hydroxy-5-nonylacetophenone oxime mixed with neodecanoic acid and bis (2-ethylhexyl) phosphoric acid: Stoichiometry and structure investigation. *Miner. Eng.* **2019**, *132*, 284–292. [CrossRef]
72. Reis, M.T.A.; Ismael, M.R.C.; Wojciechowska, A.; Wojciechowska, I.; Aksamitowski, P.; Wieszczycka, K.; Carvalho, J.M. Zinc (II) recovery using pyridine oxime-ether–novel carrier in pseudo-emulsion hollow fiber strip dispersion system. *Sep. Purif. Technol.* **2019**, *223*, 168–177. [CrossRef]
73. Xiao, B.Q.; Jiang, F.; Peng, J.H.; Ju, S.H.; Zhang, L.H.; Yin, S.H.; Zhang, L.B.; Xu, L.; Tian, S.H. Optimization study of operation parameters for extracting Cu^{2+} from sulfuric solution containing Co^{2+}. *Arab. J. Sci. Eng.* **2018**, *43*, 2145. [CrossRef]
74. Jiang, F.; Yin, S.; Zhang, L.; Peng, J.; Ju, S.; Miller, J.D.; Wang, X. Solvent extraction of Cu (II) from sulfate solutions containing Zn (II) and Fe (III) using an interdigital micromixer. *Hydrometallurgy* **2018**, *177*, 116–122. [CrossRef]
75. He, Y.; Zhang, Y.; Huang, J.; Zheng, Q.; Liu, H. Extraction of vanadium (V) from a vanadium-bearing shale leachate through bifunctional coordination in Mextral 984H extraction system. *Sep. Purif. Technol.* **2022**, *288*, 120452. [CrossRef]
76. Witt, K.; Kaczorowska, M.A.; Bożejewicz, D.; Urbaniak, W. Efficient Recovery of Noble Metal Ions (Pd^{2+}, Ag^+, Pt^{2+}, and Au^{3+}) from Aqueous Solutions Using N,N′-Bis (salicylidene) ethylenediamine (Salen) as an Extractant (Classic Solvent Extraction) and Carrier (Polymer Membranes). *Membranes* **2021**, *11*, 863. [CrossRef]
77. Elizalde, M.P.; del Sol Rúa, M.; Menoyo, B. Palladium Extraction from Nitric Acid Solutions by LIX 84 and LIX 860-I. *Solvent Extr. Ion Exc.* **2019**, *37*, 411–421. [CrossRef]
78. Torrejos, R.E.; Nisola, G.M.; Song, H.S.; Han, J.W.; Lawagon, C.P.; Seo, J.G.; Koo, S.; Kim, H.; Chung, W.J. Liquid-liquid extraction of lithium using lipophilic dibenzo-14-crown-4 ether carboxylic acid in hydrophobic room temperature ionic liquid. *Hydrometallurgy* **2016**, *164*, 362–371. [CrossRef]
79. Sun, Y.; Zhu, M.; Yao, Y.; Wang, H.; Tong, B.; Zhao, Z. A novel approach for the selective extraction of Li+ from the leaching solution of spent lithium-ion batteries using benzo-15-crown-5-ether as extractant. *Sep. Purif. Technol.* **2020**, *237*, 116325. [CrossRef]
80. Kudo, Y.; Nakamori, T.; Numako, C. Pb (II) Extraction with Benzo-18-Crown-6 Ether into Benzene under the Co-Presence of Cd (II) Nitrate in Water. *Inorganics* **2018**, *6*, 77. [CrossRef]
81. Xu, C.; Ye, G.; Wang, S.; Duan, W.; Wang, J.; Chen, J. Solvent extraction of strontium from nitric acid medium by di-tert-butyl cyclohexano-18-crown-6 in n-octanol: Extraction behavior and flowsheet demonstration. *Solvent Extr. Ion. Exc.* **2013**, *31*, 731–742. [CrossRef]
82. Wei, Z.; Gao, Y.; Zhou, Y.; Jiao, C.; Zhang, M.; Hou, H.; Liu, W. The extraction of Sr2+ with dicyclohexano-18-crown-6 in conventional organic solvent and ionic liquid diluents. *J. Serb. Chem. Soc.* **2020**, *85*, 909–922. [CrossRef]
83. Liang, Y.; Wang, X.; Zhao, S.; He, P.; Luo, T.; Jiang, J.; Cai, J.; Xu, H. A New Photoresponsive Bis (Crown Ether) for Extraction of Metal Ions. *Chem. Sel.* **2019**, *4*, 10316–10319. [CrossRef]
84. Favre-Réguillon, A.; Draye, M.; Cote, G.; Czerwinsky, K.R. Insights in uranium extraction from spent nuclear fuels using dicyclohexano-18-crown-6–Fate of rhenium as technetium homolog. *Sep. Purif. Technol.* **2018**, *209*, 338–342. [CrossRef]
85. Popova, N.N.; Tananaev, I.G.; Rovnyi, S.I.; Myasoedov, B.F. Technetium: Behaviour during processing of irradiated fuel and in environmental objects. *Rus. Chem. Rev.* **2003**, *72*, 115–137. [CrossRef]

86. Sharma, J.N.; Sinharoy, P.; Kharwandikar, B.; Thorat, V.S.; Tessy, V.; Kaushik, C.P. Process for separation of technetium from alkaline low level waste using di-tert-butyldibenzo-18-crown-6+ isodecyl alcohol/n-dodecane solvent. *Sep. Purif. Technol.* **2018**, *207*, 416–419. [CrossRef]
87. Kudo, Y.; Ikeda, S.; Morioka, S.; Tomokata, S. Silver (I) extraction with benzo-18-crown-6 ether from water into 1, 2-dichloroethane: Analyses on ionic strength of the phases and their equilibrium potentials. *Inorganics* **2017**, *5*, 42. [CrossRef]
88. Torrejos, R.E.C.; Nisola, G.M.; Min, S.H.; Han, J.W.; Lee, S.P.; Chung, W.J. Highly selective extraction of palladium from spent automotive catalyst acid leachate using novel alkylated dioxa-dithiacrown ether derivatives. *J. Ind. Eng. Chem.* **2020**, *89*, 428–435. [CrossRef]
89. Lu, X.; Zhang, D.; He, S.; Feng, J.; Tesfay, A.R.; Liu, C.; Yang, Z.; Shi, L.; Li, J. Reactive extraction of europium (III) and neodymium (III) by carboxylic acid modified calixarene derivatives: Equilibrium, thermodynamics and kinetics. *Sep. Purif. Technol.* **2017**, *188*, 250–259. [CrossRef]
90. Smirnov, I.V.; Stepanova, E.S.; Drapailo, A.B.; Kalchenko, V.I. Extraction of americium and europium with functionalized calixarenes from alkaline solutions. *Radiochemistry* **2016**, *58*, 42–51. [CrossRef]
91. Lu, Y.; Liao, W. Extraction and separation of trivalent rare earth metal ions from nitrate medium by p- phosphonic acid calix[4]arene. *Hydrometallurgy* **2016**, *165*, 300–305. [CrossRef]
92. Muravev, A.; Yakupov, A.; Gerasimova, T.; Nugmanov, R.; Trushina, E.; Babaeva, O.; Nizameeva, G.; Syakaev, V.; Katsyuba, S.; Selektor, S.; et al. Switching Ion Binding Selectivity of Thiacalix [4] arene Monocrowns at Liquid–Liquid and 2D-Confined Interfaces. *Int. J. Mol. Sci.* **2021**, *22*, 3535. [CrossRef]
93. Kim, J.Y.; Morisada, S.; Kawakita, H.; Ohto, K. Comparison of interfacial behavior and silver extraction kinetics with various types calix[4]arene derivatives at heterogeneous liquid-liquid interfaces. *J. Chromatogr. A* **2018**, *1558*, 107–114. [CrossRef]
94. Yamada, M.; Kaneta, Y.; Gandhi, M.R.; Kunda, U.; Shibayama, A. Calix [4] arene-Based Amino Extractants Containing n-Alkyl Moieties for Separation of Pd (II) and Pt (IV) from Leach Liquors of Automotive Catalysts. *Metals* **2018**, *8*, 517. [CrossRef]
95. Yamada, M.; Kaneta, Y.; Gandhi, M.R.; Kunda, U.M.R.; Shibayama, A. Recovery of Pd (II) and Pt (IV) from leach liquors of automotive catalysts with calixarene-based di-n-alkylamino extractants in saturated hydrocarbon diluents. *Hydrometallurgy* **2019**, *184*, 103. [CrossRef]
96. Kurniawan, Y.S.; Sathuluri, R.R.; Iwasaki, W.; Morisada, S.; Kawakita, H.; Ohto, K.; Miyazaki, M.; Jimina. Microfluidic reactor for Pb (II) ion extraction and removal with an amide derivative of calix [4] arene supported by spectroscopic studies. *Microchem. J.* **2018**, *142*, 377–384. [CrossRef]
97. Liu, Y.; Zhong, Z. Extraction of heavy metals, dichromate anions and rare metals by new calixarene-chitosan polymers. *J. Inorg. Organomet. Polym.* **2018**, *28*, 962–967. [CrossRef]
98. Pathak, S.; Jayabun, S.; Boda, A.; Musharaf Ali, S.; Sengupta, A. Experimental and theoretical insight into the extraction mechanism, kinetics, thermodynamics, complexation and radiolytic stability of novel calix crown ether in ionic liquid with Sr^{2+}. *J. Mol. Liq.* **2020**, *316*, 113864. [CrossRef]
99. Meng, R.; Xu, L.; Yang, X.; Sun, M.; Xu, C.; Borisova, N.E.; Zhang, X.; Lei, L.; Xiao, C. Influence of a N-Heterocyclic Core on the Binding Capability of N,O-Hybrid Diamide Ligands toward Trivalent Lanthanides and Actinides. *Inorg. Chem.* **2021**, *60*, 8754. [CrossRef]
100. Xu, L.; Pu, N.; Ye, G.; Xu, C.; Chen, J.; Zhang, X.; Lei, L.; Xiao, C. Unraveling the complexation mechanism of actinide (iii) and lanthanide (iii) with a new tetradentate phenanthroline-derived phosphonate ligand. *Inorg. Chem. Front.* **2020**, *7*, 1726. [CrossRef]
101. Xu, L.; Pu, N.; Li, Y.; Wei, P.; Sun, T.; Xiao, C.; Cheng, J.; Xu, C. Selective Separation and Complexation of Trivalent Actinide and Lanthanide by a Tetradentate Soft–Hard Donor Ligand: Solvent Extraction, Spectroscopy, and DFT Calculations. *Inorg. Chem.* **2019**, *58*, 4420. [CrossRef] [PubMed]
102. Lemport, P.S.; Matveev, P.I.; Yatsenko, A.V.; Evsiunina, M.V.; Petrov, V.S.; Tarasevich, B.N.; Roznyatovsky, V.A.; Dorovatovskii, P.V.; Khrustalev, V.N.; Zhokhov, S.S.; et al. The impact of alicyclic substituents on the extraction ability of new family of 1,10-phenanthroline-2,9-diamides. *RSC Adv.* **2020**, *10*, 26022. [CrossRef]
103. Zsabka, P.; Opsomer, T.; Van Hecke, K.; Dehaen, W.; Wilden, A.; Modolo, G.; Verwerft, M.; Cardinaels, T. Solvent extraction studies for the separation of trivalent actinides from lanthanides with a triazole-functionalized 1,10-phenanthroline extractant. *Solvent Extr. Ion Exch.* **2020**, *38*, 719. [CrossRef]
104. Zhang, X.; Yuan, L.; Chai, Z.; Shi, W. Towards understanding the correlation between UO_2^{2+} extraction and substitute groups in 2,9-diamide-1, 10-phenanthroline. *Sci. China Chem.* **2018**, *61*, 1285. [CrossRef]
105. Zhang, X.; Yuan, L.; Chai, Z.; Shi, W. A new solvent system containing N,N'-diethyl-N,N'-ditolyl-2,9-diamide-1,10-phenanthroline in 1-(trifluoromethyl)-3-nitrobenzene for highly selective UO_2^{2+} extraction. *Sep. Purif. Technol.* **2016**, *168*, 232. [CrossRef]
106. Xiao, Q.; Song, L.; Wang, X.; Xu, H.; He, L.; Li, Q.; Ding, S. Highly Efficient Extraction of Palladium (II) in Nitric Acid Solution by a Phenanthroline-Derived Diamide Ligand. *Sep. Purif. Technol.* **2021**, *280*, 119805. [CrossRef]
107. Kovalenko, O.; Baulin, V.; Baulin, D.; Tsivadze, A. Solvent-Impregnated Resins Based on the Mixture of (2-Diphenylphosphoryl)-4-ethylphenoxy) methyl) diphenylphosphine Oxide and Ionic Liquid for Nd (III) Recovery from Nitric Acid Media. *Molecules* **2021**, *26*, 2440. [CrossRef]
108. Cristiani, C.; Bellotto, M.; Dotelli, G.; Latorrata, S.; Ramis, G.; Gallo Stampino, P.; Zubiani, E.M.L.; Finocchio, E. Rare Earths (La, Y, and Nd) Adsorption Behaviour towards Mineral Clays and Organoclays: Monoionic and Trionic Solutions. *Minerals* **2021**, *11*, 30. [CrossRef]

109. Jeong, K.; Jeong, H.J.; Woo, S.M.; Bae, S. Prediction of binding stability of Pu (IV) and PuO$_2$ (VI) by nitrogen tridentate ligands in aqueous solution. *Int. J. Mol. Sci.* **2020**, *21*, 2791. [CrossRef]
110. Ahmad, H.; BinSharfan, I.I.; Khan, R.A.; Alsalme, A. 3D Nanoarchitecture of Polyaniline-MoS$_2$ Hybrid Material for Hg (II) Adsorption Properties. *Polymers* **2020**, *12*, 2731. [CrossRef]
111. Ahmad, H.; Alharbi, W.; BinSharfan, I.I.; Khan, R.A.; Alsalme, A. Aminophosphonic Acid Functionalized Cellulose Nanofibers for Efficient Extraction of Trace Metal Ions. *Polymers* **2020**, *12*, 12370. [CrossRef] [PubMed]
112. Xu, X.; Wang, M.; Wu, Q.; Xu, Z.; Tian, X. Synthesis and application of novel magnetic ion-imprinted polymers for selective solid phase extraction of cadmium (II). *Polymers* **2017**, *9*, 360. [CrossRef]
113. Kosugi, M.; Mizuna, K.; Sazawa, K.; Okazaki, T.; Kuramitz, H.; Taguchi, S.; Hata, N. Organic Ion-Associate Phase Microextraction/Back-Microextraction for Preconcentration: Determination of Nickel in Environmental Water Using 2-Thenoyltrifluoroacetone via GF-AAS. *Appl. Chem.* **2021**, *1*, 130–141. [CrossRef]
114. Han, Q.; Huo, Y.; Wu, J.; He, Y.; Yang, X.; Yang, L. Determination of ultra-trace rhodium in water samples by graphite furnace atomic absorption spectrometry after cloud point extraction using 2-(5-iodo-2-pyridylazo)-5-dimethylaminoaniline as a chelating agent. *Molecules* **2017**, *22*, 487. [CrossRef] [PubMed]
115. Radzyminska-Lenarcik, E.; Pyszka, I.; Urbaniak, W. New Polymer Inclusion Membranes in the Separation of Palladium, Zinc and Nickel Ions from Aqueous Solutions. *Polymers* **2021**, *13*, 1424. [CrossRef] [PubMed]

Article

Optimization of Iron Recovery from BOF Slag by Oxidation and Magnetic Separation

Mo Lan, Zhanwei He and Xiaojun Hu *

State Key Laboratory of Advanced Metallurgy, University of Science and Technology Beijing, Beijing 100083, China; g20199312@xs.ustb.edu.cn (M.L.); hezhanwei@ustb.edu.cn (Z.H.)
* Correspondence: huxiaojun@ustb.edu.cn

Abstract: In order to solve the problem of solid waste pollution of basic oxygen furnace (BOF) slag in the metallurgical process, this paper took BOF slag as the research object, and carried out oxidation reconstruction of BOF slag and alcohol wet magnetic separation recovery of iron phase, so as to efficiently recover and utilize BOF slag. In the early stages, the research group realized the transformation from weak magnetic iron oxide to strong magnetic magnesia-iron spinel phase in BOF slag through oxidation reconstruction experiments under different technological parameters. On this basis, different conditions in the magnetic separation process were adjusted to achieve the optimal iron recovery rate and grade in this paper. The experimental results show that, under the appropriate reconstruction temperature, with the increase of reaction time, gas flow rate and magnetic field intensity, the iron recovery will increase and the iron grade will decrease. The most suitable magnetic field intensity is 75 mT, the magnetic material yield is 46.00%, the iron grade is 29.10%, and the iron recovery is 64.12%. Compared with the initial steel slag, the iron grade increased by 8.22%, and the iron recovery increased by 46.38% compared with the direct magnetic separation without oxidation.

Keywords: BOF slag; oxidative reconstruction; magnetic separation; iron recovery

1. Introduction

After decades of rapid development, China's iron and steel industry has achieved fruitful results in all aspects. However, due to the inevitable problem of solid waste in heavy industry, iron and steel enterprises have been troubled. According to the statistics of the National Development and Reform Commission, by the end of 2020, China's crude steel output has reached 1.054 billion tons, the production rate of steel-making slag is 8–15% of crude steel, and the emissions of steel-making slag in 2020 are about 84–158 million tons; among them, BOF slag accounts for the vast majority. Due to the large output, large composition fluctuation, poor stability and other reasons, BOF slag cannot be used as cement building materials such as blast furnace slag, so its comprehensive utilization rate is low [1,2].

Stacked BOF slag not only occupies limited land resources, but also pollutes water and soil, which is also a huge waste of resources. In order to make more rational use of BOF slag, it is a key problem that metallurgical enterprises need to find a way to deal with BOF slag on a large scale and recycle it. At present, there are two common methods for recovering iron and iron oxides from BOF slag: reduction and oxidation. The reduction method is mainly to reduce the iron-containing oxides in the BOF slag to metallic iron, and then separate them by magnetic separation. Although metallic iron can be obtained, the reduction process not only consumes a significant amount of energy, but also generates greenhouse gases, which is bad for the long-term development of the environment, and it is also contrary to the goals of carbon peaking and carbon neutrality [3–5].

In recent years, the process of oxidizing steel slag has become the focus of attention. This method oxidizes the non-magnetic phase FeO inside the steel slag into a ferromagnetic phase Fe_3O_4 or $MgFe_2O_4$, and then recovers the iron-containing phase by magnetic separation. Semykina et al. [6,7] respectively analyzed the oxidation mechanism of Fe^{2+} in the FeO-CaO-SiO_2 system and the FeO-CaO-SiO_2-MnO system, and found that the required oxygen partial pressure conditions are relatively harsh and cannot be used industrially. Li et al. [8] used steam as an oxidant to oxidize and roast steel slag. After analysis, it was found that it can be transformed into strong magnetic $MgFe_2O_4$, but the method is complicated and the hydrogen generated affects the safety of the experiment. Xue et al. [9] used dry magnetic separation to separate the magnesium-iron spinel phase in the BOF slag after modification, and analyzed the formation mechanism of $MgFe_2O_4$. The addition of SiO_2 in the modification reduced the oxidizing atmosphere requirements, but also increased costs. Based on the research group's previous research on BOF slag oxidation [10], this paper explores the optimal magnetic separation process for the strongly magnetic $MgFe_2O_4$ phase in oxidized BOF slag, and solves the problem of iron resource recovery in steel slag without increasing additional cost. In addition, since the main component of the residue after the BOF slag oxidation magnetic separation is C_2S (dicalcium silicate) and other silica-containing calcium phases, it can be used as a raw material for cement and other cementitious materials after a little treatment, and it also contains a large amount of free CaO, MgO and other alkaline oxides. These alkaline substances can also be used as low-cost flue gas desulfurization and denitrification agents. Therefore, the process of steel slag oxidation and magnetic separation can provide a new way for the utilization of BOF slag solid waste, and finally realize a large-scale process application of BOF slag, so as to solve the problem of its low utilization rate [11–15].

2. Materials and Methods

2.1. Experimental Materials

The experimental slag was obtained from a steel and iron group; samples with a particle size of 140 μm were collected after jaw crushing and electromagnetic crushing, and samples larger than 140 μm were broken several times until all samples were 140 μm and used as BOF slag for the experiment. The composition was analyzed by X-ray fluorescence spectrometer (XRF, PANalytical B.V., Almelo, The Netherlands); its chemical composition is shown in Table 1.

Table 1. Chemical analysis of the tested slag.

Composition	CaO	SiO_2	MgO	Al_2O_3	MnO	P_2O_5	TiO_2	FeO	TFe
Content (wt %)	41.40	16.80	6.53	4.87	3.76	1.91	1.43	20.60	20.88

2.2. Methods

Figure 1 is the phase equilibrium diagram of the oxidized slag calculated by FactSage 7.0 (Thermfact and GTT-Technologies, Montreal and Aachen, Canada and Germany) after the initial steel slag composition in Table 1 is homogenized. The Equilib module was selected in the calculation process, and the product database was selected as Ftoxid-SLAGA, Ftoxid-SPINA, Ftoxid-MeO-A, Ftoxid-bC$_2$S and Ftoxid-aC$_2$S. The ambient atmospheric pressure was set as a standard atmospheric pressure, the partial pressure of oxygen was 0.21 atm, and the reaction temperature was 800–1500 °C. It can be seen that the spinel phase will appear at a temperature between 1050 and 1400 °C, which means that the temperature range of spinel phase formation is wide, and the FeO phase in steel slag can be transformed into a magnetic spinel phase at a lower temperature.

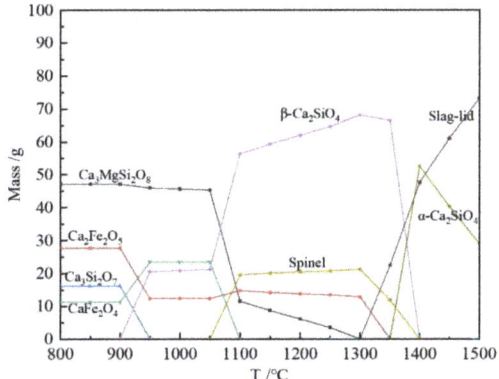

Figure 1. Phase equilibrium diagram of slag oxidized.

In this experiment, a sample weighing 50 g was placed in a corundum crucible and was heated by a vertical MoSi$_2$ high-temperature tube furnace. The experimental device was the same as that used in our previous work [10]. During the heating process, argon gas was introduced from the top of the furnace as the protective gas. We first explored the influence of different thermal insulation temperatures (950, 1000, 1050, and 1100 °C) on the experimental results. At this time, the gas flow was controlled at 1 L/min and the thermal insulation time was controlled at 60 min and the injection compressed air as the oxidizing gas. Similarly, we also explored the influence of different flow rates (0.5, 0.75, 1, and 1.25 L/min) on the experimental results. At this time, the insulation temperature was controlled at 1050 °C and the insulation time was controlled at 60 min, the oxidation gas was still selected as compressed air. Finally, we explored the influence of the heat preservation time (20, 40, 60, and 100 min) on the experimental results. At this time, the holding temperature was controlled at 1050 °C, the gas flow was controlled at 1 L/min, and other experimental conditions were the same as above. When the oxidation process was over, the sample was taken out and quickly cooled to room temperature using nitrogen blowing. The experimental conditions of oxidation are shown in Table 2.

Table 2. Experimental conditions of oxidation and magnetic separation.

Procedure		Temp. (°C)	Time (min)	Flow Rate (L/min)	Magnetic Field (mT)
Temp. (°C)	950 1000 1050 1100	-	60	1	50, 75, 100
Time (min)	20 40 60 100	1050	-	1	50, 75, 100
Flow rate (L/min)	0.5 0.75 1 1.25	1050	60	-	50, 75, 100

The oxidized slag was fully ground to 74 µm by a planetary ball mill. Five grams of the sample was used for magnetic separation, using DTCXG-ZN50 wet magnetic separator (Dongtang Electric Co., Ltd., Tangshan, China). In order to prevent the phase change of steel slag after contact with water, alcohol was selected as the solvent of steel slag. The magnetic field intensities of magnetic separation are set in Table 2, which are 50, 75 and 100 mT, respectively. After separation, the magnetic material (magnetic separation concentrate) and non-magnetic material (magnetic separation tailings) were collected. The total iron content of magnetic separation concentrate was analyzed, and the recorded quality data, the yield, iron grade and iron recovery rate of the magnetic separation concentrate were calculated, respectively.

Among them, the iron content of the magnetic separation concentrate was determined by the GB/T6730.65-2009 titanium trichloride reduction potassium dichromate titration method. The method of decomposing the sample was the hydrochloric acid-sodium fluoride decomposition method. The sample was decomposed by most of the trivalent iron in the stannous chloride reduction test solution, and then sodium tungstate was used as an indicator, titanium trichloride would reduce all the remaining trivalent iron to divalent to produce "tungsten blue", with dichromic acid potassium solution oxidizing the excess reducing agent. In a sulfuric acid-phosphoric acid medium, using sodium diphenylamine sulfonate as an indicator, potassium dichromate standard titration solution was used to titrate ferrous iron. The expression of the mass fraction of total iron in the magnetic separation concentrate is:

$$\text{Iron grade (\%)} = \frac{c \times (V - V_0) \times 55.85}{m \times 1000} \times 100. \tag{1}$$

In Formula (1): c is the concentration of potassium dichromate standard titrant (mol/L), V is the volume of potassium dichromate standard titrant consumed by the titration sample solution (mL), V_0 is the titration blank test solution consumption of potassium dichromate Standard titrant volume (mL), and m is the mass of the sample (g).

The yield expression of magnetic separation concentrate is:

$$\text{Yield (\%)} = \frac{M_1}{M_2} \times 100\%. \tag{2}$$

The expression of iron recovery rate is:

$$\text{Iron recovery rate (\%)} = \frac{TFe_1 \times M_1}{TFe_2 \times M_2} \times 100\%. \tag{3}$$

In Formulas (2) and (3), TFe_1 is the mass of total iron in the magnetic material separated by magnetic separation, M_1 is the mass of the magnetic material separated by magnetic separation, and TFe_2 is always the total mass of the magnetic material. Iron mass, M_2 is always the total material mass.

3. Results and Discussion

3.1. Phase Composition of Raw Slag, Oxide Slag and Magnetic Separation Slag

Figure 2 shows the XRD patterns of the raw slag used in the experiment and the slag under different oxidation conditions. It can be found that the mineral phase system in the raw steel slag is more complicated, including β-Ca_2SiO_4 (β-C_2S) and Ca_3SiO_5 (C_3S), $Ca_3MgSi_2O_8$ (C_3MS_2), FeO, Fe_3O_4 and a small amount of $Ca_2Fe_2O_5$ (C_2F). After oxidation under certain conditions, the number of phases is reduced, mainly composed of silicon-calcium phase and magnesium-iron spinel phase. It can also be seen from the figure that when the oxidation time is increased and the air flow rate is increased, the diffraction peak intensity of $MgFe_2O_4$ increases, indicating that, as the degree of gas-solid reaction deepens, the content of the $MgFe_2O_4$ spinel phase increases. This is mainly due to the diffusion of gas molecules from the upper sample to the lower sample during the oxidation process until the sample is completely oxidized.

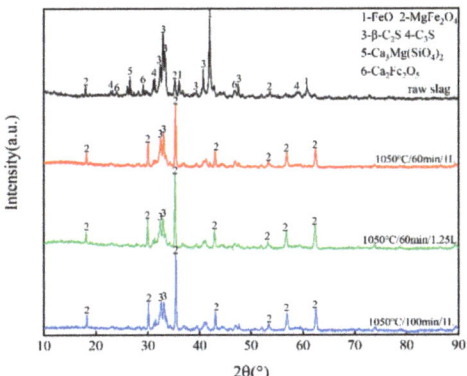

Figure 2. X-ray pattern of raw and oxidized slag under different conditions.

In the oxidation process of raw slag, FeO in slag is oxidized into Fe_3O_4 and Fe_2O_3; the Fe_2O_3 produced will form a solid solution $MgFe_2O_4$ with the MgO in the slag. Since the diameters of the two ions of Mg^{2+} and Fe^{2+} are very close, and the radius of Mg^{2+} is slightly larger than that of Fe^{2+}, in a certain temperature range, Mg^{2+} will enter the Fe_3O_4 lattice through solid-state diffusion to replace part of Fe^{2+}, and then $MgFe_2O_4$ is formed and a more stable spinel phase is precipitated. The above two methods are the main methods for oxidizing BOF slag to form strong magnetic $MgFe_2O_4$ [16–20].

The XRD patterns of the magnetic separation concentrate and magnetic separation tailings after magnetic separation are shown in Figure 3. It can be found that the magnetic separation concentrate phase is mainly composed of $MgFe_2O_4$ and β-C_2S. Compared with the diffraction peaks corresponding to $MgFe_2O_4$ and β-C_2S in the oxide slag XRD, the $MgFe_2O_4$ diffraction peak intensity is higher, while the β-C_2S diffraction peak intensity is lower, which means that the $MgFe_2O_4$ phase is the main phase in the magnetic separation concentrate. Therefore, the magnetic separation concentrate can be used for ironmaking or sintering. The phases in the magnetic separation tailings are composed of $MgFe_2O_4$, β-C_2S and FeO. Compared with the magnetic separation concentrate, the diffraction peak of $MgFe_2O_4$ has a much lower intensity, which also shows that the strong magnetic $MgFe_2O_4$ can be separated by magnetic separation.

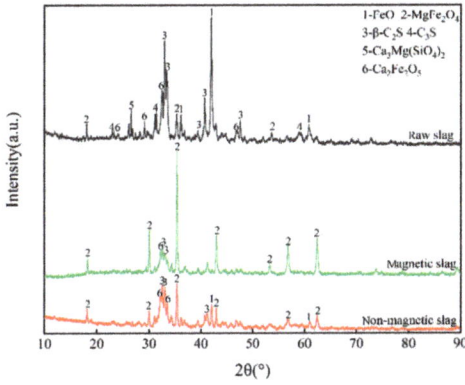

Figure 3. X-ray pattern of raw and magnetic separation slag.

3.2. The Influence of Oxidation Temperature on the Effect of Magnetic Separation

Firstly, the influence of temperature on the magnetic separation effect was investigated. When 1 L/min of air is introduced, the raw slag is oxidized at different temperatures for 60 min, then the sample is taken out and ground to 200 mesh, and magnetic separation is carried out at a magnetic field strength of 75 mT. The yield, iron grade and iron recovery of the magnetic separation concentrate of the 950–1100 °C oxide slag are shown in Figure 4. It can be found that when the oxidation temperature is greater than 1000 °C, the iron grade fluctuates between 27.32% and 27.74%, and the iron recovery fluctuates between 75.53% and 76.00%. The overall magnetic separation effect is not much different, which shows that, within the appropriate reaction temperature range, the amount of $MgFe_2O_4$ produced is not much different, and it also proves from the side that the choice of the reaction temperature range is correct.

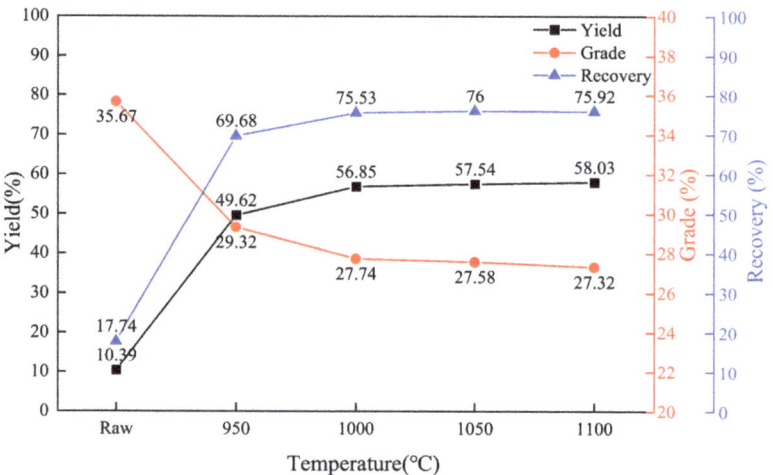

Figure 4. Magnetic separation effect between different oxidation temperatures at 75 mT.

3.3. The Influence of Gas Flow Rate on the Effect of Magnetic Separation

It can be seen from Figure 4 that the magnetic separation effect of BOF slag is the best after 60 min of oxidation at 1050 °C. Therefore, the influence of different gas flow rates (0.5, 0.75, 1, and 1.25 L/min) on the magnetic separation effect is investigated under this temperature condition. The magnetic separation effect is shown in Figure 5. The figure shows that, with the increase of the flow rate, the magnetic separation yield and iron recovery are increasing, which is far better than the magnetic separation result of the raw slag, indicating that the appropriate amount of air can oxidize the slag to a greater extent. Among them, the yield of magnetic separation concentrate is the highest at 1.25 L/min. However, due to the sharp drop in grade, the recovery is reduced. When the gas flow rate is 1 L/min, the recovery is the highest, and the iron grade is relatively high, which is a suitable gas flow rate. When the gas flow rate increases, the oxidation reaction speed is increased, and the conversion of FeO inside the sample to $MgFe_2O_4$ is intensified. The amount of formation is the largest, but the magnetic separation effect decreases. This is because with the progress of the reaction, the Fe^{2+} inside the sample is gradually oxidized to $MgFe_2O_4$, and the magnetic separation concentrate selected during the magnetic separation process increases, resulting in an increase in the yield and recovery, and $MgFe_2O_4$ is wrapped by C_2S. This can also be demonstrated by a scanning electron microscope (SEM, Carl Zeiss AG, Oberkochen, Germany) photograph of oxidized BOF slag in Figure 6. The non-magnetic phases are selected together by the magnetic field, resulting in a decline in grade.

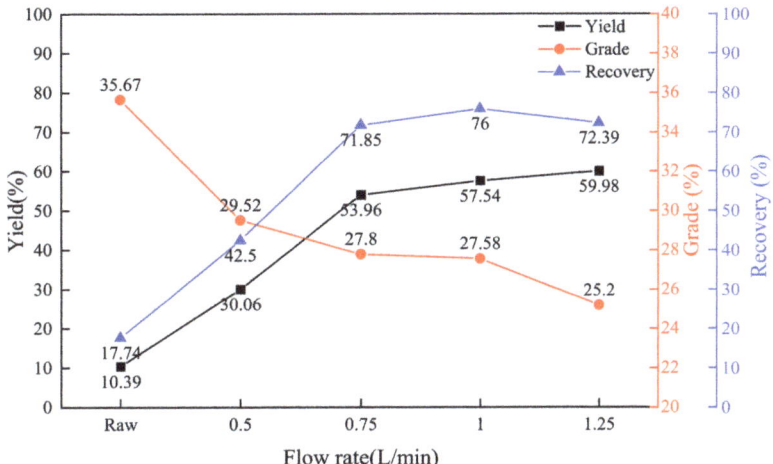

Figure 5. Magnetic separation effect between different gas inlet flow rates at 75 mT.

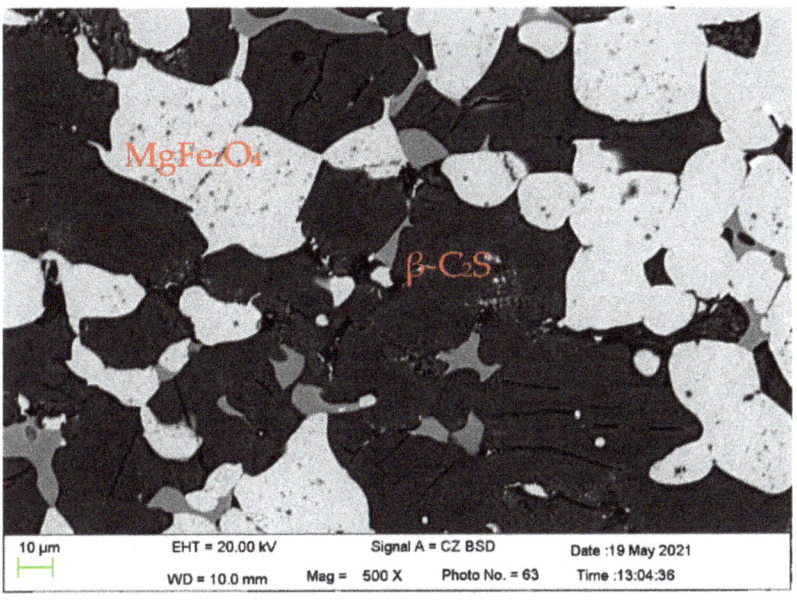

Figure 6. Scanning electron microscope photograph of oxidized BOF slag.

3.4. The Influence of Oxidation Time on the Effect of Magnetic Separation

It can be seen from Figures 4 and 5 that the magnetic separation effect is best when the oxidation temperature is 1050 °C and the gas flow rate is 1 L/min. Therefore, the comparison of the magnetic separation effect of oxidized BOF slag when holding for 20–100 min is shown in Figure 7. The effect of magnetic separation is similar to that of different flow rates. The relationship between yield and iron recovery and reaction time is positively correlated, while grade is negatively correlated. This shows that the nature of time and flow changes are the manifestation of the reaction process, which corresponds exactly to the XRD pattern of the steel oxide slag. Finally, as the reaction deepens, the magnetic separation effect first increases and then decreases.

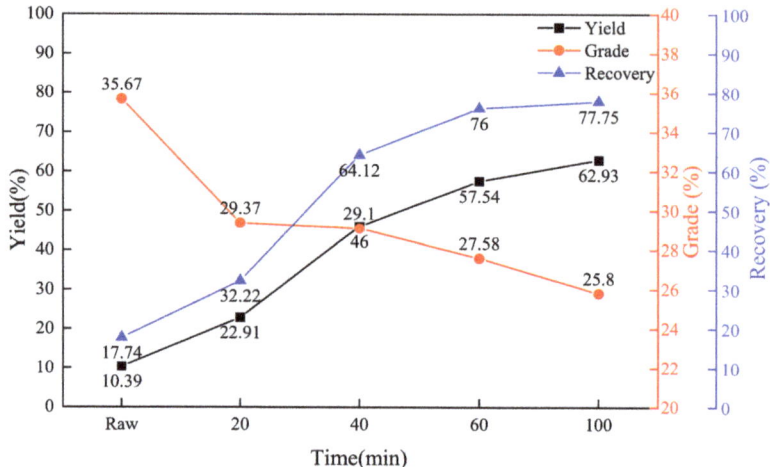

Figure 7. Magnetic separation effect under different oxidation time at 75 mT.

3.5. The Influence of Magnetic Field Strength on the Effect of Magnetic Separation

Finally, the influence of magnetic field on magnetic separation effect was explored. The optimal oxidation conditions were selected, that is the gas flow rate was 1 L/min, and the steel slag was oxidized for 60 min at 1050 °C, the oxide slag was taken out, and wet magnetic separation was used to perform magnetic separation under different magnetic fields (50, 75, 100 mT). Under different magnetic fields, the yield, iron grade and iron recovery of the magnetic separation concentrate are shown in Figure 8. It can be seen from the figure that as the magnetic field strength increases, the yield and iron recovery increase, while the iron grade is falling. $MgFe_2O_4$ is a strong magnetic phase, which can be selected in a weaker magnetic field. From the microstructure of the oxide slag, it is known that the $MgFe_2O_4$ spinel phase is embedded in the base phase β-C_2S, so it will be brought out when $MgFe_2O_4$ is selected. In the process of slag oxidation, some $MgFe_2O_4$ spinel phase particles are smaller and are covered by more β-C_2S, and cannot be selected under a smaller magnetic field. When the magnetic field is increased, this part is selected, the iron recovery was improved, and the increase of β-C_2S also led to the decrease of iron grade. When the magnetic field is increased to 100mT, almost all the magnetic phase is recovered, and the iron recovery is almost 100%. In actual production, the relationship between the yield, grade and iron recovery of the magnetic separation concentrate should be considered. That is to say, a higher iron recovery can be obtained in the case of a lower yield, and the grade is in a moderate range, so 75 mT is a more suitable magnetic separation intensity.

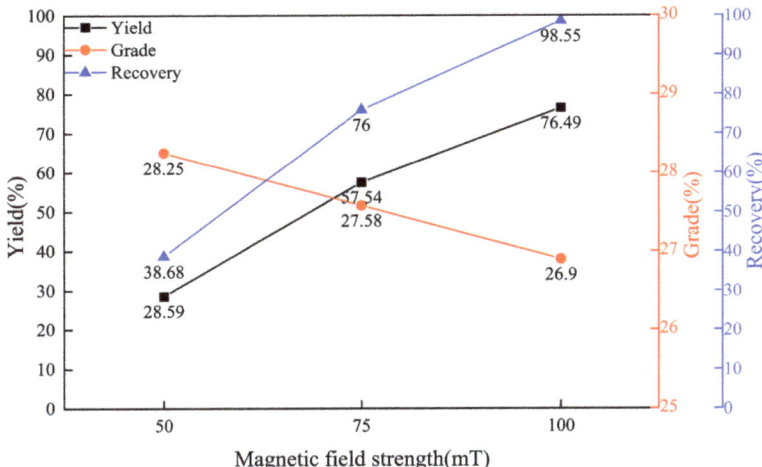

Figure 8. Magnetic separation effect under different magnetic field conditions.

4. Conclusions

(1) Through oxidation treatment in an air atmosphere, and under suitable conditions of reaction temperature, time and air flow rate, the magnetic iron oxides in the steel slag can be transformed into ferromagnetic magnesium-iron spinel;

(2) When the reaction temperature is controlled between 1050 and 1100 °C, the oxidation time is controlled between 40 and 60 min, the air flow rate is controlled at 0.75–1 L/min and the magnetic field strength is 75 mT, the yield of modified BOF slag is 46–57.54%, the iron grade can reach 27.58–29.10%, and the iron recovery can reach 64.12–76.00%, which is the best process parameter range for the magnetic separation experiment;

(3) The magnetic field strength has a great influence on the results of magnetic separation. The effect is best when the magnetic separation strength is 75 mT. After 75 mT, the magnetic separation yield will rise sharply, the grade of the concentrate will decrease, and the magnetic separation effect does not change significantly with the increase of the magnetic field strength.

Author Contributions: M.L. and Z.H. prepared and revised the manuscript under the supervision of X.H. All authors contributed to the general discussion. All authors have read and agreed to the published version of the manuscript.

Funding: This work is financially supported by State Key Laboratory of Advanced Metallurgy, University of Science and Technology Beijing (Grant No. 41620024 and 41622012).

Institutional Review Board Statement: Not applicable.

Informed Consent Statement: Not applicable.

Data Availability Statement: Not applicable.

Conflicts of Interest: The authors declare no conflict of interest.

References

1. He, K.; Wang, L.; Li, X. Review of the energy consumption and production structure of China's steel industry: Current situation and future development. *Metals* **2020**, *10*, 302. [CrossRef]
2. Guo, J.; Bao, Y.; Wang, M. Steel slag in China: Treatment, recycling, and management. *Waste Manag.* **2018**, *78*, 318–330. [CrossRef] [PubMed]
3. Das, B.; Prakash, S.; Reddy, P.S.R.; Misra, V.N. An overview of utilization of slag and sludge from steel industries. *Resour. Conserv. Recycl.* **2007**, *50*, 40–57. [CrossRef]

4. Trofimov, E.; Chumanov, I.; Dildin, A.; Samoylova, O. On expediency of the preliminary heat treatment for liquid-phase reduction of waste steelmaking slag. *Am. J. Appl. Sci.* **2015**, *12*, 952–961. [CrossRef]
5. Che Ghani, S.A.; Lim, J.W.; Chew, L.H.; Choong, T.S.Y.; Tezara, C.; Yazdi, M.H.; Alias, A.; Mamat, R.; Rahman, M.M. Overview of steel slag application and utilization. *MATEC Web Conf.* **2016**, *74*, 26.
6. Semykina, A.; Shatokha, V.; Iwase, M.; Seetharaman, S. Kinetics of oxidation of divalent iron to trivalent state in liquid FeO-CaO-SiO$_2$ slags. *Metall. Mater. Trans. B* **2010**, *41*, 1230–1239. [CrossRef]
7. Semykina, A. The kinetics of oxidation of liquid FeO-MnO-CaO-SiO$_2$ slags in air. *Metall. Mater. Trans. B* **2011**, *43*, 56–63. [CrossRef]
8. Li, P.; Guo, H.; Gao, J.; Min, J.; Yan, B.; Chen, D.; Seetharaman, S. Novel concept of steam modification towards energy and iron recovery from steel slag: Oxidation mechanism and process evaluation. *J. Clean. Prod.* **2020**, *254*, 119952. [CrossRef]
9. Xue, P.; He, D.; Xu, A.; Gu, Z.; Yang, Q.; Engström, F.; Björkman, B. Modification of industrial BOF slag: Formation of MgFe$_2$O$_4$ and recycling of iron. *J. Alloys Compd.* **2017**, *712*, 640–648. [CrossRef]
10. He, Z.; Lan, M.; Hu, X.; Xue, X.; Chou, K.-C. Study on oxidation behavior of industrial basic oxygen furnace slag and recovery of magnetic iron-containing phase. *Steel Res. Int.* **2022**, *92*, 2100558. [CrossRef]
11. Calmon, J.L.; Tristão, F.A.; Giacometti, M.; Meneguelli, M.; Moratti, M.; Teixeira, J.E.S.L. Effects of BOF steel slag and other cementitious materials on the rheological properties of self-compacting cement pastes. *Constr. Build Mater.* **2013**, *40*, 1046–1053. [CrossRef]
12. Rashad, A.M. A synopsis manual about recycling steel slag as a cementitious material. *J. Mater. Res. Technol.* **2019**, *8*, 4940–4955. [CrossRef]
13. Liu, X.; Zou, Y.; Geng, R.; Zhu, T.; Li, B. Simultaneous removal of SO$_2$ and NO$_x$ using steel slag slurry combined with ozone oxidation. *ACS Omega* **2021**, *6*, 28804–28812. [CrossRef] [PubMed]
14. Dhoble, Y.N.; Ahmed, S. Review on the innovative uses of steel slag for waste minimization. *J. Mater. Cycles Waste Manag.* **2018**, *20*, 1373–1382. [CrossRef]
15. Iizuka, A.; Yamasaki, A.; Yanagisawa, Y. Desulfurization performance of sorbent derived from waste concrete. *J. Chem. Eng. Jpn.* **2011**, *44*, 746–749. [CrossRef]
16. Blackman, L.F.C. On the formation of Fe^{2+} in the system MgO-Fe$_2$O$_3$-MgFe$_2$O$_4$ at high temperatures. *J. Am. Ceram. Soc.* **1959**, *42*, 143–145. [CrossRef]
17. Phillips, B.; Muan, A. Phase equilibria in the system CaO-iron oxide in air and at 1 atm. O$_2$ pressure. *J. Am. Ceram. Soc.* **1958**, *41*, 445–454. [CrossRef]
18. Yadav, U.S.; Pandey, B.D.; Das, B.K.; Jena, D.N. Influence of magnesia on sintering characteristics of iron ore. *Ironmak. Steelmak.* **2013**, *29*, 91–95. [CrossRef]
19. Shu, Q.F.; Liu, Y. Effects of basicity, MgO and MnO on mineralogical phases of CaO–FeO$_x$–SiO$_2$–P$_2$O$_5$ slag. *Ironmak. Steelmak.* **2017**, *45*, 363–370. [CrossRef]
20. O'Neill, H.S.C.; Annersten, H.; Virgo, D. The temperature dependence of the cation distribution in magnesioferrite (MgFe$_2$O$_4$) from powder XRD structural refinements and Mössbauer spectroscopy. *Am. Mineral.* **1992**, *77*, 725–740.

Review

Extraction of the Rare Element Vanadium from Vanadium-Containing Materials by Chlorination Method: A Critical Review

Shiyuan Liu, Weihua Xue and Lijun Wang *

Collaborative Innovation Center of Steel Technology, University of Science and Technology Beijing, Beijing 100083, China; shiyuanliu126@126.com (S.L.); 18435180612@163.com (W.X.)
* Correspondence: lijunwang@ustb.edu.cn

Abstract: Vanadium as a rare element has a wide range of applications in iron and steel production, vanadium flow batteries, catalysts, etc. In 2018, the world's total vanadium output calculated in the form of metal vanadium was 91,844 t. The raw materials for the production of vanadium products mainly include vanadium-titanium magnetite, vanadium slag, stone coal, petroleum coke, fly ash, and spent catalysts, etc. Chlorinated metallurgy has a wide range of applications in the treatment of ore, slag, solid wastes, etc. Chlorinating agent plays an important role in chlorination metallurgy, which is divided into solid (NaCl, KCl, $CaCl_2$, $AlCl_3$, $FeCl_2$, $FeCl_3$, $MgCl_2$, NH_4Cl, NaClO, and $NaClO_3$) and gas (Cl_2, HCl, and CCl_4). The chlorination of vanadium oxides (V_2O_3 and V_2O_5) by different chlorinating agents was investigated from the thermodynamics. Meanwhile, this paper summarizes the research progress of chlorination in the treatment of vanadium-containing materials. This paper has important reference significance for further adopting the chlorination method to treat vanadium-containing raw materials.

Keywords: vanadium; chlorination metallurgy; chlorination agents; NaCl roasting; carbochlorination; thermodynamics; molten salt chlorination

1. Introduction

Vanadium is located in the fourth period and fifth (VB) group of the periodic table and which occupies the 23rd position in the periodic table of elements. The symbol of the vanadium element is V. The physical characteristics of vanadium are a melting point of 1929 °C, boiling point of 3350 °C, relative atomic mass of 50.9415, and density of 5.96 (g/cm^3), and it is a silver grey metal [1,2]. The valence of vanadium in compounds can be +2, +3, +4, and +5 [3]. At present, there are known vanadium oxides such as V_2O_3, VO_2, V_2O_5, V_3O_5, V_3O_7, V_4O_7, V_5O_9, V_6O_{11}, and V_6O_{13}, among which the pentavalent vanadium compounds are the most stable [4]. The main chlorides are $VOCl_3$, VOCl, VCl_5, VCl_4, VCl_3, VCl_2, and VCl, among which $VOCl_3$ are the most stable [5]. However, the toxicity of vanadium compounds increases with the increase of vanadium valence, and the pentavalent vanadium compounds are the most toxic. Thus, compounds containing pentavalent vanadium, such as $NaVO_3$, NH_4VO_3, V_2O_5, and $VOCl_3$, are the most toxic [6]. Toxic vanadium compounds can exist in both cationic and anionic forms [7].

Vanadium as a rare element has a wide range of applications in iron and steel production, vanadium flow batteries, catalysts, etc. [8,9]. The raw materials for the production of vanadium products mainly include vanadium-titanium magnetite, vanadium slag, stone coal, petroleum coke, fly ash, and spent catalysts, etc. [1,8,9]. Salt roasting (Na_2CO_3, NaCl, NaOH, CaO, etc.) was applied to extract vanadium from vanadium-containing materials [1,8,9]. However, the salt roasting method extraction of vanadium from vanadium slag is associated with the formation of a large amount of sludge and significant losses of vanadium [10]. Previous review articles on vanadium extraction from vanadium-containing

materials mainly focused on the salt roasting process [1,8,9]. In this work, extraction of the rare element vanadium from vanadium-containing materials by chlorination method was summarized.

Table 1 shows melting temperature, boiling temperature, and sublimation temperature for vanadium compounds [11,12]. According to Table 1, the melting point and boiling point of chloride are lower than those of corresponding oxides. Thus, chloride is easier to separate and enrich than oxide [13–15]. Chlorinated metallurgy has a wide range of applications in the treatment of ore, slag, solid wastes, etc. [16–22]. In the last century, the extraction of Ti from titanium ore by chlorination method has been industrialized [23]. Chlorinating agent plays an important role in chlorination metallurgy, which is divided into solid (NaCl, KCl, $CaCl_2$, $AlCl_3$, $FeCl_2$, $FeCl_3$, $MgCl_2$, NH_4Cl, NaClO, $NaClO_3$) and gas (Cl_2, HCl, CCl_4) [24–29]. Compared with gaseous chlorinating agents, the solid chlorinating agents are easier to handle and more environmentally friendly.

Table 1. Melting temperature, boiling temperature, and sublimation temperature for vanadium compounds.

V-O-Cl Substance	Transition Temperature (°C)
VCl_2	$T_m = 1347$
	$T_s = 1407$
	$T_b = 1530$
VCl_3	$T_s = 833$
VCl_4	$T_b = 151$
$VOCl_3$	$T_b = 127$
VO_2Cl	$T_b = 177$
$VOCl_2$	$T_s = 511$
VOCl	$T_s = 1120$
VO	$T_m = 1790$
V_2O_3	$T_m = 1970$
VO_2	$T_m = 1545$
V_2O_5	$T_m = 690$

T_m, melting temperature; T_b, boiling temperature; T_s, sublimation temperature.

The traditional chlorination method of extracting vanadium with NaCl as an additive will produce $NaVO_3$ and then ammonia nitrogen wastewater will be produced in the process of preparing V_2O_5. The carbochlorination method of extracting vanadium to prepare $VOCl_3$ will not produce ammonia nitrogen wastewater. Molten salt chlorination of extracting vanadium will obtain VCl_3, and metal V will be obtained by molten salt electrolysis. In this work, these two new processes will be introduced.

2. Vanadium Reserves and the Major Vanadium Producers

Table 2 shows the world's vanadium ore reserves in 2018. More than 99% of the world's vanadium ore reserves are concentrated in China, Russia, South Africa and Australia [30]. Meanwhile, China has the largest vanadium reserves. According to statistics, in 2018, about 16% of the world's vanadium products directly came from vanadium-titanium magnetite, about 68% of the vanadium products came from the vanadium-rich steel slag (and a small amount of phosphorus-rich vanadium slag) obtained by vanadium-titanium magnetite after iron and steel metallurgical processing, and approximately 16% of vanadium products were produced from recovered vanadium-containing by-products (vanadium-containing fuel ash, waste chemical catalysts) and vanadium-containing stone coal [30]. Table 3 shows the overview of major vanadium producers in the world in 2018 [30]. In 2018, the world's total vanadium output calculated in the form of metal vanadium was 91,844 t [30]. The global market share of vanadium products in 2018 was approximately 90.8% ferroalloy products (FeV, VN, ferrovanadium nitride, etc), approximately 4.2% non-ferrous metals such as Ti, and about 5% of vanadium compounds (vanadium oxide, ammonium vanadate, $VOSO_4$, etc.) for the chemical industry, energy storage and other fields [30]. Like the consumption pattern of the global vanadium market, more than 90% of China's vanadium is used in the steel industry in the form of vanadium alloys [30].

Table 2. Global vanadium ore reserve calculated by metallic vanadium in 2018 (10 kt) [30].

China	Russia	South Africa	Australia	United States	Brazil
950	500	350	210	4.5	13

Table 3. Overview of major vanadium producers in the world in 2018 [30].

Company Name	Production Capacity (V_2O_5)/t	Products	Raw Material
Ansteel Pangang Group Co., Ltd.	40,000	FeV, VN, vanadium oxide, V-Al alloy	Vanadium slag
Russian (Evraz) company	30,000	FeV, vanadium oxide, V-Al alloy, catalyst	Vanadium slag, fly ash, spent catalyst
HBIS Group Chengsteel company	25,000	FeV, VN, ferrovanadium nitride, vanadium oxide	Vanadium slag
Beijing Jianlong Heavy Industry Group Co., Ltd.	15,000	VN, vanadium oxide	Vanadium slag
Austria Treibacher Industrie AG	13,000	V_2O_3, V_2O_5, FeV	Vanadium slag
Glencore (Xstrata)	12,000	FeV, vanadium oxide	vanadium-titanium magnetite
Sichuan Chuanwei Group Chengyu Vanadium Titanium Technology Co., Ltd.	12,000	V_2O_5	Vanadium slag
Sichuan Desheng Group Vanadium and Titanium Co., Ltd.	12,000	Vanadium slag	Vanadium slag
Largo Resources Ltd. Brazil Maracás Menchen Mine	11,000	V_2O_5	Vanadium-titanium magnetite
Bushveld Vametco, South Africa	6000	VN, vanadium oxide	Vanadium-titanium magnetite
Australia Atlantic Vanadium PTY Ltd.	12,000	FeV, vanadium oxide	Vanadium-titanium magnetite
Vanchem Vanadium Product (Pty) Ltd.	10,000	FeV, vanadium oxide, catalyst	Vanadium-titanium magnetite, vanadium slag
Czech Republic, Germany, Canada, Japan, India, Taiwan, Thailand, etc.	12,000	V_2O_5, V-Al alloy, FeV, etc.	Slag, waste catalyst, fuel ash, etc.
Other Chinese manufacturers	37,000	V_2O_5, V-Al alloy, VN, FeV, etc.	Vanadium slag, waste catalyst, stone coal

3. Chlorination Thermodynamics of Vanadium Oxides

Vanadium in vanadium-titanium magnetite, vanadium slag, and stone coal mainly exists in trivalent form. Meanwhile, the V^{5+} compounds are the very stable. Thus, V_2O_3 and V_2O_5 were selected as the reactants for thermodynamic calculation by HSC Chemistry 6.4. The possibilities of V_2O_3 reacting with different chlorinating agents are calculated from the thermodynamic viewpoint as shown in Equations (1)–(10). Figure 1 shows the standard Gibbs free energies of reactions between V_2O_3 and chlorination agents at 0–1300 °C. V_2O_3 can be chlorinated to VCl_3 by $AlCl_3$, CCl_4 and $COCl_2$. Gibbs free energies of reaction between V_2O_3 and $AlCl_3$ increases with increasing temperature. Thermodynamically, increasing temperature is not conducive to $AlCl_3$ chlorination. However, V_2O_3 cannot be chlorinated to VCl_3 by the NaCl, $CaCl_2$, $FeCl_2$, $FeCl_3$, $MgCl_2$, HCl or Cl_2 at 0–1300 °C.

$$V_2O_3 + 6NaCl = 2VCl_3 + 3Na_2O \tag{1}$$

$$V_2O_3 + 3CaCl_2 = 2VCl_3 + 3CaO \tag{2}$$

$$V_2O_3 + 3FeCl_2 = 2VCl_3 + 3FeO \tag{3}$$

$$V_2O_3 + 2FeCl_3 = 2VCl_3 + Fe_2O_3 \quad (4)$$

$$V_2O_3 + 2AlCl_3 = 2VCl_3 + Al_2O_3 \quad (5)$$

$$V_2O_3 + 3MgCl_2 = 2VCl_3 + 3MgO \quad (6)$$

$$V_2O_3 + 6HCl\ (g) = 2VCl_3 + 3H_2O\ (g) \quad (7)$$

$$V_2O_3 + 3Cl_2\ (g) = 2VCl_3 + 1.5O_2\ (g) \quad (8)$$

$$V_2O_3 + 1.5CCl_4\ (g) = 2VCl_3 + 1.5CO_2\ (g) \quad (9)$$

$$V_2O_3 + 3COCl_2\ (g) = 2VCl_3 + 3CO_2\ (g) \quad (10)$$

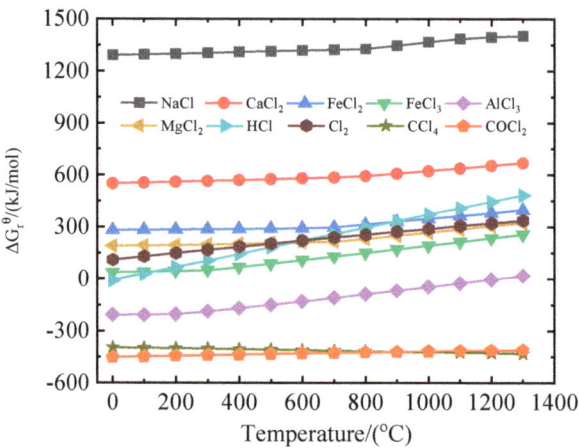

Figure 1. The standard Gibbs free energies of reactions between V_2O_3 and chlorination agents (reactions 1–10).

The V_2O_5 reacting with different chlorinating agents are as follows: Equations (11)–(20). Figure 2 shows the standard Gibbs free energies of reactions between V_2O_5 and chlorination agents. V_2O_5 can be chlorinated to $VOCl_3$ by $FeCl_3$, $AlCl_3$, CCl_4 and $COCl_2$ at 0–1300 °C. However, V_2O_3 cannot be chlorinated to $VOCl_3$ by the NaCl, $CaCl_2$, $FeCl_2$, $MgCl_2$, HCl and Cl_2 at 0–1300 °C.

$$V_2O_5 + 6NaCl = 2VOCl_3\ (g) + 3Na_2O \quad (11)$$

$$V_2O_5 + 3CaCl_2 = 2VOCl_3 + 3CaO \quad (12)$$

$$V_2O_5 + 3MgCl_2 = 2VOCl_3 + 3MgO \quad (13)$$

$$V_2O_5 + 3FeCl_2 = 2VOCl_3 + 3FeO \quad (14)$$

$$V_2O_5 + 2FeCl_3 = 2VOCl_3 + Fe_2O_3 \quad (15)$$

$$V_2O_5 + 2AlCl_3 = 2VOCl_3 + Al_2O_3 \quad (16)$$

$$V_2O_5 + 6HCl\ (g) = 2VOCl_3 + 3H_2O\ (g) \quad (17)$$

$$V_2O_5 + 3COCl_2\ (g) = 2VOCl_3 + 3CO_2\ (g) \quad (18)$$

$$V_2O_5 + 1.5CCl_4\ (g) = 2VOCl_3 + 1.5CO_2\ (g) \quad (19)$$

$$V_2O_5 + 3Cl_2\ (g) = 2VOCl_3 + 1.5O_2\ (g) \quad (20)$$

The V_2O_5 and V_2O_3 reacting with C and Cl_2 in the temperature range from 0 °C to 1300 °C are expressed as follows in Equations (21) and (22). Figure 3 shows standard Gibbs free energies of reactions 21–22 at 0–1300 °C. Adding C realizes the chlorination of V_2O_5

and V_2O_3 to $VOCl_3$ by Cl_2 at 0–1300 °C. However, the effect of increasing temperature on the chlorination of V_2O_3 and V_2O_5 is opposite.

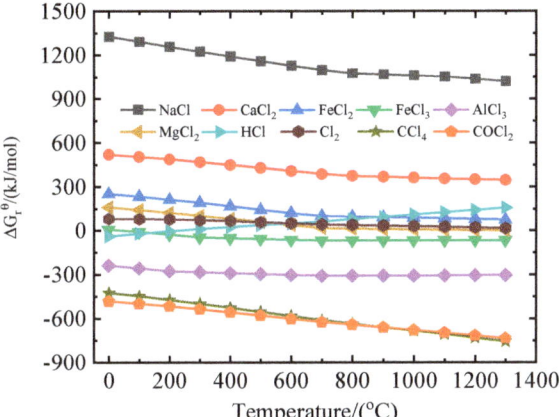

Figure 2. The standard Gibbs free energies of reactions between V_2O_5 and chlorination agents (reactions 11–20).

$$2V_2O_3 + 6Cl_2 \text{ (g)} + C = 4VOCl_3 + CO_2 \text{ (g)} \qquad (21)$$

$$V_2O_5 + 3Cl_2 \text{ (g)} + 1.5C = 2VOCl_3 + 1.5CO_2 \text{ (g)} \qquad (22)$$

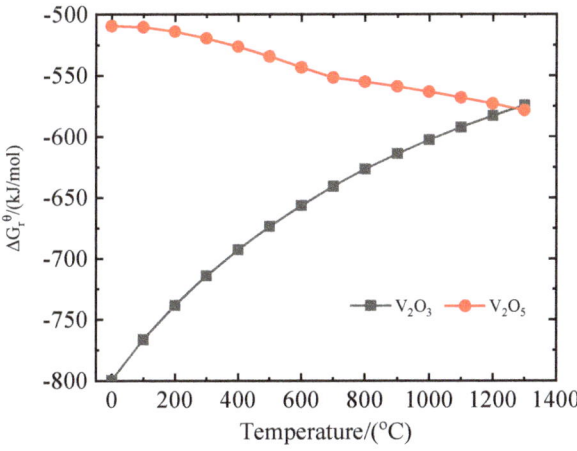

Figure 3. Variation of standard Gibbs free energy of reactions 21–22 with temperature.

Under an oxygen atmosphere, the equations for the NaCl roasting reaction of V_2O_3 and V_2O_5 are as shown in Equations (23) and (24). Figure 4 shows the variation of standard Gibbs free energy of reactions 23–24 with temperature. Under the same conditions, V_2O_3 is more easily chlorinated. Thermodynamically, increasing temperature is not conducive to NaCl chlorination of V_2O_3. The reaction of V_2O_3 and different chlorinating agents (FeCl$_2$ and FeCl$_3$) are as shown in Equations (25) and (26). It can be seen from Figure 4 that V_2O_3 can be chlorinated by $FeCl_2$ and $FeCl_3$ under an oxygen atmosphere.

$$V_2O_3 + 2NaCl + 1.5O_2 \text{ (g)} = 2NaVO_3 + Cl_2 \text{ (g)} \qquad (23)$$

$$2V_2O_5 + 4NaCl + O_2 (g) = 4NaVO_3 + 2Cl_2 (g) \qquad (24)$$

$$V_2O_3 + 2FeCl_3 + O_2 (g) = 2VOCl_3 (g) + Fe_2O_3 \qquad (25)$$

$$V_2O_3 + 3FeCl_2 + 1.75O_2 (g) = 2VOCl_3 (g) + 1.5Fe_2O_3 \qquad (26)$$

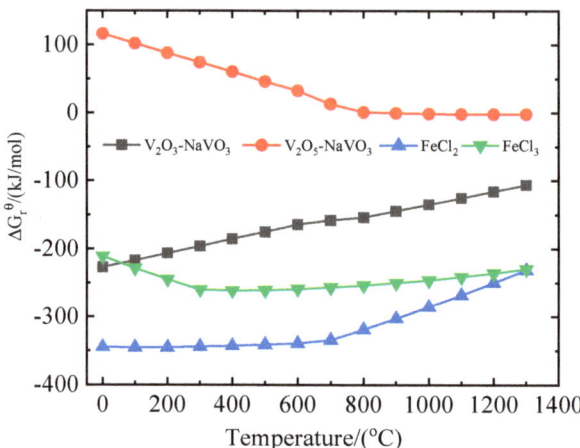

Figure 4. Variation of standard Gibbs free energy of reactions 23–26 with temperature.

According to the above thermodynamic analysis, the valence state of vanadium, reaction temperature, atmosphere and chlorinating agent play a very important role in the chlorination of vanadium. Thus, the chlorination of vanadium can be achieved by selecting appropriate conditions. The following will introduce the progress of chlorination of vanadium-containing materials.

4. Application of Chlorination Method
4.1. Chlorination Extraction of Vanadium from Vanadium Titanomagnetite

Vanadium-titanium magnetite is mainly composed of iron (Fe), vanadium (V) and titanium (Ti) elements, which is multi-element symbiotic iron ore containing a small amount of cobalt (Co), nickel (Ni), chromium (Cr), scandium (Sc) and gallium (Ga) [31,32]. The reserves of vanadium-titanium magnetite in the Panzhihua-Xichang regions in China amount to about 9.66 billion tons [33]. The content of vanadium pentoxide in vanadium-titanium magnetite is 0.1 wt%–2 wt% [34]. Jena et al. [23] proposed that under the action of oxygen and water, NaCl as an additive reacts with the vanadium in the vanadium bearing titaniferous magnetite. The roasted samples were leached with hot water. More than 90% of V was extracted. The reaction Equations are as follows in Equations (27)–(31).

$$SiO_2 + 2NaCl + H_2O = Na_2SiO_3 + 2HCl \qquad (27)$$

$$Na_2SiO_3 + O_2 + V_2O_3 = 2NaVO_3 + SiO_2 \qquad (28)$$

$$2NaCl + 3/2O_2 + V_2O_3 = 2NaVO_3 + Cl_2 \qquad (29)$$

$$3Cl_2 + 3V_2O_3 = 2VOCl_3 + 2V_2O_5 \qquad (30)$$

$$4VOCl_3 + 3O_2 = 2V_2O_5 + 6Cl_2 \qquad (31)$$

To some extent, the presence of SiO_2 and the formation of HCl and Cl_2 can promote the extraction of vanadium [23,35,36]. Zheng et al. [37] first calculated the feasibility of extraction vanadium from vanadium-rich resources with $FeCl_2$ and $FeCl_3$. Thermodynamic calculations show that the higher the valence of vanadium in vanadium titanomagnetite, the easier it is to extract vanadium. Therefore, the chlorinated atmosphere was selected as the oxygen atmosphere. The reaction Equations are as follows in Equations (32)–(36). Under the optimal experimental conditions (827 °C, reactant (vanadium titanomagnetite)—chlorination agent (FeCl$_3$) molar ratio of 1:2, 2 h, oxygen atmosphere), the extraction ratio of vanadium is 32%.

$$V_2O_5 + 2FeCl_3 = VOCl_3 + Fe_2O_3 \quad (32)$$

$$2V_2O_4 + 4FeCl_3 + O_2 = 4VOCl_3 + 2Fe_2O_3 \quad (33)$$

$$2V_2O_3 + 4FeCl_3 + (2x-1)O_2 = 4VOCl_3 + 4FeO_x \quad (34)$$

$$V_2O_4 + FeCl_3 + (x-1)O_2 = VOCl_3 + FeO_x \quad (35)$$

$$12FeCl_2 + 4V_2O_5 + 3O_2 = 6Fe_2O_3 + 8VOCl_3 \quad (36)$$

Chloride extraction of vanadium from vanadium-titanium magnetite has long been used. However, the content of vanadium in vanadium-titanium magnetite is low, and the cost of directly extracting vanadium in vanadium slag using chlorination method is high. Thus, it is not recommended to extract vanadium directly from vanadium-titanium magnetite by chlorination method.

4.2. Chlorination Extraction of Vanadium from Vanadium Slag

Vanadium slag is produced from vanadium-titanium magnetite by blast furnace smelting and the vanadium extraction process in a converter [38,39]. Vanadium slags contain 30–40 wt% total Fe, 6.9–14.4 wt% TiO$_2$, 13.5–19.0 wt% V$_2$O$_3$, 0.9–4.6 wt% Cr$_2$O$_3$, and 7.4–10.7 wt% MnO. The main phases of vanadium slag consist of (Fe,Mn)(V,Cr)$_2$O$_4$, (Fe,Mn)$_2$SiO$_4$ and Fe$_2$TiO$_4$. According to the phases of vanadium slag, vanadium is present in the form of V^{3+}, from which it is difficult to extract vanadium by direct leaching [40,41].

In order to extract vanadium, the traditional method is to oxidize insoluble low-valent vanadium to soluble high-valent vanadium in aqueous solution [42,43]. Figure 5 shows a flow chart of extracting vanadium from vanadium slag by NaCl roasting. The roasting temperature is about 800 °C. After roasting, vanadium in the solid exists in the form of NaVO$_3$, and then dissolves to obtain NaVO$_3$ solution. Vanadium is precipitated in the form of ammonium vanadate by adding ammonium salt (NH$_4$Cl, NH$_4$HCO$_3$, (NH$_4$)$_2$SO$_4$, (NH$_4$)$_2$CO$_3$). Ammonium vanadate is calcined to obtain V$_2$O$_5$ at about 550 °C. Under the action of oxygen, NaCl as an additive reacts with the vanadium spinel in the vanadium slag. The reaction Equation is as follows in Equation (37). The conversion rate of vanadium can reach 85% [44,45].

$$4FeV_2O_4 + 8NaCl + 7O_2 = 8NaVO_3 + 4Cl_2 + 2Fe_2O_3 \quad (37)$$

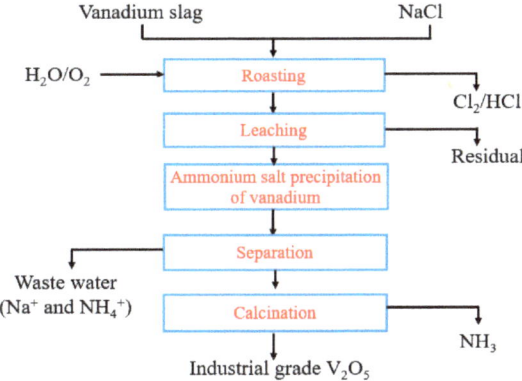

Figure 5. Flow chart of extracting vanadium from vanadium slag by NaCl roasting.

A total of 85.8% of V in vanadium slag was extracted by acidic sodium chlorate solution. V^{3+} in vanadium slag was oxidized by $NaClO_3$ as a chlorinating agent. The reaction Equation is as follows in Equation (38) [46].

$$6FeV_2O_4 + 5NaClO_3 + 15H_2SO_4 = 5NaCl + 6(VO_2)_2SO_4 + 3Fe_2(SO_4)_3 + 15H_2O \quad (38)$$

Sun et al. [47] proposed chlorination of vanadium slag by $FeCl_3$. Under the optimal experimental conditions (827 °C, reactant (vanadium slag)—chlorination agent ($FeCl_3$) molar ratio of 1:2, 2 h, oxygen atmosphere), the extraction ratio of vanadium in vanadium slag is 57%. Du [48] investigated carbochlorination of pre-oxidized vanadium slag. The flow chart of extracting vanadium from vanadium slag by chlorination is shown in Figure 6. The carbochlorination temperature is about 650 °C. Vanadium is volatile in the form of $VOCl_3$. $VOCl_3$ was oxidized to V_2O_5 by O_2. The equations of the main reactions involved are (39)–(41). The effect of time, temperature, petroleum coke and chlorine pressure fraction were studied. Under optimal process conditions (650 °C,120 min, $P(Cl_2)/P(Cl_2 + N_2)$ = 0.5, 10% of petroleum coke mass fraction), 18.8% of Fe and 87.5% of V were extracted. Wastewater containing high Na^+ and NH_4^+ is scarcely produced in whole process.

$$(Fe, Mn)(V, Cr, Ti)_2O_4\,(s) + O_2\,(g) \rightarrow Fe_2O_3\,(s) + MnO\,(s) + Cr_2O_3\,(s) + V_2O_5\,(s) + TiO_2\,(s) \quad (39)$$

$$1/3V_2O_5\,(s/l) + 1/2C\,(s) + Cl_2\,(g) \rightarrow 2/3VOCl_3\,(g) + 1/2CO_2\,(g) \quad (40)$$

$$1/3Fe_2O_3\,(s) + 1/2C\,(s) + Cl_2\,(g) \rightarrow 2/3FeCl_3\,(g) + 1/2CO_2\,(g) \quad (41)$$

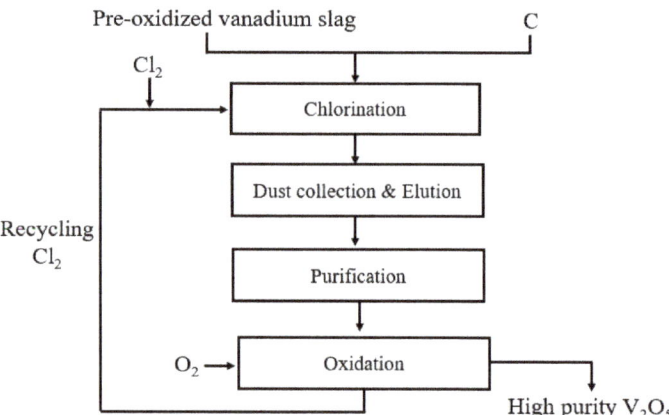

Figure 6. Flow chart of extracting vanadium from vanadium slag by carbochlorination.

In order to extract vanadium from vanadium slag, Liu et al. [49–55] proposed to use selective chlorination method to extract vanadium. Because to the existence form and value of valuable metal elements (Fe, Mn, V, Cr and Ti) in vanadium slag, NH_4Cl was selected to chlorinate Fe and Mn in vanadium slag. Thermodynamic calculations show that the iron and manganese in vanadium slag could be chlorinated by hydrogen chloride, but the V, Cr and Ti could not be chlorinated in the temperature range from 0 to 1000 °C. Under optimal chlorination conditions, the chlorination ratio of iron and manganese were 72% and 95%, respectively. Meanwhile, the enrichment ratio of V, Cr and Ti was obtained as 48%. In addition, $AlCl_3$ was selected to chlorinate V, Cr and Ti in vanadium slag. Figure 7 shows a flow chart of extracting vanadium from vanadium slag by $AlCl_3$ chlorination. The chlorination temperature is about 900 °C. Vanadium after chlorination exists in the form of VCl_3. Metal V was obtained by molten salt electrolysis at 900 °C. The effects of reaction temperature, reaction time, mass ratio of $AlCl_3$/slag and mass ratio of salt/$AlCl_3$ on the chlorination ratio of valuable elements were investigated. Under optimal chlorination conditions (AlCl$_3$—slag mass ratio of 1.5:1, (NaCl-KCl)-AlCl$_3$ mass ratio of 1.66:1, at 900 °C, 8 h.), the chlorination ratio of iron, vanadium, chromium and manganese were 90.3%, 76.5%, 81.9% and 97.3%. The volatilization ratio of titanium was 79.9%. The results of kinetic study indicate that the rate-control step of vanadium chlorination process was the surface chemical reaction. The vanadium and chromium in vanadium slag after AlCl$_3$ chlorination were present in the form of VCl$_3$ and CrCl$_3$ in molten salt. The main reaction was as follow (42).

$$8AlCl_3 + 3FeV_2O_4 = 3FeCl_2 + 4Al_2O_3 + 6VCl_3 \qquad (42)$$

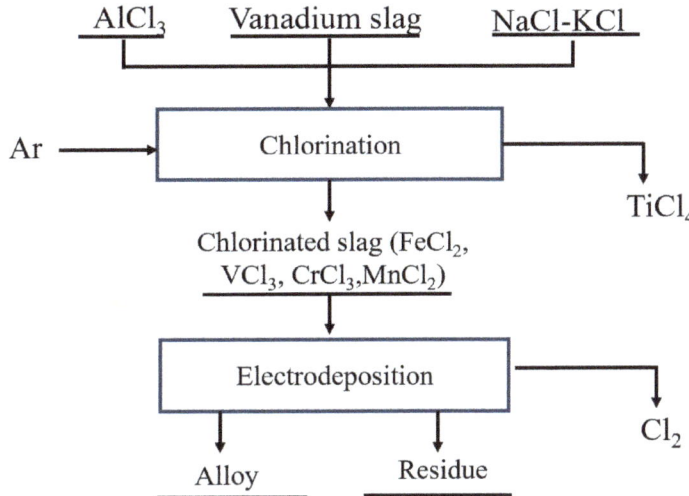

Figure 7. Flow chart of extracting vanadium from vanadium slag by AlCl$_3$ chlorination.

4.3. Chlorination Extraction of Vanadium from BOF-Slag

The basic oxygen furnace (BOF)-slags contains 31–56% CaO, 10–27% SiO$_2$, 1–4.5% Al$_2$O$_3$, 5–35% Fe compounds, and less than 1% of vanadium [56]. Seron et al. investigated the recovery of vanadium from BOF-slags by oxy-carbochlorination. Under specific conditions (900 °C, chlorine partial pressure 0.2, 90 min, 50% carbon content), the recovery ratio of vanadium in slag can reach 95% [57].

4.4. Chlorination Extraction of Vanadium from Stone Coal

Black shale is one of China's most important vanadium resources, accounting for more than 87% of domestic vanadium reserves [58,59]. It is estimated that the reserves of vanadium in the form of V$_2$O$_5$ in stone coal are 118 million tons [60]. However, the ordinary grade of vanadium in black shale is usually below 2 wt% [58,59]. In China, vanadium in most of the stone coal replaces trivalent aluminum in mica minerals in a quasi-homogeneous form. The chemical formula of vanadium-containing illite is K(Al,V)$_2$(OH)$_2$[Si$_3$Al]O$_{10}$. The mica mineral structure is very stable. It is difficult to destroy the lattice structure by general concentration of acid and alkali. Thus, in order to extract vanadium from vanadium-containing mica, the lattice structure of vanadium containing mica first needs to be destroyed [61].

Under the action of oxygen and water, NaCl as an additive reacts with the pre-decarburized stone coal [62,63]. The reaction Equation is expressed as follow (43):

$$K(Al,V)_2(OH)_2[Si_3Al]O_{10} + 2NaCl + 3(2-m)SiO_2 + (m-1/2)O_2 = (3-m)(K,Na)AlSi_3O_8 + NaVO_3 + 2HCl + Cl_2 \qquad (43)$$

where m is the number of vanadium ions replacing aluminum ions in hydromica octahedron.

NaCl has a melting point of 801 °C, which tends to keep the structure stable and does not decompose at high temperature. However, due to the presence of V, Al, Fe, and other oxides in the stone coal, NaCl can be decomposed at lower temperature to generate Cl_2 with high chemical reactivity. The reaction is described as follows in Equations (44)–(47). Cl_2 can react with low-valent vanadium to form $VOCl_3$, and $VOCl_3$ is an intermediate product that can be further oxidized to V_2O_5. The presence of Cl_2 promotes the high temperature roasting to destroy the crystal structure of illite. The oxidation of the exposed trivalent vanadium changes to a higher valence state. Cl_2 is more active than oxygen at high temperature and is more easily adsorbed on the surface of minerals. The promotion of Cl_2 on the oxidation of low-cost vanadium cannot be ignored. Thus, NaCl as an additives agent was selected for extracting vanadium from stone coal [63–66].

$$4NaCl + O_2 = 2Na_2O + 2Cl_2 \tag{44}$$

$$3Cl_2 + 3V_2O_3 = 2VOCl_3 + 2V_2O_5 \tag{45}$$

$$4VOCl_3 + 3O_2 = 2V_2O_5 + 6Cl_2 \tag{46}$$

$$xNa_2O + yV_2O_5 = xNa_2O \cdot yV_2O_5 \tag{47}$$

The possible chemical reaction between vanadium oxide (V_2O_3, VO_2, and V_2O_5) and the solid chlorinating agent (NaCl, $CaCl_2$ and $FeCl_3$) was calculated by FactSage 7.1 (Montreal, Canada) using the database of FactPS, FToxid and FT salt. The results show that vanadium oxide cannot be directly chlorinated thermodynamically by NaCl and $CaCl_2$ as solid chlorinating agents. However, V_2O_4 and V_2O_5 can be chlorinated by $FeCl_3$. Meanwhile, V can be separated from black shale by controlled roasting temperature of chlorination volatilization [67]. In the air, the structure of illite and muscovite in stone coal is hard to be destroyed by roasting without additives. Zhang et al. [68] studied that the vanadium-bearing stone coal was roasted in chlorine, and 90% of V in the form of $VOCl_3$ was extracted at 1000 °C for 1 h. Li et al. [69] investigated extraction of vanadium by leaching. Under the optimal leaching conditions (liquid-to-solid ratio of 2, oxygen partial pressure of 1200 kPa, 90 °C, 6 h, 1.5 g/L NaClO, 15 g/L HF, 100 g/L H_2SO_4), 91% of V in vanadium slag was extracted by NaClO-H_2SO_4-HF system under atmospheric pressure. V^{3+} in stone was oxidized by NaClO as a chlorinating agent and oxidant.

4.5. Chlorination Extraction of Vanadium from Spent Catalysts

Catalysts are extensively used in sulfuric acid production and petroleum refining [70,71]. More than 100,000 tons of spent hydrodesulphurization catalysts are produced every year, which usually contain the valuable elements V, Mo, Ni, and Co [72]. Vanadium in spent catalyst is present in the form of sulfide (V_2S_3 or V_3S_4) [73]. Oxidation roasting of spent catalyst and subsequent NaCl/H_2O roasting of oxide were proposed by Biswas et al. [74], and 81.9% of V was extracted. The reactions were as follows in Equations (48) and (49):

$$4V_3S_4 + 31O_2 = 6V_2O_5 + 16SO_2 \tag{48}$$

$$V_2O_5 + 2NaCl + 2H_2O = 2NaVO_3 + 2HCl \tag{49}$$

There are two processes (direct chlorination and roasting chlorination) for recovering vanadium from spent hydrodesulphurization catalysts by Cl_2 chlorination.

In addition to metal elements such as vanadium and molybdenum, spent catalysts also contain elemental carbon and sulfur. Direct chlorination of spent catalysts was investigated by Gaballah et al. [75]. In order to recover Mo, V, Ni and Co, Cl_2/N_2, Cl_2/air, and Cl_2/CO/N_2, gas mixtures were used to chloride spent catalysts. Vanadium sulphide was chlorinated to vanadium chloride as expressed in the reaction Equations (50)–(52). A total of 75% of V in the form of VCl_4 and/or $VOCl_3$ was recovered by Cl_2/air gas mixture at less than 600 °C.

$$1/7V_2S_3 + Cl_2 = 2/7VCl_4 \text{ (l,g)} + 3/7SCl_2 \tag{50}$$

$$1/4V_2S_3 + 3/4O_2 + Cl_2 = 1/2VCl_4 \text{ (l,g)} + 3/4SO_2 \tag{51}$$

$$1/3V_2S_3 + 3/4O_2 + Cl_2 = 2/3VOCl_3 \text{ (l,g)} + SO_2 \tag{52}$$

Vanadium sulfide in the spent catalyst is first oxidized to oxide at 300–500 °C. Vanadium oxide was chlorinated by Cl_2/N_2, Cl_2/O_2, or Cl_2/CO in the temperature range 300 °C to 600 °C. Finally, V was volatilized in the form of VCl_4 or $VOCl_3$ to achieve separation from other elements (Co, Ni). A total of 65% of V from oxidized V sulfide can be recovered by $Cl_2/N_2 = 1$ at 500 °C for 19 h. Meanwhile, V sulfide is directly chlorinated without roasting, and 80% of V can be recovered by $Cl_2/N_2 = 1$ at 500 °C for 0.5 h [76].

The affinity of metal to oxide is stronger than that of metal to sulfur. Under the same conditions, sulfides are easier to chlorinate than oxides. Thus, direct chlorination of vanadium sulfide is better than chlorination after oxidation of vanadium sulfide [77].

4.6. Chlorination of V_2O_5

Mink et al. [78] reported that CCl_4 reversibly dissociate and adsorbs on the two exposed vanadium atoms of the basic (001) plane of V_2O_5 before the chlorination reaction. The mechanism of chlorination of V_2O_5 by CCl_4 was analyzed by MS and XPS. Before the formation of the volatile final product $VOCl_3$, the surface vanadium atoms gradually acquire two chlorine atoms [79]. The kinetics of chlorination of V_2O_5 by CCl_4 was investigated by Jean et al. [80]. A total of 87% of V_2O_5 could be chlorinated by CCl_4 at 480 °C in 30min. Chlorination reaction conforms to topochemical reaction model. According to analysis of kinetics results, the following mechanisms, Equations (53)–(58), at different temperatures, were proposed.

a. 280–370 °C

$$V_2O_5 + CCl_4 \xrightarrow{slow} VOCl_3 + VO_2Cl + CO_2 \tag{53}$$

$$VO_2Cl + CCl_4 \xrightarrow{fast} VCl_5 + CO_2 \tag{54}$$

$$VCl_5 \xrightarrow{fast} VCl_4 + 1/2Cl_2 \tag{55}$$

$$VOCl_3 + CCl_4 + 1/2Cl_2 \xrightarrow{fast} VCl_4 + COCl_2 + Cl_2 \tag{56}$$

b. 410–515 °C

$$CCl_4 \xrightarrow{fast} C + 4Cl \tag{57}$$

$$V_2O_5 + C + 4Cl \xrightarrow{slow} VOCl_3 + VO_2Cl + CO_2 \tag{58}$$

The whole reaction can be expressed by the following Formula (59)

$$V_2O_5 + 3CCl_4 = 2VCl_4 + 2CO_2 + COCl_2 + Cl_2 \tag{59}$$

Gaballah et al. [81] studied kinetics of chlorination of V_2O_5 with Cl_2-CO-N_2, Cl_2-N_2, and Cl_2-air gas mixtures. Thermodynamic calculation showed that chlorinated product was mainly $VOCl_3$ during the Cl_2 chlorination of V_2O_5. However, VCl_4 may be formed during the carbochlorination of vanadium pentoxide. The results of kinetics indicated that the rate-control step of V_2O_5 chlorination process between 500 °C and 570 °C with Cl_2-N_2 was a chemical reaction. Pore diffusion and chemical reaction were the limiting step for the V_2O_5 chlorination in the temperature range of 570 °C to 650 °C. In Cl_2-CO-N_2 atmosphere, the limiting step of carbochlorination of V_2O_5 at 400–620 °C was the chemical reaction. Brocchi et al. [6] systematically studied the carbon-chlorination of V_2O_5 from thermodynamics between 627 °C and 1327 °C. In carbon-chlorination reaction of V_2O_5, the most stable vanadium oxychloride and vanadium chloride are $VOCl_3$ and VCl_4, respectively. E. Mccarley et al. [82] reported a process for preparing high-purity V_2O_5 (maximum of 100 ppm impurities) by carbon chlorination using V red cake (88 wt% of V_2O_5) as raw materials. Pap et al. [83] reported that the chlorination of V_2O_5 by three chlorinating agents (Cl_2, $COCl_2$ and CCl_4) is compared from the aspects of thermodynamics and kinetics. The results showed that V_2O_5 can be chlorinated thermodynamically by $COCl_2$ and CCl_4 at 127 °C, and the chlorinated products were $VOCl_3$ and CO_2. However, the reaction of V_2O_5 with Cl_2 can occur obviously when the temperature exceeds 477 °C. The chlorinated product was $VOCl_3$ and O_2. The chlorination kinetics showed that the apparent activation energy of V_2O_5 chlorinated by Cl_2, CCl_4, and $COCl_2$ were 126 kJ/mol, 77 kJ/mol, and 48 kJ/mol, respectively. High-purity V_2O_5 (99.95 wt%) is prepared by chlorinating industrial grade V_2O_5 (96.7 wt%) with $AlCl_3$. The reactions involved are as follows in Equations (60)–(64). Under the protection of purity Ar, the chlorination ratio of V_2O_5 at a V_2O_5:$AlCl_3$ mole ratio of 1:6, 180 °C, and 3.5 h was 62%, and a large amount of $VOCl_3$ was collected. When NaCl is added to the chlorination reaction system, the chlorination ratio of vanadium can reach 83.4% at mole ratio of $AlCl_3$:V_2O_5 of 6:1 and mole fraction of NaCl of 0.6 in the NaCl-$AlCl_3$ system [84,85].

$$2AlCl_3 + V_2O_5 = Al_2O_3 + 2VOCl_3 \text{ (g)} \quad (60)$$

$$6VOCl_3 + 20NH_3 \cdot H_2O = (NH_4)_2V_6O_{16} + 18NH_4Cl + 10H_2O \quad (61)$$

$$VOCl_3 + 4NH_4OH = NH_4VO_3 + 3NH_4Cl + 2H_2O \quad (62)$$

$$(NH_4)_2V_6O_{16} = 3V_2O_5 + 2NH_3 + H_2O \quad (63)$$

$$NH_4VO_3 = V_2O_5 + 2NH_3 + H_2O \quad (64)$$

4.7. Chlorination Extraction of Vanadium from Other Vanadium-Containing Materials

Petroleum coke, fly ash and carbonaceous gold ore also contain a certain amount of vanadium. 0.6 Mt/year of petroleum coke was produced from Syrian petroleum refineries. The extraction ratio of vanadium can reach 60% by NaCl-roasting [86]. The vanadium content in fly ash is as low as 1–7%. The fly ash was treated by acid leaching, oxidation of $NaClO_3$, and precipitation [87]. Murase et al. [88] investigated extraction and separation of vanadium from a fly ash of Orimulsion. Air-Cl_2 or N_2-Cl_2 gas mixture were used to chlorinate valuable elements (V, Ni and Mg). The separation of V and Fe were achieved by controlled temperature of chlorination. V and Fe was selectively extracted by chlorination at 400 °C and 500 °C, respectively. Mg and Ni in residue were extracted by chlorination of N_2-Cl_2-Al_2Cl_6 (g) at 600 °C. The extraction and separation of V, Ni and Mg were successfully achieved by the method of chlorination. The content of vanadium in refractory carbonaceous gold ore is 1.1 wt%. Wang et al. [89] investigated extraction and separation of vanadium from carbonaceous gold ore by NaCl roasting. After NaCl roasting, Au volatilizes in the form of $AuCl_3$ and V in the form of $NaVO_3$ remains in the roasted solid. The reactions were as follows in Equations (65)–(69).

$$4FeS_2 + 11O_2 = Fe_2O_3 + 8SO_2 \quad (65)$$

$$SO_2 + 2NaCl + O_2 = Na_2SO_4 + Cl_2 \quad (66)$$

$$4V_xO_y + (5x-2y)O_2 = 2xV_2O_5 \ (1 \leq x \leq 2; 2 \leq y \leq 4) \quad (67)$$

$$2V_2O_5 + 4NaCl + O_2 = 4NaVO_3 + 2Cl_2 \quad (68)$$

$$2Au + 3Cl_2 = 2AuCl_3 \quad (69)$$

The influences of experiment conditions including NaCl dosage, time and temperature were studied. Under optimal process conditions, (air gas flow rate 1 L/min, 4 h, 800 °C, NaCl 10%), the extraction ratios of V and Au are 85.3% and 92%.

4.8. Treatments of Chlor-Containing Compounds in Gas, Solid and Solution

Regarding the Cl_2 and HCl off-gas generated in the chlorination process, the first method is recycling, and the second method is the absorption of alkaline solution. The chloride in the solid can be washed to remove chloride ions. The chlor-containing wastewater can be treated by solvent extraction, the electrochemical method, separation interception method, the principle of precipitation, and ion exchange [90].

5. Conclusions and Outlook

The research progress on the treatment of vanadium-containing materials with various chlorinating agents (solid and gas) is summarized in terms of thermodynamics and kinetics.

The NaCl roasting method is used to treat vanadium titanomagnetite, vanadium slag, stone, spent catalysts, petroleum coke, and carbonaceous gold ore. The NaCl roasting method has the characteristics of short process, less investment and less equipment, etc. In the 1970s, in China, the price of vanadium was very high and hundreds of small-scale vanadium extraction plants adopted the NaCl roasting method to extract vanadium from vanadium-containing materials (stone coal) [63]. However, Cl_2 and HCl generated during the NaCl roasting process makes it highly demanding for the equipment's anti-corrosion performance. The environmental pollution caused by Cl_2 and HCl gas and the threat of Cl_2 and HCl gas to workers' health are also fatal defects of NaCl roasting. Due to increasingly strict environmental protection policies, the NaCl roasting method has become outdated and has gradually been replaced by other roasting methods.

The demand for high-purity vanadium pentoxide is increasing in all-vanadium flow batteries and high-purity metal vanadium. Therefore, efficient preparation of high-purity vanadium pentoxide is urgently needed [91]. Due to the low melting point and boiling point of vanadium chloride, vanadium chloride has a greater advantage than vanadium oxide in separation and enrichment. The advantages of chlorination method in the preparation of high-purity vanadium are very obvious, and it has very good development prospects.

Trivalent vanadium oxide is difficult to leach. Thus, the traditional vanadium extraction method is to oxidize vanadium to pentavalent vanadium for extraction. However, the toxicity of vanadium compounds increases with the increase of vanadium valence, and the pentavalent vanadium compounds are the most toxic. More than 90% of vanadium produced in industry is added to steel in the form of vanadium alloys. Trivalent vanadium oxide can be chlorinated to VCl_3 by $AlCl_3$. Metal V can be obtained by reduction or electrolysis of VCl_3. Thus, the direct chlorination of low-valent vanadium is also a very promising process.

Author Contributions: Conceptualization, S.L. and L.W.; methodology, L.W.; validation, S.L., W.X., and L.W.; formal analysis, S.L.; investigation, S.L.; resources, W.X.; data curation, W.X.; writing—original draft preparation, S.L. and L.W.; writing—review and editing, L.W.; visualization, S.L. and W.X.; supervision, L.W.; funding acquisition, S.L. and L.W. All authors have read and agreed to the published version of the manuscript.

Funding: The authors are grateful for the financial support of this work from the National Natural Science Foundation of China (No. 51904286, 51922003, 51734002).

Data Availability Statement: The data presented in this study are available from the corresponding author, upon reasonable request.

Conflicts of Interest: The authors declare no conflict of interest.

References

1. Peng, H. A literature review on leaching and recovery of vanadium. *J. Environ. Chem. Eng.* **2019**, *7*, 103313. [CrossRef]
2. Bauer, G.; Güther, V.; Hess, H.; Otto, A.; Roidl, O.; Roller, H.; Sattelberger, S. *Vanadium and Vanadium Compounds*; Excerpt from Ullmann's; Wiley-VCH Verlag GmbH: Weinheim, Germany, 2002.
3. Del Carpio, E.; Hernández, L.; Ciangherotti, C.; Coa, V.V.; Jiménez, L.; Lubes, V.; Lubes, G. Vanadium: History, chemistry, interactions with α-amino acids and potential therapeutic applications. *Coord. Chem. Rev.* **2018**, *372*, 117–140. [CrossRef] [PubMed]
4. Zhao, Q.S.; Li, Z.J. *Vanadium Metallurgy(fine)/Nonferrous Metals Theory and Technology Frontier Series*; Central South University Press: Changsha, China, 2015.
5. Hildenbrand, D.L.; Lau, K.H.; Perez-Mariano, J.; Sanjurjo, A. Thermochemistry of the Gaseous Vanadium Chlorides VCl, VCl_2, VCl_3, and VCl_4. *J. Phys. Chem. A* **2008**, *112*, 9978–9982. [CrossRef] [PubMed]
6. Aregay, G.G.; Ali, J.; Shahzad, A.; Lfthikar, J.; Oyekunle, D.T.; Chen, Z.Q. Application of layered double hydroxide enriched with electron rich sulfide moieties ($S_2O_4^{2-}$) for efficient and selective removal of vanadium (V) from diverse aqueous medium. *Sci. Total Environ.* **2021**, *792*, 148543. [CrossRef]
7. Zhang, R.C.; Leiviskä, T. Surface modification of pine bark with quaternary ammonium groups and its use for vanadium removal. *Chem. Eng. J.* **2020**, *385*, 123987. [CrossRef]
8. Petranikova, M.; Tkaczyk, A.; Bartl, A.; Amato, A.; Lapkovskis, V.; Tunsu, C. Vanadium sustainability in the context of innovative recycling and sourcing development. *Waste Manag.* **2020**, *113*, 521–544. [CrossRef]
9. Gilligan, R.; Nikoloski, A.N. The extraction of vanadium from titanomagnetites and other sources. *Miner. Eng.* **2020**, *146*, 106106. [CrossRef]
10. Kologrieva, U.; Volkov, A.; Zinoveev, D.; Krasnyanskaya, I.; Stulov, P.; Wainstein, D. Investigation of vanadium-containing sludge oxidation roasting process for vanadium extraction. *Metals* **2021**, *11*, 100. [CrossRef]
11. Brocchi, E.; Navarro, R.; Moura, F. A chemical thermodynamics review applied to V_2O_5 chlorination. *Thermochim. Acta* **2013**, *559*, 1–16. [CrossRef]
12. Zhou, X.J.; Cui, X.M.; Peng, F.C. *Thermodynamics of Ti, V and Their Chemical Compounds*; Metallurgical Industry Press: Beijing, China, 2019. (In Chinese)
13. Wang, H.-J.; Feng, Y.-L.; Li, H.-R.; Kang, J.-X. Simultaneous extraction of gold and zinc from refractory carbonaceous gold ore by chlorination roasting process. *Trans. Nonferr. Met. Soc. China* **2020**, *30*, 1111–1123. [CrossRef]
14. Fan, C.; Xu, J.; Yang, H.; Zhu, Q. High-purity, low-Cl V_2O_5 via the gaseous hydrolysis of $VOCl_3$ in a fluidized bed. *Particuology* **2020**, *49*, 9–15. [CrossRef]
15. Kim, J.Y.; Lee, M.S.; Jung, E.J. A study of formation behavior of porous structure induced by selective chlorination of ilmenite. *Mater. Chem. Phys.* **2020**, *241*, 122433. [CrossRef]
16. Xing, Z.; Cheng, G.; Yang, H.; Xue, X.; Jiang, P. Mechanism and application of the ore with chlorination treatment: A review. *Miner. Eng.* **2020**, *154*, 106404. [CrossRef]
17. Jena, S.K.; Dash, N.; Angadi, S.I. A novel application of Linz-Donawitz Slag for potash recovery from waste mica scrap using chlorination roasting coupled water leaching process. *Sep. Sci. Technol.* **2021**, *56*, 2310–2326. [CrossRef]
18. Mochizuki, Y.; Tsubouchi, N.; Sugawara, K. Separation of valuable elements from steel making slag by chlorination. *Resour. Conserv. Recycl.* **2020**, *158*, 104815. [CrossRef]
19. Long, H.L.; Li, H.Y.; Pei, J.N.; Srinivasakannan, C.; Yin, S.H.; Zhang, L.B.; Ma, A.Y.; Li, S.W. Cleaner recovery of multiple valuable metals from cyanide tailings via chlorination roasting. *Sep. Sci. Technol.* **2021**, *56*, 2113–2123. [CrossRef]
20. Guo, X.; Zhang, B.; Wang, Q.; Li, Z.; Tian, Q. Recovery of Zinc and Lead from Copper Smelting Slags by Chlorination Roasting. *JOM* **2021**, *73*, 1861–1870. [CrossRef]
21. Kim, J.; Lee, Y.R.; Jung, E.J. A Study on the Roasting Process for Efficient Selective Chlorination of Ilmenite Ores. *JOM* **2021**, *73*, 1495–1502. [CrossRef]
22. Kang, J.; Okabe, T.H. Thermodynamic Consideration of the Removal of Iron from Titanium Ore by Selective Chlorination. *Met. Mater. Trans. A* **2014**, *45*, 1260–1271. [CrossRef]
23. Jena, P.K.; Brocchi, E.A. Metal extraction through chlorine metallurgy. *Miner. Process. Extr. Metall. Rev.* **1997**, *16*, 211–237. [CrossRef]
24. Li, H.Y.; Li, S.W.; Ma, P.C.; Zhou, Z.F.; Long, H.L.; Peng, J.H.; Zhang, L.B. Evaluation of a cleaner production for cyanide tailings by chlorination thermal treatments. *J. Clean. Prod.* **2021**, *281*, 124195. [CrossRef]
25. Chen, S.; Guan, J.; Yuan, H.; Wu, H.C.; Gu, W.X.; Gao, G.L.; Guo, Y.G.; Dia, J.; Su, R.J. Behavior and Mechanism of Indium Extraction from Waste Liquid-Crystal Display Panels by Microwave-Assisted Chlorination Metallurgy. *JOM* **2021**, *73*, 1290–1300. [CrossRef]
26. Cui, F.; Mu, W.; Zhai, Y.; Guo, X. The selective chlorination of nickel and copper from low-grade nickel-copper sulfide-oxide ore: Mechanism and kinetics. *Sep. Purif. Technol.* **2020**, *239*, 116577. [CrossRef]
27. Kumari, A.; Raj, R.; Randhawa, N.; Sahu, S.K. Energy efficient process for recovery of rare earths from spent NdFeB magnet by chlorination roasting and water leaching. *Hydrometallurgy* **2021**, *201*, 105581. [CrossRef]
28. Okabe, P.; Newton, M.; Rappleye, D.; Simpson, M.F. Gas-solid reaction pathway for chlorination of rare earth and actinide metals using hydrogen and chlorine gas. *J. Nucl. Mater.* **2020**, *534*, 152156. [CrossRef]

29. Xie, K.; Hu, H.; Xu, S.; Chen, T.; Huang, Y.; Yang, Y.; Yang, F.; Yao, H. Fate of heavy metals during molten salts thermal treatment of municipal solid waste incineration fly ashes. *Waste Manag.* **2020**, *103*, 334–341. [CrossRef]
30. Chen, D.H. Annual evaluation for vanadium industry in 2018. *HeBei Metall.* **2019**, *284*, 5–15. (In Chinese)
31. Fan, G.; Wang, M.; Dang, J.; Zhang, R.; Lv, Z.; He, W.; Lv, X. A novel recycling approach for efficient extraction of titanium from high-titanium-bearing blast furnace slag. *Waste Manag.* **2021**, *120*, 626–634. [CrossRef] [PubMed]
32. Zhang, R.; Dang, J.; Liu, D.; Lv, Z.; Fan, G.; Hu, L. Reduction of perovskite-geikielite by methane–hydrogen gas mixture: Thermodynamic analysis and experimental results. *Sci. Total Environ.* **2020**, *699*, 134355. [CrossRef] [PubMed]
33. Zheng, F.; Chen, F.; Guo, Y.; Jiang, T.; Travyanov, A.Y.; Qiu, G. Kinetics of Hydrochloric Acid Leaching of Titanium from Titanium-Bearing Electric Furnace Slag. *JOM* **2016**, *68*, 1476–1484. [CrossRef]
34. Yang, S.L. *Non-Blast Furnace Smelting Technology of Vanadium-Titanium Magnetite*; Metallurgical Industry Press: Beijing, China, 2012. (In Chinese)
35. Bhaskarasarma, P.; Rao, P.; Jena, P. Extraction of vanadium from titaniferrous vanadium bearing magnetites by salt roasting. *Trans. Ind. Inst. Met.* **1980**, *33*, 166–169.
36. Jena, P.K.; Brocchi, E.A. Halide Metallurgy of Refractory Metals. *Miner. Process. Extr. Met. Rev.* **1992**, *10*, 29–40. [CrossRef]
37. Zheng, H.Y.; Sun, Y.; Lu, J.W.; Dong, J.H.; Zhang, W.L.; Shen, F.M. Vanadium extraction from vanadium-bearing titano-magnetite by selective chlorination using chloride wastes (FeClx). *J. Cent. South Univ.* **2017**, *24*, 311–317. [CrossRef]
38. Lee, J.-C.; Kim, E.-Y.; Chung, K.W.; Kim, R.; Jeon, H.-S. A review on the metallurgical recycling of vanadium from slags: Towards a sustainable vanadium production. *J. Mater. Res. Technol.* **2021**, *12*, 343–364. [CrossRef]
39. Liu, S.Y. Fundamental Studies on Selective Chlorination of Valuable Elements (Fe, Mn, V, Cr and Ti) from Vanadium Slag and Utilizations towards High Value. Ph.D. Thesis, University of Science and Technology Beijing, Beijing, China, 2019.
40. Xiang, J.; Huang, Q.; Lv, X.; Bai, C. Extraction of vanadium from converter slag by two-step sulfuric acid leaching process. *J. Clean. Prod.* **2018**, *170*, 1089–1101. [CrossRef]
41. Liu, S.Y.; Wang, L.J.; Chou, K.C. Viscosity measurement of $FeO–SiO_2–V_2O_3–TiO_2$ slags in the temperature range of 1644–1791 K and modelling by using ion-oxygen parameter. *Ironmak. Steelmak.* **2018**, *45*, 641–647. [CrossRef]
42. Ji, Y.; Shen, S.; Liu, J.; Xue, Y. Cleaner and effective process for extracting vanadium from vanadium slag by using an innovative three-phase roasting reaction. *J. Clean. Prod.* **2017**, *149*, 1068–1078. [CrossRef]
43. Zhang, G.Q.; Hu, T.; Liao, W.J.; Ma, X.D. An energy-efficient process of leaching vanadium from roasted tablet of ammonium sulfate, vanadium slag and silica. *J. Environ. Chem. Eng.* **2021**, *9*, 105332. [CrossRef]
44. Li, H.; Wang, X.H.; Zhou, R.Z. Experimental study on chloridizing volatilization of Ga during sodiumination roasting of V-slag. *Iron Steel Vanadium Titan.* **1993**, *14*, 44–50. (In Chinese)
45. Wang, H.G.; Wang, M.; Wang, X.W. Leaching behaviour of chromium during vanadium extraction from vanadium slag. *Miner. Process. Extr. Met.* **2015**, *124*, 127–131. [CrossRef]
46. Liu, S.; Shen, S.; Chou, K. An effective process for simultaneous extraction of valuable metals (V, Cr, Ti, Fe, Mn) from vanadium slag using acidic sodium chlorate solution under water bath conditions. *J. Min. Met. Sect. B Met.* **2018**, *54*, 153–159. [CrossRef]
47. Sun, Y. Comprehensive Utilization of Vanadium-Bearing Titanomagnetite with Extracting Vanadium by Selective Chlorination technology. Ph.D. Thesis, Northeastern University, Shenyang, China, 2015. (In Chinese).
48. Du, G.; Fan, C.; Yang, H.; Zhu, Q. Selective extraction of vanadium from pre-oxidized vanadium slag by carbochlorination in fluidized bed reactor. *J. Clean. Prod.* **2019**, *237*, 117765. [CrossRef]
49. Liu, S.-Y.; Li, S.-J.; Wu, S.; Wang, L.-J.; Chou, K.-C. A novel method for vanadium slag comprehensive utilization to synthesize Zn-Mn ferrite and Fe-V-Cr alloy. *J. Hazard. Mater.* **2018**, *354*, 99–106. [CrossRef] [PubMed]
50. Liu, S.; Wang, L.; Chou, K. Selective Chlorinated Extraction of Iron and Manganese from Vanadium Slag and Their Application to Hydrothermal Synthesis of $MnFe_2O_4$. *ACS Sustain. Chem. Eng.* **2017**, *5*, 10588–10596. [CrossRef]
51. Liu, S.; Wang, L.; Chou, K.-C.; Kumar, R.V. Electrolytic preparation and characterization of VCr alloys in molten salt from vanadium slag. *J. Alloy. Compd.* **2019**, *803*, 875–881. [CrossRef]
52. Liu, S.; Wang, L.; Chou, K. A Novel Process for Simultaneous Extraction of Iron, Vanadium, Manganese, Chromium, and Titanium from Vanadium Slag by Molten Salt Electrolysis. *Ind. Eng. Chem. Res.* **2016**, *55*, 12962–12969. [CrossRef]
53. Liu, S.; He, X.; Wang, Y.; Wang, L. Cleaner and effective extraction and separation of iron from vanadium slag by carbothermic reduction-chlorination-molten salt electrolysis. *J. Clean. Prod.* **2021**, *284*, 124674. [CrossRef]
54. Liu, S.Y.; Wang, L.J.; Chou, K.C. Innovative method for minimization of waste containing Fe, Mn and Ti during comprehensive utilization of vanadium slag. *Waste Manag.* **2021**, *127*, 179–188. [CrossRef]
55. Liu, S.-Y.; Zhen, Y.-L.; He, X.-B.; Wang, L.-J.; Chou, K.-C. Recovery and separation of Fe and Mn from simulated chlorinated vanadium slag by molten salt electrolysis. *Int. J. Miner. Met. Mater.* **2020**, *27*, 1678–1686. [CrossRef]
56. Menad, N.-E.; Kanari, N.; Save, M. Recovery of high grade iron compounds from LD slag by enhanced magnetic separation techniques. *Int. J. Miner. Process.* **2014**, *126*, 1–9. [CrossRef]
57. Seron, A.; Menad, N.; Galle-Cavalloni, P.; Bru, K. Selective Recovery of Vanadium by Oxy-carbochlorination of Basic Oxygen Furnace Slag: Experimental Study. *J. Sustain. Met.* **2020**, *6*, 478–490. [CrossRef]
58. Bin, Z.Y. Progress of the research on extraction of vanadium pentoxide from black shale and the market of the V_2O_5. *Hu-Nan Nonferr. Met.* **2006**, *22*, 16–20. (In Chinese)

59. Xiao, W.D. Mineralogy of black shale from Shanglin of Guangxi and vanadium extraction with hydrometallurgical process. *Nonferr. Met.* **2007**, *59*, 85–90.
60. He, D.; Feng, Q.; Zhang, G.; Ou, L.; Lu, Y. An environmentally-friendly technology of vanadium extraction from stone coal. *Miner. Eng.* **2007**, *20*, 1184–1186. [CrossRef]
61. Hu, K.L.; Liu, X.H. Review of roasting processing of vanadium-bearing carbonaceous shale. *Rare Met. Cem. Carbides* **2015**, *43*, 1–7.
62. Lin, H.L.; Fan, B.W. Study on mechanism of phase transformation during roasting and extracting vanadium from Fangshankou bone coal. *Chin. J. Rare Met.* **2001**, *25*, 273–277. (In Chinese)
63. Zhang, Y.M.; Bao, S.X.; Liu, T.; Chen, T.J.; Huang, J. The technology of extracting vanadium from stone coal in China: History, current statues and future prospects. *Hydrometallurgy* **2011**, *109*, 116–124. [CrossRef]
64. Wang, M.; Huang, S.; Chen, B.; Wang, X. A review of processing technologies for vanadium extraction from stone coal. *Miner. Process. Extr. Met.* **2020**, *129*, 290–298. [CrossRef]
65. Bie, S.; Wang, Z.J.; Li, Q.H.; Zhang, Y.G. Review of vanadium extraction from stone coal by roasting technique with sodium chloride and calcium oxide. *Chin. J. Rare Met.* **2010**, *34*, 291–297. (In Chinese)
66. Wang, H.S. Extraction of vanadtum from stone coal by roasting in the presence of sodium salts. *Min. Metall. Eng.* **1994**, *14*, 49–52.
67. Wan, J.Y.; Chen, T.J.; Han, J.; Feng, Y. Thermodynamics analysis of chlorination volatilization enrichment process of black shale. *Nonferr. Met. (Extr. Metall.)* **2018**, *5*, 50–55. (In Chinese)
68. Zhang, Y.; Hu, Y.; Bao, S. Vanadium emission during roasting of vanadium-bearing stone coal in chlorine. *Miner. Eng.* **2012**, *30*, 95–98. [CrossRef]
69. Li, C.-X.; Wei, C.; Deng, Z.-G.; Li, M.-T.; Li, X.-B.; Fan, G. Recovery of vanadium from black shale. *Trans. Nonferr. Met. Soc. China* **2010**, *20*, s127–s131. [CrossRef]
70. Erust, C.; Akcil, A.; Bedelova, Z.; Anarbekov, K.; Baikonurova, A.; Tuncuk, A. Recovery of vanadium from spent catalysts of sulfuric acid plant by using inorganic and organic acids: Laboratory and semi-pilot tests. *Waste Manag.* **2016**, *49*, 455–461. [CrossRef]
71. Li, Z.; Chen, M.; Zhang, Q.; Liu, X.; Saito, F. Mechanochemical processing of molybdenum and vanadium sulfides for metal recovery from spent catalysts wastes. *Waste Manag.* **2017**, *60*, 734–738. [CrossRef] [PubMed]
72. Beolchini, F.; Fonti, V.; Dell'Anno, A.; Rocchetti, L.; Vegliò, F. Assessment of biotechnological strategies for the valorization of metal bearing wastes. *Waste Manag.* **2012**, *32*, 949–956. [CrossRef]
73. Akcil, A.; Vegliò, F.; Ferella, F.; Okudan, M.D.; Tuncuk, A. A review of metal recovery from spent petroleum catalysts and ash. *Waste Manag.* **2015**, *45*, 420–433. [CrossRef]
74. Biswas, R.; Wakihara, M.; Taniguchi, M. Recovery of vanadium and molybdenum from heavy oil desulphurization waste catalyst. *Hydrometallurgy* **1985**, *14*, 219–230. [CrossRef]
75. Gaballah, I.; Djona, M. Recovery of Co, Ni, Mo and V from unroasted spent hydrorefining catalysts by selective chlorination. *Metall. Mater. Trans. B* **1995**, *26*, 41–50. [CrossRef]
76. Gaballah, I.; Djona, M.; Mugica, J.; Solozobal, R. Valuable metals recovery from spent catalysts by selective chlorination. *Resour. Conserv. Recycl.* **1994**, *10*, 87–96. [CrossRef]
77. Zhang, J.Y. *Metallurgical Physical Chemistry*; Metallurgical Industry Press: Beijing, China, 2004. (In Chinese)
78. Mink, G.; Bertóti, I.; Székely, T. Chlorination of V_2O_5 by CCl_4. Adsorption and steady state reaction. *React. Kinet. Catal. Lett.* **1985**, *27*, 33–38. [CrossRef]
79. Mink, G.; Bertóti, I.; Székely, T. Chlorination of V_2O_5 by CCl_4, the proposed reaction mechanism. *React. Kinet. Catal. Lett.* **1985**, *27*, 39–45. [CrossRef]
80. Jena, P.; Brocchi, E.; González, J. Kinetics of Low-Temperature Chlorination of Vanadium Pentoxide by Carbon Tetrachloride Vapor. *Metall. Mater. Trans. B* **2005**, *36*, 195–199. [CrossRef]
81. Gaballah, I.; Djona, M.; Allain, E. Kinetics of chlorination and carbochlorination of vanadium pentoxide. *Met. Mater. Trans. A* **1995**, *26*, 711–718. [CrossRef]
82. McCarley, R.E.; Roddy, J.W. The preparation of high purity vanadium pentoxide by a chlorination procedure. *J. Less Common Met.* **1960**, *2*, 29–35. [CrossRef]
83. Pap, I.S.; Mink, G.; Bertóti, I.; Székely, T.; Babievskaya, I.Z.; Karmazsin, E. Comparative kinetic and thermodynamic study on the chlorination of V_2O_5 with CCl_4, $COCl_2$ and Cl_2. *J. Therm. Anal. Calorim.* **1989**, *35*, 163–173. [CrossRef]
84. Jiang, D.-D.; Zhang, H.-L.; Xu, H.-B.; Zhang, Y. A novel method to prepare high-purity vanadium pentoxide by chlorination with anhydrous aluminum chloride. *Chem. Lett.* **2017**, *46*, 669–671. [CrossRef]
85. Jiang, D.D.; Zhang, H.L.; Xu, H.B.; Zhang, Y. Chlorination and purification of vanadium pentoxide with anhydrous aluminum chloride. *J. Alloy. Compd.* **2017**, *709*, 505–510. [CrossRef]
86. Shlewit, H.; Alibrahim, M. Extraction of sulfur and vanadium from petroleum coke by means of salt-roasting treatment. *Fuel* **2006**, *85*, 878–880. [CrossRef]
87. Vitolo, S.; Seggiani, M.; Filippi, S.; Brocchini, C. Recovery of vanadium from heavy oil and Orimulsion fly ashes. *Hydrometallurgy* **2000**, *57*, 141–149. [CrossRef]
88. Murase, K.; Nishikawa, K.I.; Ozaki, T.; Machida, K.I.; Adachi, G.Y.; Suda, T. Recovery of vanadium, nickel and magnesium from a fly ash of bitumen-in-water emulsion by chlorination and chemical transport. *J. Alloy. Compd.* **1998**, *264*, 151–156. [CrossRef]

89. Wang, H.; Feng, Y.; Li, H.; Kang, J. The separation of gold and vanadium in carbonaceous gold ore by one-step roasting method. *Powder Technol.* **2019**, *355*, 191–200. [CrossRef]
90. Zhang, L.J.; Lv, P.; He, Y.; Li, S.W.; Chen, K.H.; Yin, S.H. Purification of chlorine-containing wastewater using solvent extraction. *J. Clean. Prod.* **2020**, *273*, 122863. [CrossRef]
91. Fan, C.; Yang, H.; Zhu, Q. Selective hydrolysis of trace $TiCl_4$ from $VOCl_3$ for preparation of high purity V_2O_5. *Sep. Purif. Technol.* **2017**, *185*, 196–201. [CrossRef]

Article

Fluoride Leaching of Titanium from Ti-Bearing Electric Furnace Slag in [NH$_4^+$]-[F$^-$] Solution

Fuqiang Zheng, Yufeng Guo *, Feng Chen, Shuai Wang *, Jinlai Zhang, Lingzhi Yang and Guanzhou Qiu

School of Minerals Processing and Bioengineering, Central South University, Changsha 410083, China; f.q.zheng@csu.edu.cn (F.Z.); csuchenf@csu.edu.cn (F.C.); zhangjinlai@csu.edu.cn (J.Z.); yanglingzhi@csu.edu.cn (L.Y.); qgz@csu.edu.cn (G.Q.)
* Correspondence: yfguo@csu.edu.cn (Y.G.); wang_shuai@csu.edu.cn (S.W.)

Abstract: The effects of F$^-$ concentration, leaching temperature, and time on the Ti leaching from Ti-bearing electric furnace slag (TEFS) by [NH$_4^+$]-[F$^-$] solution leaching process was investigated to reveal the leaching mechanism and kinetics of titanium. The results indicated that the Ti leaching rate obviously increased with the increase of leaching temperature and F$^-$ concentration. The kinetic equation of Ti leaching was obtained, and the activation energy was 52.30 kJ/mol. The fitting results of kinetic equations and calculated values of activation energy both indicated that the leaching rate of TEFS was controlled by surface chemical reaction. The semi-empirical kinetics equation was consistent with the real experimental results, with a correlation coefficient (R^2) of 0.996. The Ti leaching rate reached 92.83% after leaching at 90 °C for 20 min with F$^-$ concentration of 14 mol/L and [NH$_4^+$]/[F$^-$] ratio of 0.4. The leaching rates of Si, Fe, V, Mn, and Cr were 94.03%, 7.24%, 5.36%, 4.54%, and 1.73%, respectively. The Ca, Mg, and Al elements were converted to (NH$_4$)$_3$AlF$_6$ and CaMg$_2$Al$_2$F$_{12}$ in the residue, which can transform into stable oxides and fluorides after pyro-hydrolyzing and calcinating.

Keywords: Ti-bearing electric furnace slag; fluorination method; leaching; kinetics; titanium dioxide

Citation: Zheng, F.; Guo, Y.; Chen, F.; Wang, S.; Zhang, J.; Yang, L.; Qiu, G. Fluoride Leaching of Titanium from Ti-Bearing Electric Furnace Slag in [NH$_4^+$]-[F$^-$] Solution. *Metals* **2021**, *11*, 1176. https://doi.org/10.3390/met11081176

Academic Editor: Jean François Blais

Received: 30 June 2021
Accepted: 21 July 2021
Published: 24 July 2021

Publisher's Note: MDPI stays neutral with regard to jurisdictional claims in published maps and institutional affiliations.

Copyright: © 2021 by the authors. Licensee MDPI, Basel, Switzerland. This article is an open access article distributed under the terms and conditions of the Creative Commons Attribution (CC BY) license (https://creativecommons.org/licenses/by/4.0/).

1. Introduction

Titanium dioxide is an irreplaceable high-grade functional white pigment and widely used in coatings [1], rubber [2], papermaking [3], printing [4] and other chemical industries [5–10], due to its characteristics, stable chemical properties, good color covering capability, high tinting strength and dispersion, nontoxicity, and electronic properties. Its application is a key indicator for evaluating the national modernization [11]. The titanium reserves in China account for about 38.8% of that worldwide, providing an important material basis for the development of titanium dioxide production and the promotion of China's modernization process. Statistically, titanium dioxide production consumes 90% of titanium raw materials all over the world [12–14]. Nowadays, about 62.7% of global titanium dioxide is produced by chlorination process, and the other is produced by sulfuric acid process [15].

The chlorination process requires high quality titanium-rich feedstock containing TiO$_2$ ≥ 90% and CaO + MgO < 1.5%, because the non-volatile CaCl$_2$ and MgCl$_2$ will hinder the chlorination reaction and obstruct the reaction bed [16–18]. The chlorination temperature of TiO$_2$ reaches 900–1000 °C, and the oxidation temperature of TiCl$_4$ reaches 1800–2000 °C. A large number of toxic and harmful, flammable, and explosive media, such as Cl$_2$, TiCl$_4$, and CO, participate in the high temperature reaction, which is not conducive to safe production [19]. The chlorination process can only produce rutile titanium dioxide, which belongs to high-end titanium dioxide product [20]. The hazardous wastes, including CaCl$_2$, MgCl$_2$, unreacted titanium-rich feedstocks, and petroleum coke, cannot be recycled except for by deep burial.

Ilmenite and titanium slag can be used as the feedstocks of the sulfuric acid process. The reaction temperature of sulfuric acid and titanium-bearing feedstock is about 250 °C, and the calcination temperature for TiO$_2$ transition is 700–900 °C [21]. The sulfuric acid process can produce the anatase and rutile titanium dioxide products belonging to low-end titanium dioxide products [21,22]. The iron in ilmenite is transformed into ferrous sulfate which causes the waste of iron resources and environmental pollution [23]. A large amount of waste acid and acid wastewater produced by the sulfuric acid process cannot be recycled. Environmental pollution has become an insurmountable problem in the production of titanium dioxide. Therefore, the method for the clean production of titanium dioxide is the focus of the titanium dioxide industry.

The fluorination method was proposed to prepare titanium dioxide, ferric oxide, or calcium fluoride from ilmenite or perovskite [24,25]. The leaching agent of fluorination method is NH$_4$HF$_2$ or NH$_4$F. The reaction temperature of fluoride leaching process is about 100 °C, which is much lower than that of chlorination process and sulfuric acid process. Fluorine and ammonia in the fluorination method can be recycled in the thermal hydrolysis process of fluorinated ammonium salt [26]. The solution containing fluorine and ammonia is returned to the leaching or impurity removal process for recycling, which ensures the cleanliness of the fluorination method. The fluorination reaction of ilmenite or perovskite should be carried out under high temperature and high pressure. The Ti leaching rate is only about 78% at 160 °C for 120 min in the leaching of perovskite by NH$_4$HF$_2$ [25].

The titanium resources contained in the vanadium titanomagnetite concentrate in the Panxi region account for more than 50% of China's total titanium resources. At present, iron and vanadium in vanadium titanomagnetite concentrate are mainly recovered by the blast furnace process. Titanium in the blast furnace titanium slag cannot be effectively recovered [27,28]. The direct reduction-electric furnace smelting process is an effective method to utilize iron and enrich titanium from vanadium titanomagnetite concentrate, and has been commercialized in South Africa and New Zealand [29]. The titanium in vanadium titanomagnetite concentrate is enriched in Ti-bearing electric furnace slag (TEFS) containing high SiO$_2$, Al$_2$O$_3$, CaO, and MgO, which cannot be the feedstock of chlorination process and sulfuric acid process. TEFS can be used to prepare titanium dioxide by fluorination method due to its low requirements for impurities in feedstock. However, the main Ti-bearing mineral phase in TEFS is a complex solid solution with a stable structure and is more difficult to decompose by NH$_4$HF$_2$ or NH$_4$F than ilmenite and perovskite. Our team studied the pressurized fluoride leaching of TEFS, including the leaching of the main elements and the transformation of mineral phase and particle morphology [30]. Based on the results, we found that there are some shortcomings in pressurized fluoride leaching in large-scale leaching equipment. During the leaching process, the F-bearing solution and steam cause serious corrosion to the equipment. The outer layer of the heating parts in the fluoride leaching equipment needs to be sprayed with fluorine-resistant materials which are easy to deform and fall off under high temperature. The leaching pressure and temperature reach the designated values too slowly. It is more harmful to fluorine-resistant materials, which makes it difficult to enlarge the pressurized fluoride leaching reactor. The completion of fluoride leaching reaction at a temperature under normal pressure was more conducive to the industrial production. However, compared with pressurized leaching, the rate of the chemical reaction was lower during normal pressure leaching. Therefore, it is necessary to study the fluoride leaching mechanism and kinetics of titanium from TEFS under a normal pressure system, in order to reveal the key factors affecting the Ti leaching rate. It has important practical significance for increasing the leaching speed and realizing the industrialization.

In this paper, the effects of F$^-$ concentration and temperature on the TEFS fluoride leaching mechanism and kinetics were investigated. The transformation of mineral phase and microstructure during leaching can reveal the fluoride leaching mechanism of TEFS. The kinetics study can determine the rate-controlling step and key factors in order to provide a basis for regulating the leaching reaction rate. The findings will provide a technical basis for the preparation of titanium dioxide by fluorination method with TEFS and is of great significance for improving the utilization of titanium resources in the Panxi region of China.

2. Experimental methods

2.1. Experimental Procedure

The leaching experiment was carried out in a 500 mL plastic three-necked flask fitted with an agitator and a reflux condenser. The leaching reaction of fine TEFS particles was completed in a very short time, which made it impossible to accurately study the leaching kinetics. Too wide particle size range is not conducive to the study of leaching kinetics. Therefore, the TEFS used in this study were crushed and screened with standard Taylor sieves from 200 mesh to 250 mesh, and the volume average particle size was 67.35 μm. The mass transfer rate between the solution and TEFS particles was accelerated by strengthening the stirring. The influence of stirring on the external diffusion can be ignored, as the rotating speed was 200 r/min. During the leaching process, the flask was heated in a thermostatically controlled water bath. The $[NH_4^+]/[F^-]$ molar ratio of $[NH_4^+]$-$[F^-]$ solution was 0.4. First, 300 g of $[NH_4^+]$-$[F^-]$ solution with required F^- concentration was loaded into the flask and heated to the designated temperature. After that, 15 g TEFS sample was put into the $[NH_4^+]$-$[F^-]$ solution. There was a large excess of leaching solution, with the L/S ratio of 20:1, which meant the change of F^- concentration can be neglected in the whole leaching process. The F^- concentration range was chosen to be 8 mol/L to 14 mol/L in order to ensure that the concentration of leaching agent was much larger than the theoretical consumption and lower than the maximum solubility. When the leaching time reaches the experimental required time, the slurry was filtered and washed with distilled water immediately. Then the washed residue was dried for 4 h at 110 °C for the analyses of chemical composition, mineral phase, and microstructure.

2.2. Definition of Parameters

The Ti leaching rate is defined as follows:

$$X = \left[1 - \left(\frac{m \times w_{Ti}}{m_0 \times w_0}\right)\right] \times 100\% \tag{1}$$

where X demotes the Ti leaching rate (%), m demotes the mass of dry leaching residue (g), w_{Ti} denotes the Ti content of dry leaching residue (wt%), m_0 demotes the initial mass of dry TEFS sample (g), and w_0 denotes the Ti content of dry TEFS sample (wt%).

2.3. Analysis and Characterization

The chemical compositions of TEFS sample and leaching residue were quantitatively analyzed by the fusion method of X-ray fluoroscopy (XRF, Axios mAX, Holand PANalytical Co., Ltd., Almelo, The Netherlands).

The chemical compositions of the leaching solution were determined by inductively coupled plasma atomic emission spectrometry (ICP-MS, Optima 5300DV, PerkinElmer, Waltham, MA, USA).

The mineral phases of the TEFS sample and leaching residue were characterized by X-ray diffraction (XRD, Cu Kα radiation, λ = 0.154056 nm, 40 kV, 300 mA, SCAN: 10.0/80.0/0.0085/0.15 sec, D/max2550PC, Rigaku Co., Ltd., Tokyo, Japan).

The scanning electron microscope equipped with an energy diffraction spectrum analyzer (SEM, JEOL JSM-6490LV, Tokyo, Japan) was used to study the microstructure and element analysis of TEFS sample and leaching residue.

The particle size distribution of TEFS sample was characterized by laser practical size analyzer (Mastersize 2000, Malvern Panalytical, Malvern, England).

3. Results and Discussion

3.1. Ti-Bearing Electric Furnace Slag

The TEFS used in this investigation was provided by the Chongqing Iron and Steel (group) Co., Ltd. (Chongqing, China). The TEFS was produced from vanadium titano-magnetite concentrate by the direct reduction-electric furnace smelting process. The main chemical composition of TEFS was listed in Table 1. The X-ray diffraction (XRD) pattern of TEFS was shown in Figure 1. The SEM microstructure and EDS analysis of TEFS was shown in Figure 2, which indicates that the columnar $M_xTi_{3x}O_5$ particles embedded in the gangue phase of augite. The width of columnar $M_xTi_{3x}O_5$ particles is less than 50 μm. All reagents used in the study were of analytical grade.

Table 1. The chemical composition of Ti-bearing electric furnace slag (TEFS) (wt%).

TiO_2	SiO_2	Al_2O_3	MgO	CaO	Total Fe	MnO	V_2O_5	Cr_2O_3
51.50	19.01	14.38	7.42	3.22	1.88	1.10	0.63	0.03

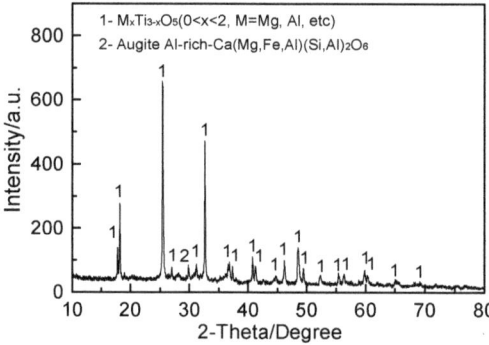

Figure 1. XRD pattern of Ti-bearing electric furnace slag.

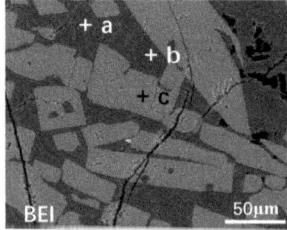

Point	O	Ti	Ca	Mg	Al	Si	Fe	Mn	V	Cr
a	34.07	3.83	8.26	9.55	13.68	27.29	0.61	2.4	—	0.32
b	34.78	3.57	8.01	8.28	16.14	26.34	0.53	2.11	—	0.24
c	27.39	65.72	—	3.34	3.54	—	—	—	—	—

Figure 2. SEM microstructure and EDS analysis of Ti-bearing electric furnace slag (wt%).

3.2. Thermodynamic Analysis of Fluoride Leaching Reactions

The major Ti-bearing mineral phase in TEFS is $M_xTi_{3x}O_5$, which is the solid solution of Ti_3O_5, $MgTi_2O_5$, and Al_2TiO_5. The reactions of Ti-bearing mineral phases in TEFS during fluoride leaching can be simply expressed as follows.

$$2Ti_3O_{5(s)} + 24H^+_{(aq)} + 36F^-_{(aq)} + O_{2(g)} = 6TiF_6^{2-}{}_{(aq)} + 12H_2O_{(aq)} \qquad (2)$$

$$Al_2TiO_{5(s)} + 10H^+_{(aq)} + 18F^-_{(aq)} = 2AlF_6^{3-}{}_{(aq)} + TiF_6^{2-}{}_{(aq)} + 5H_2O_{(aq)} \qquad (3)$$

$$MgTi_2O_{5(s)} + 10H^+_{(aq)} + 14F^-_{(aq)} = MgF_{2(s)} + 2TiF_6^{2-}_{(aq)} + 5H_2O_{(aq)} \quad (4)$$

The major gangue in TEFS is augite, which can be simply regarded as $CaSiO_3$, $MgSiO_3$, Al_2SiO_5, $CaAl_2O_4$, $MgAl_2O_4$, $CaO \cdot MgO \cdot 2SiO_2$, and $CaO \cdot Al_2O_3 \cdot 2SiO_2$. The reactions of gangue in TEFS during fluoride leaching can be expressed as follows.

$$CaSiO_{3(s)} + 6H^+_{(aq)} + 8F^-_{(aq)} = CaF_{2(s)} + SiF_6^{2-}_{(aq)} + 3H_2O_{(aq)} \quad (5)$$

$$MgSiO_{3(s)} + 6H^+_{(aq)} + 8F^-_{(aq)} = MgF_{2(s)} + SiF_6^{2-}_{(aq)} + 3H_2O_{(aq)} \quad (6)$$

$$Al_2SiO_{5(s)} + 10H^+_{(aq)} + 18F^-_{(aq)} = 2AlF_6^{3-}_{(aq)} + SiF_6^{2-}_{(aq)} + 5H_2O_{(aq)} \quad (7)$$

$$CaAl_2O_{4(s)} + 8H^+_{(aq)} + 14F^-_{(aq)} = CaF_{2(s)} + 2AlF_6^{3-}_{(aq)} + 4H_2O_{(aq)} \quad (8)$$

$$MgAl_2O_{4(s)} + 8H^+_{(aq)} + 14F^-_{(aq)} = MgF_{2(s)} + 2AlF_6^{3-}_{(aq)} + 4H_2O_{(aq)} \quad (9)$$

$$CaO \cdot MgO \cdot 2SiO_{2(s)} + 12H^+_{(aq)} + 16F^-_{(aq)} = 2SiF_6^{2-}_{(aq)} + CaF_{2(s)} + MgF_{2(s)} + 6H_2O_{(aq)} \quad (10)$$

$$CaO \cdot Al_2O_3 \cdot 2SiO_{2(s)} + 16H^+_{(aq)} + 26F^-_{(aq)} = 2SiF_6^{2-}_{(aq)} + 2AlF_6^{3-}_{(aq)} + CaF_{2(s)} + 8H_2O_{(aq)} \quad (11)$$

Figure 3 shows the ΔG^θ-T relationship lines of the possible reactions during fluoride leaching. According to the calculated results, the ΔG^θ of Equations (2)–(11) at the temperature range of 298 K to 448 K are less than zero, and all the ΔG^θ decreases with increasing leaching temperature. The calculation results illustrate that Equations (2)–(11) can proceed spontaneously, and the reactions occurred easier as the reaction temperature increased. Comparing the ΔG^θ of Equations (2)–(4), the ΔG^θ line of Equation (2) was below the ΔG^θ lines of Equations (3) and (4), which means the Ti_3O_5 was easier to decompose in the leaching process than Al_2TiO_5 and $MgTi_2O_5$. According to Equations (5)–(7), the $MgSiO_3$ was easier to decompose in the leaching process than Al_2SiO_5 and $CaSiO_3$. The ΔG^θ lines of Equations (8) and (9) were the fluoride leaching reactions of aluminate, which shows that the fluoride leaching occurring process of $MgAl_2O_4$ was easier than that of $CaAl_2O_4$. In addition, the ΔG^θ lines of Equations (10) and (11) were below that of other silicates and aluminates. The solubility of $(NH_4)_3AlF_6$ was only 1.03 g/100g H_2O. Thus, the AlF_6^{3-} may react with the NH_4^+ in the leaching solution as follows.

$$3NH_4^+_{(aq)} + AlF_6^{3-}_{(aq)} = (NH_4)_3AlF_{6(s)} \quad (12)$$

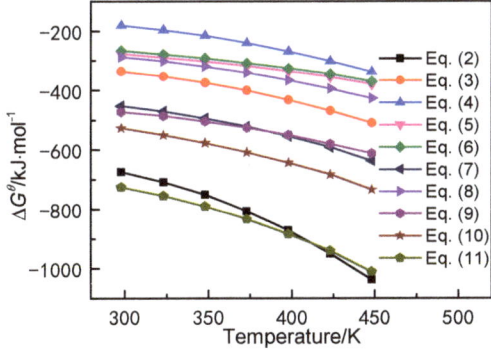

Figure 3. ΔG^θ–T relationship lines of the possible reactions during fluoride leaching.

According to the results of thermodynamic analysis, the Ti and Si elements were converted to TiF_6^{2-} and SiF_6^{2-} in the leaching solution. The Ca, Mg, and Al elements were converted to CaF_2, MgF_2, and $(NH_4)_3AlF_6$ in the residue.

3.3. Effects of Experimental Parameters on Ti Leaching Rates

The leaching experiments were carried out in the F^- concentration range of 8 mol/L to 14 mol/L and temperature range of 60 °C to 90 °C at L/S ratio of 20:1 with rotating speed of 200 r/min. The effect of temperature and F^- concentration on Ti leaching rates are shown in Figure 4.

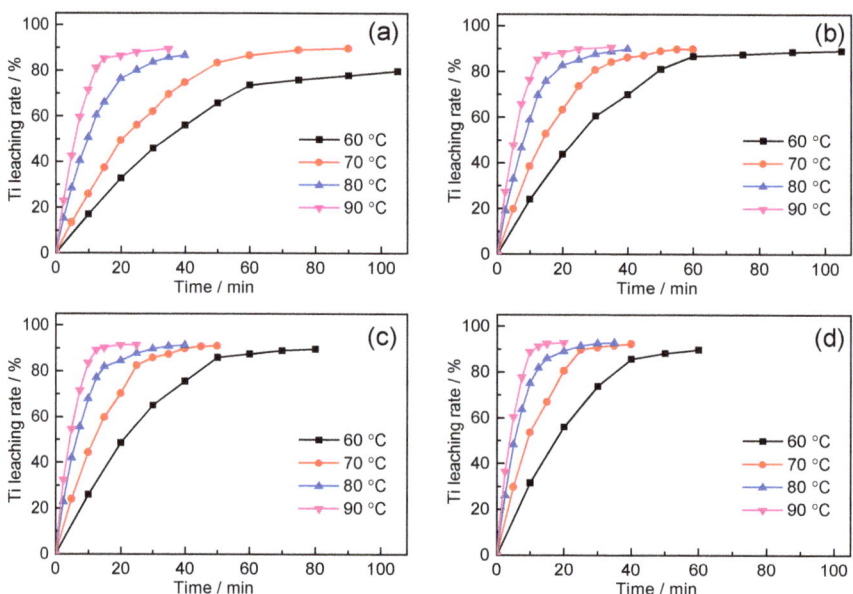

Figure 4. Effects of temperature on the Ti leaching rate with L/S ratio of 20:1 and rotating speed of 200 r/min. (**a**) C_{F^-} = 8 mol/L; (**b**) C_{F^-} = 10 mol/L; (**c**) C_{F^-} = 12 mol/L; (**d**) C_{F^-} = 14 mol/L.

When the Ti leaching rate reaches a certain value, prolonging the leaching time has no significant effect on the Ti leaching rate. The Ti leaching rate obviously increased with the increase of F^- concentration at the leaching temperatures of 60 °C and 70 °C. The F^- concentration had no obvious effect on Ti leaching rate when the leaching temperature exceeded 80 °C. With the increase of F^- concentration, the effect of leaching temperature on Ti leaching rate decreased. Increased leaching temperature and F^- concentration could increase the effective collision of TEFS particles and reactive ions, which promote the leaching reactions.

3.4. Transformation of Mineral Phase

The effect of leaching temperature on the mineral phase of leaching residue was shown in Figure 5. The other leaching conditions were F^- concentration of 10 mol/L, leaching time of 30 min, L/S ratio of 20:1, and rotating speed of 200 r/min. When the leaching temperature reached 60 °C, the peaks of $(NH_4)_3AlF_6$ and $CaMg_2Al_2F_{12}$ were detected in the leaching residue and the peaks of augite disappeared, as shown in Figure 5b. The peaks of $M_xTi_{3x}O_5$ became very weak at the leaching temperature of 90 °C, as shown in Figure 5c. The intensity of $(NH_4)_3AlF_6$ peaks increased sharply as the leaching temperature rose from 60 °C to 90 °C. The reactions can be expressed as follows.

$$Al_2TiO_{5(s)} + 10H^+_{(aq)} + 18F^-_{(aq)} + 6NH_4^+_{(aq)} = 2(NH_4)_3AlF_{6(s)} + TiF_6^{2-}_{(aq)} + 5H_2O_{(aq)} \quad (13)$$

$$CaSiO_{3(s)} + 2MgSiO_{3(s)} + Al_2SiO_{5(s)} + 28H^+_{(aq)} + 36F^-_{(aq)} = CaMg_2Al_2F_{12(s)} + 4SiF_6^{2-}_{(aq)} + 14H_2O_{(aq)} \quad (14)$$

$$CaAl_2O_{4(s)} + 2MgAl_2O_{4(s)} + 24H^+_{(aq)} + 36F^-_{(aq)} + 12NH_4^+_{(aq)} = CaMg_2Al_2F_{12(s)} + 4(NH_4)_3AlF_{6(s)} + 12H_2O_{(aq)} \quad (15)$$

$$CaO \cdot MgO \cdot 2SiO_{2(s)} + CaO \cdot Al_2O_3 \cdot 2SiO_{2(s)} + 3MgTi_2O_{5(s)} + Al_2TiO_{5(s)} + 68H^+_{(aq)} + 90F^-_{(aq)}$$
$$= 2CaMg_2Al_2F_{12(s)} + 4SiF_6^{2-}_{(aq)} + 7TiF_6^{2-}_{(aq)} + 34H_2O_{(aq)} \quad (16)$$

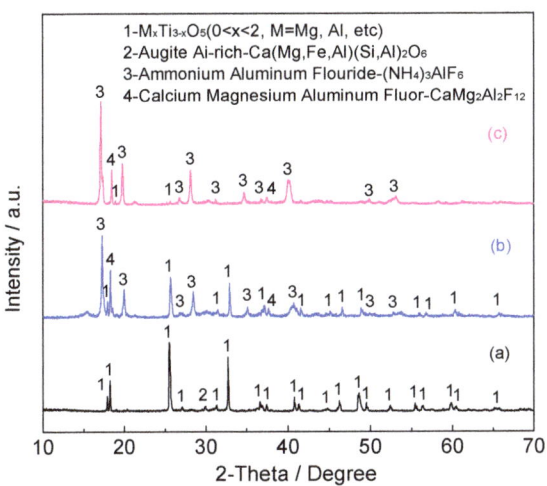

Figure 5. XRD patterns of the leaching residue at different leaching temperatures with F$^-$ concentration of 10 mol/L, leaching time of 30 min, L/S ratio of 20:1, and rotating speed of 200 r/min. (**a**) Ti-bearing electric furnace slag; (**b**) Leaching temperature: 60 °C; (**c**) Leaching temperature: 90 °C.

3.5. Transformation of Microstructure

The SEM images and EDS analysis of TEFS and leaching residue are shown in Figure 6. Before the fluoride leaching, the TEFS particle was a dense irregular particle, as shown in Figure 6a. It was observed from the EDS analysis that point 1 was $M_xTi_{3x}O_5$ and point 2 was augite. Figure 6b illustrates that the outer layer of the particle is the product layer, and the inside of particle is still unreacted TEFS. Many cracks appeared on the surface of residue. The granular grain on the surface of residue was $CaMg_2Al_2F_{12}$ (point 3). The octahedral grain was $(NH_4)_3AlF_6$, as shown in point 4. Point 5 was the surface of unreacted TEFS. When the fluoride leaching reaction progressed, the surface of the TEFS particle reacts with the leaching reagent and forms the F-bearing ions and F-bearing compounds. The TiF_6^{2-} and SiF_6^{2-} ions dissolved in the leaching solution. The $CaMg_2Al_2F_{12}$ and $(NH_4)_3AlF_6$ grains adsorb on the surface of unreacted TEFS.

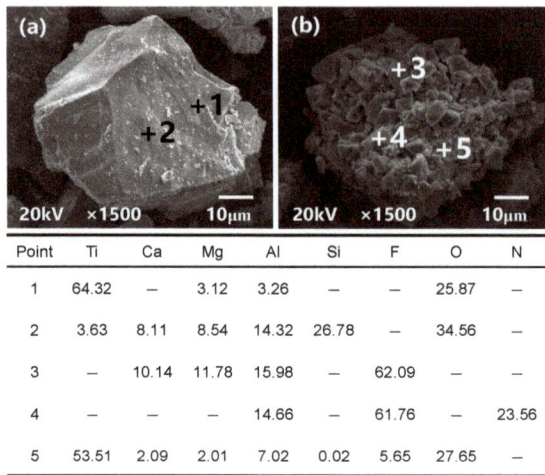

Point	Ti	Ca	Mg	Al	Si	F	O	N
1	64.32	—	3.12	3.26	—	—	25.87	—
2	3.63	8.11	8.54	14.32	26.78	—	34.56	—
3	—	10.14	11.78	15.98	—	62.09	—	—
4	—	—	—	14.66	—	61.76	—	23.56
5	53.51	2.09	2.01	7.02	0.02	5.65	27.65	—

Figure 6. SEM images and EDS analysis of TEFS and fluoride leaching residue (wt%). (**a**) TEFS; (**b**) fluoride leaching residue at 60 °C with F^- concentration of 10 mol/L, leaching time of 30 min, L/S ratio of 20:1, and rotating speed of 200 r/min.

3.6. Kinetics of Fluoride Leaching of Titanium

3.6.1. Selection of Leaching Kinetics Model

When the fluid reactant reacted with the dense and non-porous solid particles, in addition to forming a fluid product, the resulting solid product or residual inert material was loose and porous. This reaction model is the shrinking core model. The SEM microstructure showed that the TEFS particles were essentially dense grains. The non-porous TEFS particles gradually shrank during leaching. The product layer composed of $CaMg_2Al_2F_{12}$ and $(NH_4)_3AlF_6$ was loose and porous. Thus, the reaction model of TEFS leaching process accords with the shrinking core model [12]. The rate-controlling steps of shrinking core model contain surface chemical reaction and diffusion control through the ash or production layer [31,32]. The rate equations for shrinking core model are listed in Table 2.

Table 2. The rate equations for shrinking core model.

Rate-Controlling Step	Rate Equation
Surface chemical reaction	$1-(1-X)^{1/3} = k \cdot t$
Diffusion control through the ash or production layer	$1-\frac{2}{3}X-(1-X)^{2/3} = k' \cdot t$

X: Ti leaching rate (%); t: time (min); k: apparent rate constant (min^{-1}); k': apparent rate constant (min^{-1}).

In order to establish the control step, the Ti leaching rates obtained at different conditions in this study were analyzed by the rate equations for shrinking core model. When the Ti leaching rate reached a certain value at the determining F^- concentration and temperature, there was no further increase with prolonged time. In this study, the Ti leaching rate increased significantly was used to analyze kinetics. The plots of $1-(1-X)^{1/3}$ versus time and plots of $1-(2/3)X-(1-X)^{1/3}$ versus time for different conditions are shown in Figures 7 and 8, respectively. The slope of the fitted straight line denoted the apparent rate constant of rate-controlling step. The correlation coefficients (R^2) of the fitted straight line could be used to determine the rate-controlling step. Comparison of R^2 between plots of $1-(1-X)^{1/3}$ and $1-(2/3)X-(1-X)^{2/3}$ versus time is listed in Table 3. The plots of $1-(1-X)^{1/3}$ obtained a higher linear relationship with time. The surface chemical reaction step with the highest kinetics resistance is the rate-controlling step. Therefore, the fluoride leaching rate of TEFS was controlled by surface chemical reaction.

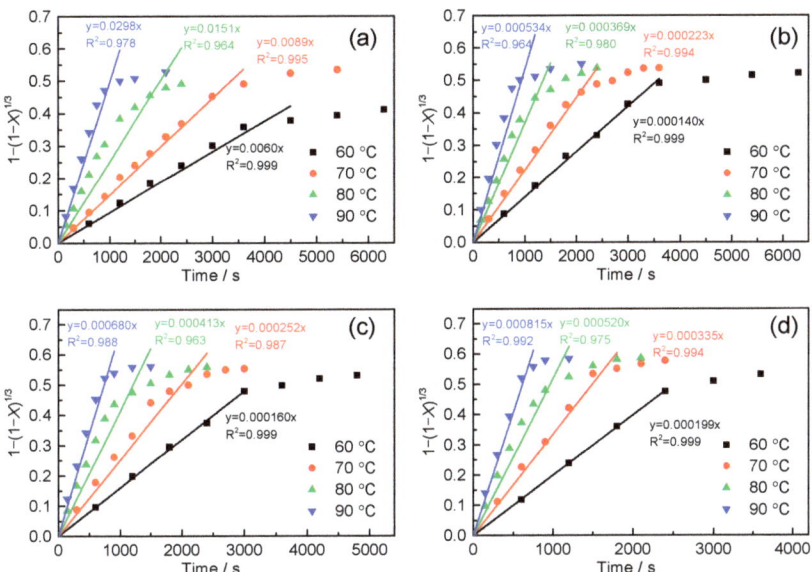

Figure 7. Plots of $1 - (1 - X)^{1/3}$ versus time for different leaching temperatures with L/S ratio of 20:1 and rotating speed of 200 r/min. (**a**) C_{F^-} = 8 mol/L; (**b**) C_{F^-} = 10 mol/L; (**c**) C_{F^-} = 12 mol/L; (**d**) C_{F^-} = 14 mol/L.

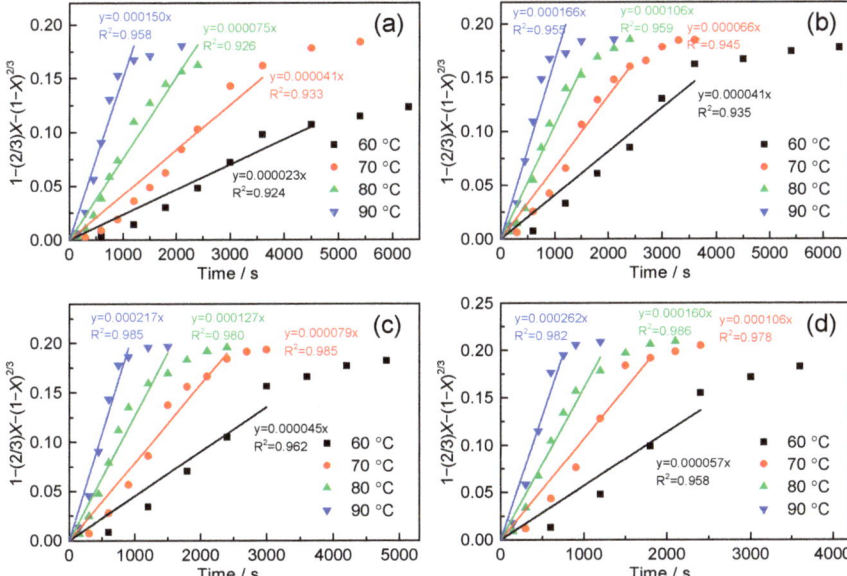

Figure 8. Plots of $1 - (2/3)X - (1 - X)^{1/3}$ versus time for different leaching temperature with L/S ratio of 20:1 and rotating speed of 200 r/min. (**a**) C_{F^-} = 8 mol/L; (**b**) C_{F^-} = 10 mol/L; (**c**) C_{F^-} = 12 mol/L; (**d**) C_{F^-} = 14 mol/L.

Table 3. Comparison of R^2 between plots of $1 - (1 - X)^{1/3}$ and $1 - (2/3)X - (1 - X)^{2/3}$ versus time for different F$^-$ concentrations and temperatures with L/S ratio of 20:1 and rotating speed of 200 r/min.

C_{F^-} (mol/L)	Temperature (°C)	Correlation Coefficient R^2	
		$1 - (1 - X)^{1/3}$	$1 - (2/3)X - (1 - X)^{2/3}$
8	60	0.999	0.924
	70	0.995	0.933
	80	0.964	0.926
	90	0.978	0.958
10	60	0.999	0.935
	70	0.994	0.945
	80	0.980	0.959
	90	0.964	0.955
12	60	0.999	0.962
	70	0.987	0.985
	80	0.963	0.980
	90	0.988	0.985
14	60	0.999	0.958
	70	0.994	0.978
	80	0.975	0.986
	90	0.992	0.982

3.6.2. Determination of Rate Constant

F$^-$ Concentration Index

The kinetic equation of the fluoride leaching rate of TEFS can be expressed as Equation (17).

$$1 - (1 - X)^{1/3} = k \cdot t \quad (17)$$

where X is Ti leaching rate (%), and k is apparent rate constant (min^{-1}).

The apparent rate constant (k) is affected by F$^-$ concentration and leaching temperature, as indicated in Equation (18).

$$k = k_0 \cdot C_{F^-}^b \cdot e^{-E/RT} \quad (18)$$

where k_0 is the preexponential factor of the kinetic equation, C_{F^-} is F$^-$ concentration (mol/L), b is F$^-$ concentration index, E is activation energy (kJ/mol), R is molar gas constant (8.314 J·mol^{-1}·K^{-1}), and T is leaching temperature (K).

The semi-empirical kinetics equation can be established from Equation (17) and Equation (18) as:

$$1 - (1 - X)^{1/3} = k_0 \cdot C_{F^-}^b \cdot e^{-E/RT} \cdot t \quad (19)$$

While $k_0 \cdot e^{-E/RT}$ can be regarded as k_1 in order to calculate the F$^-$ concentration index (b), Equation (19) can be transformed into:

$$1 - (1 - X)^{1/3} = k_1 \cdot C_{F^-}^b \cdot t \quad (20)$$

Equation (20) can be transformed into:

$$\frac{1 - (1 - X)^{1/3}}{t} = k_1 \cdot C_{F^-}^b \quad (21)$$

Equation (21) can be transformed into Equation (22) after logarithmic calculation on both sides of the equation.

$$\ln\left[\frac{1 - (1 - X)^{1/3}}{t}\right] = \ln k_1 + b \ln C_{F^-} \quad (22)$$

The plots of $\ln[[1-(1-X)^{1/3}]/t]$ vs. $\ln C_{F^-}$ at different leaching temperatures are presented in Figure 9. The slopes of fitted straight lines were the F^- concentration indexes at different temperatures. The four slopes were calculated to be 1.17, 1.43, 1.00, and 0.81 with R^2 of 0.967, 0.972, 0.959, and 0.946, respectively. The statistical average of b was calculated to be 1.10 according to Figure 9.

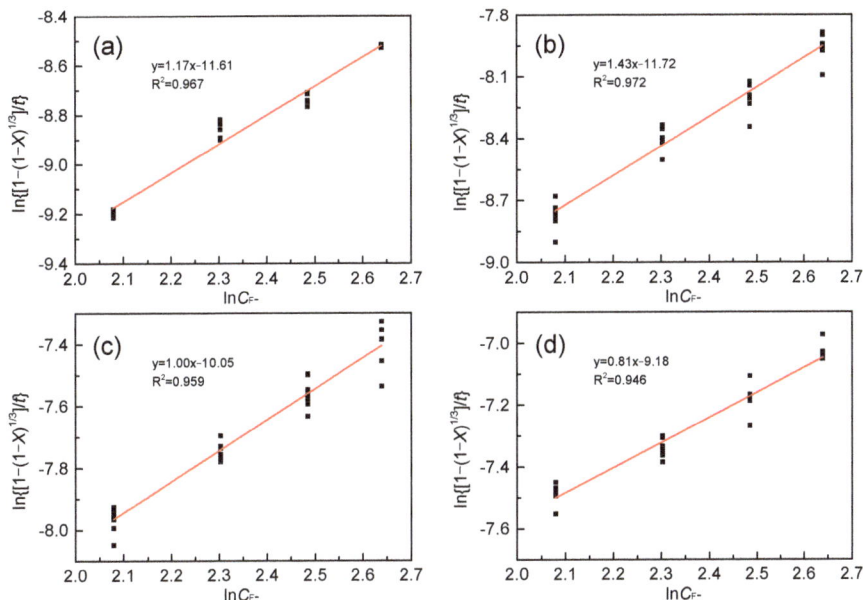

Figure 9. Plot of $\ln[[1-(1-X)^{1/3}]/t]$ vs. $\ln C$ with L/S ratio of 20:1 and rotating speed of 200 r/min. (**a**) 60 °C, (**b**) 70 °C, (**c**) 80 °C, (**d**) 90 °C.

Equation (19) can be written as:

$$1-(1-X)^{1/3} = k_0 \cdot C_{F^-}^{1.10} \cdot e^{-E/RT} \cdot t \tag{23}$$

Activation Energy and Preexponential Factor

In order to calculate the activation energy and preexponential factor, the Equation (18) can be transformed into:

$$\frac{1-(1-X)^{1/3}}{C_{F^-}^{1.10} \cdot t} = k_0 \cdot e^{-E/RT} \tag{24}$$

Equation (24) can be transformed into Arrhenius equation after logarithmic calculation on both sides of the equation.

$$\ln\left[\frac{1-(1-X)^{1/3}}{C_{F^-}^{1.10} \cdot t}\right] = -\frac{E}{RT} + \ln k_0 \tag{25}$$

The Arrhenius plot, which describes the relationship of $\ln\left\{\left[1-(1-X)^{1/3}\right]/\left(C_{F^-}^{1.10} \cdot t\right)\right\}$ and $1/T \times 1000$, is shown in Figure 10. The correlation coefficient (R^2) of the fitted straight line in Figure 10 is 0.946.

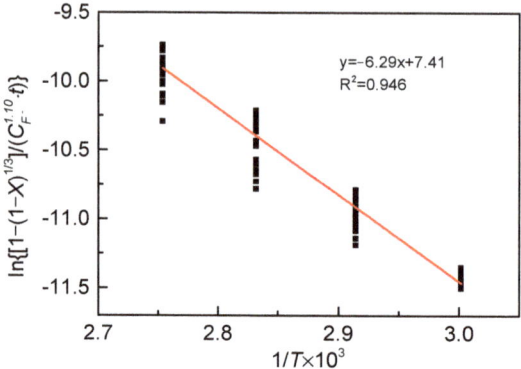

Figure 10. Plot of $\ln\{[1-(1-X)^{1/3}]/C_{F^-}^{1.10} \cdot t\}$ versus $1/T \times 1000$ with L/S ratio of 20:1 and rotating speed of 200 r/min.

The slope of the fitted straight line denoted the $-E/R$. The intercept of the fitted straight line and vertical axis represented $\ln k_0$.

The slope of fitted straight line in Figure 10 was calculated to be -6.29.

$$-\frac{E}{1000 \times R} = -6.29 \qquad (26)$$

Thus, the activation energy, E, can be expressed as:

$$E = -6.29 \times (-1000 \times R) = 6290 \times 8.314 = 52.30 \text{ kJ/mol} \qquad (27)$$

The activation energy of the fluoride leaching reaction is 52.30 kJ/mol. The activation energy of the reaction controlled by surface chemical reaction was in excess of 40.0 kJ/mol. The value of the activation energy confirmed that the fluoride leaching rate of TEFS was controlled by surface chemical reaction [31,33].

The intercept of fitted straight line and vertical axis in Figure 10 was calculated to be 7.41.

$$\ln k_0 = 7.41 \qquad (28)$$

Thus, the preexponential factor k_0 can be expressed as:

$$k_0 = e^{7.41} = 1652.43 \qquad (29)$$

The Ti leaching kinetic equation preexponential factor of fluoride leaching reaction was 1652.43.

3.6.3. Determination of Leaching Kinetics

In the conditions of F^- concentration range from 8 mol/L to 14 mol/L, leaching temperature range from 60 °C to 90 °C, L/S ratio of 20:1, and rotating speed of 200 r/min, the final semi-empirical kinetics equation can be obtained by synthesizing the Equations (19), (23), (27), and (29).

$$1-(1-X)^{1/3} = 1652.43 \times C_{F^-}^{1.10} \times e^{-52300/RT} \times t \qquad (30)$$

where X is Ti leaching rate (%), C_{F^-} is F^- concentration (mol/L), R is molar gas constant (8.314 J·mol^{-1}·K^{-1}), and T is leaching temperature (K).

3.6.4. Comparison of Experimental and Calculated Results

Comparison of the calculated Ti leaching rate with Equation (30) and all the experimental results in this work is shown in Figure 11. The R^2 of the fitted straight line is 0.996, which indicates that the calculated results by Equation (30) agree with the experimental ones.

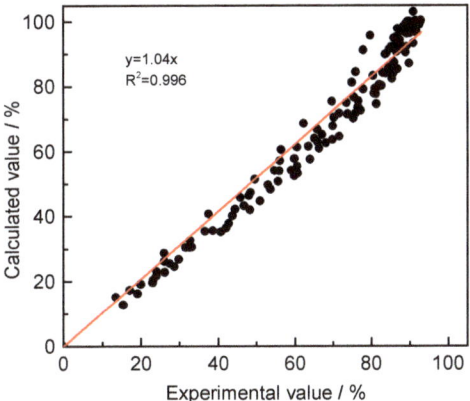

Figure 11. Comparison between the calculated Ti leaching rate using Equation (30) and all the experimental results in this work.

3.7. Leaching Solution and Leaching Residues

Figure 12 shows the leaching rates of elements in TEFS under the conditions of F^- concentration at 14 mol/L and leaching temperature of 90 °C for 20 min. Almost of the Ti and Si elements dissolved in the leaching solution. Most of Fe, V, Cr, and Mn exist in the leaching residue. The mineral phase, containing iron, vanadium, chromium, and manganese, was not detected in the leaching residue because their contents were too low to be detected in the X-ray diffraction analysis. All the Al, Mg, and Ca elements exist in the leaching residue.

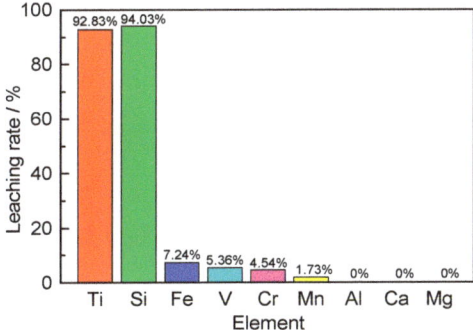

Figure 12. Leaching rates of elements in TEFS under the conditions of F^- concentration at 14 mol/L, leaching temperature of 90 °C for 20 min with L/S ratio of 20:1, and rotating speed of 200 r/min.

The composition of leaching solution is listed in Table 4. The Si is the main impurity in the leaching solution. Although the Fe, V, Mn, and Cr contents in the leaching solution were not high, they were all harmful elements which will seriously deteriorate the whiteness of titanium dioxide. Non-silicon impurities can be selectively precipitated and removed by adjusting the pH and NH_4^+ concentration of the leaching solution. Then, the recovery of titanium can be achieved by selective precipitation of fluorinated titanate. Finally, silicon can be precipitated from the solution to prepare Si-bearing product.

Table 4. Composition of leaching solution under the conditions of F^- concentration at 14 mol/L, leaching temperature of 90 °C for 20 min with L/S ratio of 20:1, rotating speed of 200 r/min (g/L).

Ti	Si	Fe	V	Mn	Cr	Ca	Mg	Al
26.64	7.81	0.13	0.02	0.016	0.001	0	0	0

The main mineral phases in leaching residue are $(NH_4)_3AlF_6$ and $CaMg_2Al_2F_{12}$, which have poor thermal stability. The whole NH_4^+ and most of the F element in residue can be recycled in the form of NH_4F by pyro-hydrolyzing at 250 °C for 120 min in the presence of water vapor and ammonia gas. Then, the residue is calcined at 900 °C for 60 min to become more stable. The composition and mineral phases of the calcinated residues are shown in Table 5 and Figure 13. The main mineral phases in the calcinated residue are aluminum oxide, sellaite, and fluorite. The calcinated residue has been identified for the leaching toxicity of solid waste according to HJ/T 298-2007 and GB 5085.3-2007. The inorganic fluoride content in the leaching solution was 59 mg/L, which was less than the hazard limit of 100 mg/L (GB 5085.3-2007). Therefore, the calcinated residue was not hazardous waste and can be used as fluxes of steelmaking.

Table 5. Composition of calcinated residue under the conditions of F^- concentration at 14 mol/L, leaching temperature of 90 °C for 20 min with L/S ratio of 20:1, and rotating speed of 200 r/min (wt%).

Ti	Al	Mg	Ca	Fe	Mn	Si	V	Cr	F
1.81	6.22	3.64	1.88	1.42	0.68	0.43	0.27	0.02	44.54

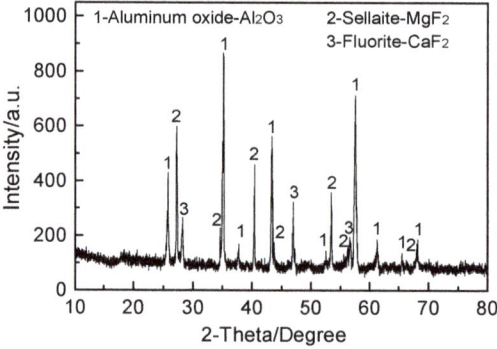

Figure 13. XRD patterns of the calcinated residue under the conditions of F^- concentration at 14 mol/L, leaching temperature of 90 °C for 20 min with L/S ratio of 20:1, and rotating speed of 200 r/min.

4. Conclusions

The fluoride leaching mechanism and kinetics of TEFS under a normal pressure system was studied to achieve efficient leaching of titanium and provide the foundation for industrialization. The TEFS is mainly composed of $M_xTi_{3-x}O_5$ (0 < x < 2, M = Mg, Al) and augite, which can both react with $[NH_4^+]$-$[F^-]$ solution. The TiF_6^{2-} and SiF_6^{2-} were dissolved in the leachate, and the $CaMg_2Al_2F_{12}$ and $(NH_4)_3AlF_6$ grains were deposited on the surface of the unreacted TEFS particles. The Ti leaching rate was controlled by surface chemical reaction, with the activation energy of 52.30 kJ/mol. The semi-empirical kinetics equation can effectively predict the result of Ti leaching rate and provide a basis for the determination of actual leaching process parameters.

In the fluoride leaching process of TEFS, the Ti leaching rate can be increased by increasing the temperature and F^- concentration. Within the condition scope of this study, increasing the leaching temperature had a more significant increase in the leaching rate. In addition, the Ti leaching rate was also improved by reducing the TEFS particle size. Higher specific surface area of TEFS particles can increase the reaction interface and thus achieve an increase in the Ti leaching rate.

Author Contributions: Conceptualization, F.Z., Y.G. and G.Q.; methodology, F.Z., Y.G. and F.C.; investigation, S.W., J.Z. and L.Y.; resources, F.Z., S.W. and L.Y.; data curation, S.W., J.Z. and L.Y.; writing—original draft preparation, F.Z.; writing—review and editing, Y.G. and S.W.; supervision, G.Q.; project administration, F.Z.; funding acquisition, F.Z. and F.C. All authors have read and agreed to the published version of the manuscript.

Funding: This research was funded by the Natural Science Foundation of Hunan Province, China (Grant No.2019JJ50816), the National Natural Science Foundation of China (Grant No. 51904348) and the State Key Laboratory of Vanadium and Titanium Resources Comprehensive Utilization.

Data Availability Statement: The data presented in this study are available in this article.

Conflicts of Interest: The authors declare no conflict of interest.

References

1. Noormohammed, S.; Sarkar, D.K. Rf-sputtered teflon-modified superhydrophobic nanostructured titanium dioxide coating on aluminum alloy for Icephobic applications. *Coatings* **2021**, *11*, 432. [CrossRef]
2. Kumar, V.; Kumar, A.; Song, M.; Lee, D.J.; Park, S.S. Properties of silicone rubber-based composites reinforced with few-layer graphene and iron oxide or titanium dioxide. *Polymers* **2021**, *13*, 1550. [CrossRef]
3. Zhang, D.; Pan, L.; Li, C.; Xia, L.; Gao, W. Fabrication of flower-like TiO_2 on Bucky paper with enhanced photocatalytic activity. *Int. J. Mod. Phys. B* **2019**, *33*, 195–207. [CrossRef]
4. Turner, J.; Aspinall, H.C.; Rushworth, S.; Black, K. A hybrid nanoparticle/alkoxide ink for inkjet printing of TiO_2: A templating effect to form anatase at 200 °C. *RSC Adv.* **2019**, *9*, 39143–39146. [CrossRef]
5. Mahdi, B.S.; Hasson, S.A. Linear optical characterization of novel dye doped poly methyl methacrylate with titanium dioxide nanoparticale. *NeuroQuantology* **2021**, *19*, 16–21. [CrossRef]
6. Wan, L.; Deng, C.; Zhao, Z.Y.; Zhao, H.B.; Wang, Y.Z. A titanium dioxide–carbon nanotube hybrid to simultaneously achieve the mechanical enhancement of natural rubber and its stability under extreme frictional conditions. *Mater. Adv.* **2021**, *2*, 2408–2418. [CrossRef]
7. Zhang, M.M.; Chen, J.Y.; Li, H.; Wang, C.R. Recent progress in Li-ion batteries with TiO_2 nanotube anodes grown by electrochemical anodization. *Rare Met.* **2020**, *40*, 249–271. [CrossRef]
8. Li, F.; Wang, C.L.; Ding, S.; Yang, K.; Tian, F. Photoelectrochemical performance of TiO_2 nanotube arrays by in situ decoration with different initial states. *Rare Met.* **2021**, *40*, 720–727. [CrossRef]
9. Liu, H.; Zhang, R.; Yan, S.; Li, J.; Yang, S. Effect of La_2O_3 on the microstructure and grain refining effect of novel Al-TiO_2-C-XLa_2O_3 refiners. *Metals* **2020**, *10*, 182. [CrossRef]
10. Rahmani, R.; Rosenberg, M.; Ivask, A.; Kollo, L. Comparison of mechanical and antibacterial properties of TiO_2/Ag ceramics and Ti_6Al_4V-TiO_2/Ag composite materials using combined SLM-SPS techniques. *Metals* **2019**, *9*, 874. [CrossRef]
11. Bordbar, H.; Yousef, A.A.; Abedini, H. Production of titanium tetrachloride ($TiCl_4$) from titanium ores: A review. *Polyolefns J.* **2017**, *4*, 150–169. [CrossRef]
12. Zheng, F.; Chen, F.; Guo, Y.; Jiang, T.; Travyanov, A.Y.; Qiu, G. Kinetics of hydrochloric acid leaching of titanium from titanium-bearing electric furnace slag. *JOM* **2016**, *68*, 1476–1484. [CrossRef]
13. Yang, J.; Tang, Y.; Yang, K.; Rouff, A.A.; Elzinga, E.J.; Huang, J.H. Leaching characteristics of vanadium in mine tailings and soils near a vanadium titanomagnetite mining site. *J. Hazard. Mater.* **2014**, *264*, 498–504. [CrossRef] [PubMed]
14. Fan, S.G.; Dou, Z.H.; Zhang, T.A.; Yan, J.S. Self-propagating reaction mechanism of Mg–TiO_2 system in preparation process of titanium powder by multi-stage reduction. *Rare Metals* **2021**, *40*, 2645–2656. [CrossRef]
15. Wang, X.; Liu, W.; Liang, B.; Lv, L.; Li, C. Combined oxidation and 2-octanol extraction of iron from a synthetic ilmenite hydrochloric acid leachate. *Sep. Purif. Technol.* **2016**, *158*, 96–102. [CrossRef]
16. Zhang, W.; Zhu, Z.; Cheng, C. A literature review of titanium metallurgical processes. *Hydrometallurgy* **2011**, *108*, 177–188. [CrossRef]
17. El-Sadek, M.H.; Fouad, O.A.; Morsi, M.B.; El-Barawy, K. Controlling conditions of fluidized bed chlorination of upgraded titania slag. *Trans. Indian Inst. Met.* **2019**, *72*, 423–427. [CrossRef]
18. Zheng, F.; Guo, Y.; Liu, S.; Qiu, G.; Chen, F.; Jiang, T.; Wang, S. Removal of magnesium and calcium from electric furnace titanium slag by H_3PO_4 oxidation roasting–leaching process. *Trans. Nonferr. Metal. Soc.* **2018**, *28*, 356–366. [CrossRef]

19. Middlemas, S.; Fang, Z.Z.; Peng, F. Life cycle assessment comparison of emerging and traditional titanium dioxide manufacturing processes. *J. Clean. Prod.* **2015**, *89*, 137–147. [CrossRef]
20. Li, Y.; Gang, X.; Lin, T.; Zhan, L.; Yan, C.; Yan, Q. Model on titanium tetrachloride gas phase oxidation process. *Mater. Sci. Forum* **2015**, *833*, 56–60. [CrossRef]
21. Zhou, Q.; Liu, Q.; Yao, L.; Ren, S.; Zhu, B.; Meng, F.; Lu, R. Influence of different conditions on the precipitation of metatitanic acid to synthesis rutile TiO_2 via sulfuric acid process. *Mater. Sci. Forum* **2016**, *850*, 851–856. [CrossRef]
22. Mehta, S.M.; Patel, S.R. The behavior of solutions of titanium dioxide in sulfuric acid in the presence of metallic sulfates. *JACS* **1951**, *73*, 224–226. [CrossRef]
23. Sasikumar, C.; Rao, D.S.; Srikanth, S.; Mukhopadhyay, N.K.; Mehrotra, S.P. Dissolution studies of mechanically activated Manavalakurichi ilmenite with HCl and H_2SO_4. *Hydrometallurgy* **2007**, *88*, 154–169. [CrossRef]
24. Anderrv, A.A.; Diachenko, A.N. Conditions for the production of pigmentary titanium dioxide of rutile and anatase modifications by ilmenite processing with ammonium fluoride. *Theor. Found. Chem. Eng.* **2009**, *43*, 707–712. [CrossRef]
25. Krysenko, G.F.; Epov, D.G.; Medkov, M.A.; Sitnik, P.V.; Nikolaev, A.I. Processing of perovskite concentrate by ammonium hydrodifluoride. *Theor. Found. Chem. Eng.* **2016**, *50*, 588–592. [CrossRef]
26. Bakeeva, N.G.; Gordienko, P.S.; Pashnina, E.V. Preparation of high-purity titanium salts from the system $(NH_4)_2TiF_6$-$(NH_4)_3FeF_6$-NH_4F-H_2O. *Russ. J. Gen. Chem.* **2009**, *79*, 1–6. [CrossRef]
27. Zhao, Y.; Sun, T.; Zhao, H.; Chen, C.; Wang, X. Effect of reductant type on the embedding direct reduction of beach titanomagnetite concentrate. *Int. J. Miner. Metall. Mater.* **2019**, *26*, 152–159. [CrossRef]
28. Li, T.; Sun, C.; Song, S.; Wang, Q. Roles of MgO and Al_2O_3 on the viscous and structural behavior of blast furnace primary slag, part 1: C/S = 1.3 containing TiO_2. *Metals* **2019**, *9*, 866. [CrossRef]
29. Steinberg, W.S.; Geyser, W.; Nell, J. The history and development of the pyrometallurgical processes at Evraz Highveld steel & vanadium. *J. S. Afr. Inst. Min. Metall.* **2011**, *111*, 705–710.
30. Zheng, F.; Guo, Y.; Qiu, G.; Chen, F.; Wang, S.; Sui, Y.; Jiang, T.; Yang, L. A novel process for preparation of titanium dioxide from Ti-bearing electric furnace slag: NH_4HF_2-HF leaching and hydrolyzing process. *J. Hazard. Mater.* **2018**, *344C*, 490–498. [CrossRef]
31. Sohn, H.Y.; Wadsworth, M.E. *Rate Processes of Extractive Metallurgy*, 1st ed.; Plenum Press: New York, NY, USA, 1979; pp. 135–143.
32. Levenspiel, O. *Chemical Reaction Engineering*, 3rd ed.; John Wiley and Sons Inc.: New York, NY, USA, 1999; pp. 566–570.
33. Habashi, F. *Principles of Extractive Metallurgy, General Principles*, 2nd ed.; Gordon and Breach Science Publishers, Inc.: New York, NY, USA, 1969; Volume 1, pp. 111–169.

MDPI
St. Alban-Anlage 66
4052 Basel
Switzerland
www.mdpi.com

Metals Editorial Office
E-mail: metals@mdpi.com
www.mdpi.com/journal/metals

Disclaimer/Publisher's Note: The statements, opinions and data contained in all publications are solely those of the individual author(s) and contributor(s) and not of MDPI and/or the editor(s). MDPI and/or the editor(s) disclaim responsibility for any injury to people or property resulting from any ideas, methods, instructions or products referred to in the content.

www.ingramcontent.com/pod-product-compliance
Lightning Source LLC
LaVergne TN
LVHW070503100526
838202LV00014B/1779